Spectroelectrochemistry

Spectroelectrochemistry

Edited by

Wolfgang Kaim
Institute of Inorganic Chemistry, University of Stuttgart, Stuttgart, Germany

Axel Klein
Department of Chemistry, University of Cologne, Cologne, Germany

RSCPublishing

ISBN: 978-0-85404-550-1

A catalogue record for this book is available from the British Library

© Royal Society of Chemistry, 2008

Published by The Royal Society of Chemistry,
Thomas Graham House, Science Park, Milton Road,
Cambridge CB4 0WF, UK

Registered Charity Number 207890

For further information see our web site at www.rsc.org

Preface

As a truly interdisciplinary field of science, electrochemistry is involved in the transformation of materials, in the transfer of information (especially in living systems), and in the conversion and storage of energy. Furthermore, electrochemical processes constitute a major class of chemical reactions in the laboratory and on large industrial scales. Conventional analytical electrochemistry continues to be an excellent method to determine concentrations (sensor technology), to yield energy data in the form of redox potentials, and to elucidate formal reaction mechanisms *via* kinetic analysis. However, these techniques are not by themselves suitable to identify unknown species that are formed as intermediates or as products in a redox reaction. The combination of *reaction-oriented electrochemistry* with *species-focused spectroscopy*, in what is known as spectroelectrochemistry (SEC), can solve this problem and thus allow for a more complete analysis of electron-transfer processes and complex redox reactions. While the technique has been well developed during the last few decades, its application in various fields of chemistry has only recently become more widespread. Readily accessible, inexpensive equipment and lower barriers to application have contributed to this situation and, at the same time, it is becoming less and less acceptable in chemical research to assign redox transformations without spectral evidence. Yet, while the method has become more commonplace, there are still aspects to be considered that require sound knowledge and experience. This book is meant to serve as a guide and as an illustration of the kind of research where SEC can make a difference in the understanding of redox reactions through identification of their intermediates and products. The examples covered include organometallics, coordination compounds (mixed-valence complexes, metalloporphyrins), and compounds of biochemical interest such as iron-containing proteins. Solutions available from spectroelectrochemical investigation not only provide simultaneous reaction analysis and species identification but also an assessment of electronic situations, of intra- and intermolecular electron transfer. We hope that this presentation helps to familiarise the scientific

Spectroelectrochemistry
Edited by Wolfgang Kaim and Axel Klein
© Royal Society of Chemistry, 2008

community with the method by describing the experimental approaches possible and by pointing out under what diverse circumstances this technique can be useful. We sincerely thank Mrs. Angela Winkelmann for her contribution in preparing this presentation.

Wolfgang Kaim
Axel Klein

Contents

Spectroelectrochemistry
Edited by Wolfgang Kaim and Axel Klein
© Royal Society of Chemistry, 2008

**Chapter 3 Mixed-Valence Intermediates as Ideal Targets
for Spectroelectrochemistry (SEC)**
Wolfgang Kaim, Biprajit Sarkar and Goutam Kumar Lahiri

Chapter 4 Spectroelectrochemistry of Metalloporphyrins
Axel Klein

Contents

Chapter 7   EPR Spectroelectrochemistry
P. R. Murray and L. J. Yellowlees

CHAPTER 1
Infrared Spectroelectrochemistry

STEPHEN P. BEST,[1] STACEY J. BORG[1] AND
KYLIE A. VINCENT[2]

[1] School of Chemistry, University of Melbourne, 3010, Victoria, Australia;
[2] Inorganic Chemistry Laboratory, University of Oxford, UK

1.1 Introduction

A set of electrochemical measurements may, with the aid of simulations, provide skeletal details of the redox-coupled reactions of a system, although the extent of the detail or uniqueness of the description depends on the complexity of the system and the relative rates of reaction. Since such an approach can, at best, yield only limited insight into the structure of intermediate species there is a clear need to supplement the electrochemical measurements by spectroscopic investigations. This need has spawned a number of approaches designed to provide the spectroscopic details of the electrogenerated intermediates/products. Spectroelectrochemical (SEC) techniques allow *in-situ* spectroscopic interrogation of electrogenerated complexes and this may permit the study of shorter-lived species and also establish the chemical reversibility of these reactions. This allows the building, testing and refinement of the mechanism and, crucially, provides insights into the structures of the intermediates.

The structure of the intermediate implicitly encompasses molecular, electronic, and vibrational components where the molecular structure is most commonly deduced by X-ray crystallography. More limited structural data may also be obtained from solute species through analysis of the X-ray absorption fine structure (XAFS) spectra and this will be discussed briefly in Section 1.6. Clearly the electronic and vibrational structure must be obtained from analysis of the spectra. The interconnection between these aspects of the structure is reinforced by *in-silico* techniques, where advances in DFT (density-functional theory) have greatly expanded the range of transition-metal compounds and

Spectroelectrochemistry
Edited by Wolfgang Kaim and Axel Klein
© Royal Society of Chemistry, 2008

smaller clusters that are amenable to study. For systems of moderate complexity a combination of structural, spectroscopic and *in-silico* approaches is required in order to achieve a satisfactory understanding of the intermediates formed during reaction. This chapter focuses on the use of IR spectroscopy to delineate the chemistry following redox activation and the integration of these results with a range of electrochemical, spectroscopic and computational methods to characterise the charge state and structure of intermediate species. While the vibrational structure of a species would ideally be determined through the examination of both its IR and Raman spectra, in most cases the complementary nature of the physical constraints associated with the two techniques results in studies concentrating on one or the other approaches. In cases where the system under investigation incorporates strongly IR absorbing chromophores, such as CO or CN⁻, IR spectroscopy can be both effective and easily implemented.

Since the objective of the studies described herein is the characterisation of the solute species formed following redox reaction the very extensive research dealing with characterisation of the electrode/solute interface will not be discussed, excellent overviews of the experimental aspects of this subject are available.[1] While this contribution focuses on applications involving IR, Raman spectroscopy has proved to be invaluable to many SEC studies where surface-enhanced Raman spectroscopy (SERS) and resonance Raman spectroscopy dominate. Reviews and recent studies attest to the value of these approaches.[2]

In this contribution we aim to illustrate the impact of IR-SEC techniques on the elucidation of the chemistry following a redox reaction. The most effective experimental approach will depend on the stability of the redox products together with the rates or nature of the following reactions. In Section 1.5 we show the experimental results obtained from several systems chosen so as to highlight the different experimental approaches that can be applied to good effect. We have limited the discussion to studies of solute species and to concentrating on examples drawn from our own research, published and unpublished. This is driven, in large part, by the availability of the raw experimental data and the opportunity that this provides to recast the figures in a self-consistent form. As a result, there is an overrepresentation of studies conducted using external reflectance SEC cells.

1.2 Overview of IR-SEC Techniques for the Study of Solute Species

The marriage between the spectroscopic and electrochemical requirements of the SEC experiment necessarily involves compromise, the nature of which will be dictated by the objectives of the study. For thin-layer cells with large surface area electrodes uncompensated solution resistance will generally present problems and these will be accentuated for studies conducted in highly resistive solvents. In many cases it is not practicable to use a conventional reference electrode and in these instances a pseudoreference consisting of a silver or platinum wire or foil is used. While such electrodes are susceptible to a drift in potential the impact of this deficiency may be minimised if the duration of the experiment is short relative to a change in the concentration of the species near the reference electrode. Several different experimental approaches have proved to be effective for the collection of IR-SEC results from electrogenerated solute species and these may be distinguished in terms of the characteristics of the working electrode. These include (i) optically transparent electrodes, (ii) perforated electrodes and (iii) reflective electrodes. To these may be added approaches in which a probe beam is brought close to the working electrode of an electrosynthesis cell by means of a waveguide or optical fibre. The sampling element may consist either of a pair of launch and collection fibres or include an optical element that is arranged so as to give near-total internal reflection (attenuated total reflection, ATR). In the latter case the spectrum of the solution in contact with the ATR crystal is sampled through its interaction with the evanescent wave that propagates beyond the reflecting surface. Depending on the cell geometry, and volume of solution, the time required for electrosynthesis can be substantial (>1 h) in which case the distinction between *in-situ* and *ex-situ* spectroscopic interrogation is not clear cut. With careful attention to the design it is possible to reduce the volume of solution subject to electrosynthesis and thereby reduce the response time. An ATR IR-SEC cell featuring a sample chamber with a volume of 20 µL has recently been reported, although even in this case the cell requires *ca.* 6 min for redox equilibration following a potential step.[3]

1.3 Transmission Cells Using Optically Transparent or Perforated Electrodes

In optical terms the simplest SEC approach involves the use of transparent electrodes. For visible spectroscopy doped tin oxide affords a useful spectral window.[4] More recently boron-doped diamond has been shown to be suitable for UV-Vis and IR spectroscopy.[5,6] This material offers a wide potential range, inertness in chemically aggressive environments and biocompatability.[7] The spectroscopic range and conductivity depends on the boron doping level. Diamond films with moderate levels of boron doping (0.5 to $0.05 \,\Omega\,cm^{-1}$, 1–10×10^{19} B cm^{-3}) retain a high transmittance in the region below 1500 cm^{-1}.[6]

An alternative strategy to the use of transparent electrodes is to use a conductor in the form of a grid or fine gauze. Cells constructed using this approach date back to the work of Murray *et al.*[8] and have been used for UV-Vis and IR spectroscopic studies. In the latter case the solution pathlength must be minimised in order to avoid problems associated with strong solvent absorption. The principle for construction of the cell is straightforward, generally involving modification of a solution IR cell by incorporation of a fine metal gauze working electrode located in the path of the IR beam. A good example of a compact, airtight cell based on this approach has been reported by Hartl and

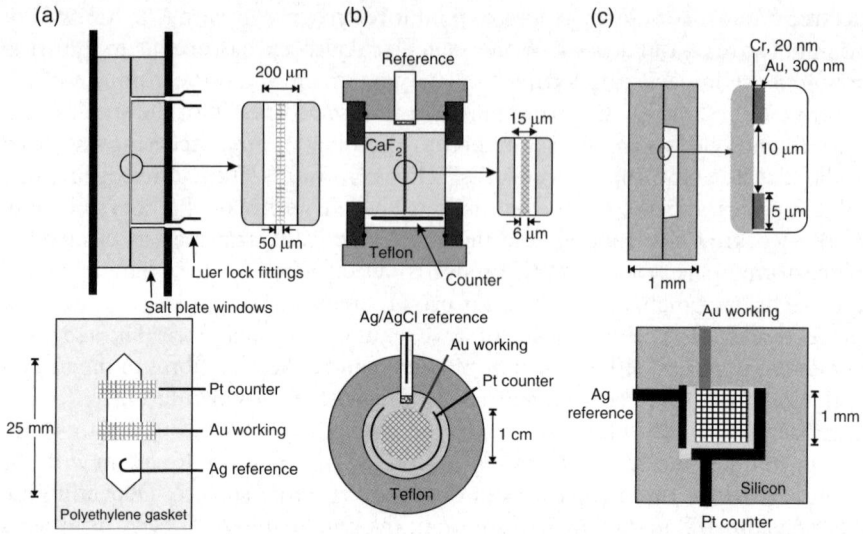

Figure 1.1 Schematic of transmission SEC cells (a) after Hartl,[9] (b) after Moss[10] and (c) after Yun.[11]

coworkers.[9] In this case the working, counter, and reference electrodes occupy the space between the salt plate windows as shown in Figure 1.1a. The working and counter gauze electrodes and silver pseudoreference electrode are melt-sealed into a single 200-μm thick polyethylene spacer. The working electrode has a dimension slightly larger than the IR beam (6 × 5 mm) and is formed from a highly transmitting (80%) Au minigrid (200 lines/mm). Platinum gauze of similar surface area serves as the counter electrode. The approach is well suited to the study of air-sensitive species since solutions may be transferred to the cell anaerobically using gas-tight syringes, however, the relatively long electro-synthesis time (up to 4 min for 5–10 mM solutions) limits the suitability of the approach for products that undergo following reaction.

A variation of this approach has been described by Moss *et al.*[10] who have reported a cell design that is optimised for IR-SEC studies of proteins. A schematic cross-sectional view of the Moss cell is shown in Figure 1.1.(b) The working electrode consists of a 6-μm thick, 70% transparent Au mini-grid and this, together with the Pt foil counter and Ag/AgCl reference electrodes, completes the three-electrode geometry. An important feature of the cell for applications using proteins is the low sample requirement, where a 3–5 μL drop of solution is sufficient to fill the 20 mm diameter × 15 μm thick space formed by the CaF$_2$ windows. The outer cavity of the cell is filled with buffer/salt solution. Despite the contact between the sample solution and the surrounding medium the rate of dilution was reported to be ∼5% over 24 h. In order to improve the rate of heterogeneous electron transfer and reduce the effects of protein denaturation, electrode-surface modifiers or electron-transfer reagents may be added to the sample solution. Owing to the high electrode surface area to volume ratio the system rapidly equilibrates following a change in applied potential (1–2 min). Although factors such as the solution resistance within the cell may lead to an offset to the cell potential, excellent Nernstian concentration/potential plots were obtained for model studies of cytochrome *c*.[10]

More recently, Yun and coworkers[11] have reported the design of a transmission IR/UV-Vis cell that is micromachined from single-crystal silicon wafers (Figure 1.1(c)). In this case the Au minigrid, Au counter and Ag/AgCl pseudo-reference electrodes are deposited directly onto the silicon wafer. The cover plate is also etched from a single crystal, allowing fine control of the pathlength of the solution. The sample solution (*ca.* 0.1 mL) is placed between the two silicon wafers and these are clamped together in the cell mount. This arrangement was reported to work well for aqueous solutions but in the case of nonaqueous solvents it was necessary to enclose the edge of the thin-layer cell with Teflon tape so as to prevent solvent leakage. While transmission through silicon cells is lowered as a result of high reflectance losses at the interfaces this is offset by the control over the solution volume and electrode geometry provided by the microfabrication techniques.

1.4 External Reflection-Absorption SEC Cells

IR reflectance spectroscopy has proved to be highly effective for the examination of both the solute/electrode interface and the electrogenerated products. Two general approaches, designated internal and external reflection, may be employed in these studies.[12] For internal reflection the radiation passes through a transparent substrate and is reflected from a thin film of metal deposited thereon. The metal film also serves as the working electrode. The penetration of the electric vector of the radiation into the solution in contact with the reflecting surface provides a means of obtaining selective information related to the solute/electrode interface. For external reflectance the radiation passes through a suitable window and is specularly reflected from a solid electrode. The solution trapped between the electrode and window is interrogated by the IR beam. External reflectance approaches have been used to study both the electrode/electrolyte interface and the electrogenerated species. Since the study of monolayer or submonolayers of adsorbed species introduces significant challenges in terms of sensitivity, a number of different approaches have been developed that has spawned an extraordinarily diverse range of acronyms (EMIRS, IRRAS, SNIFTIRS, PM-FTIRS, LPSIRS). Briefly, discrimination between molecules bound to the electrode and those in the bulk solution may be achieved by examining the spectral changes that result from either potential or polarisation modulation. A good review of the techniques, and their physical basis has been outlined by Beden and Lamy.[12]

Examination of the solute species by external reflectance presents far fewer challenges and a variety of cell designs have been reported. These include strategies that permit the study of air-sensitive compounds[13] over a range of temperatures[14,15] and/or pressures.[16] A schematic diagram of a basic external reflection SEC cell is shown in Figure 1.2(a).[14] If the object of the investigation

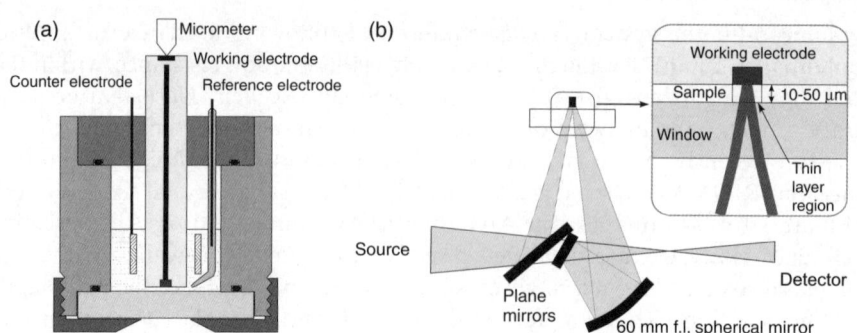

Figure 1.2   (a) Cross section of an external reflectance SEC cell and (b) an example of
an optical arrangement that permits its use in a conventional FTIR
spectrometer.

is the study of solute species then the incident beam strikes the working
electrode near to normal incidence, this minimises reflection losses from the air/
IR-transmitting window interface and reduces the cross-sectional area of the
focus of the IR beam when projected onto the electrode. Since the layer of
solution under investigation is typically 10–20 µm the cells inevitably suffer
from a substantial uncompensated solution resistance and this may best be
controlled by minimising the surface area of the working electrode. Thus,
the diameter of the working electrode should match the IR beam diameter.
Depending on the instrument this typically translates into an electrode diameter
of 3–5 mm. Naturally, the electrode must be fashioned from a material highly
reflecting to radiation in the energy range of interest. While this is best achieved
with metals such as Au and Pt, it is important to note that vitreous carbon has
sufficient reflectivity in the IR to be suitable for such measurements. In its
simplest form the working electrode is encased in an inert insulator, usually
KelF or glass (Figure 1.3(a)). The thickness of the layer of solution under
examination is determined by the position of the working electrode relative to
the front window and this is adjusted under micrometer control. In most cases
the counter and reference electrodes are located outside the thin-layer region
where diffusion between the bulk solution and the solution in contact with the
working electrode is negligibly small over the timescale of the SEC experiments.

In view of the need to restrict the size of the IR beam at the electrode plane
the use of reflection absorption cells is generally limited to FTIR instru-
mentation where it is often necessary to use a high-sensitivity photovoltaic
detector (*e.g.* liquid-nitrogen-cooled mercury cadmium telluride, MCT). Since
in these cases the detector element is small, in addition to providing a flat,
highly polished working electrode, it is necessary to provide an optical train
that does not increase the size of the IR beam at the detector focus. The optical
design used in our laboratory is shown in Figure 1.2(b). The IR beam is dir-
ected to the working electrode located at the beam focus using a plane mirror.
The central ray of the IR beam makes an angle of 14.3° to the normal to the
CaF$_2$ window and 10° to that of the working electrode. The reflected beam is

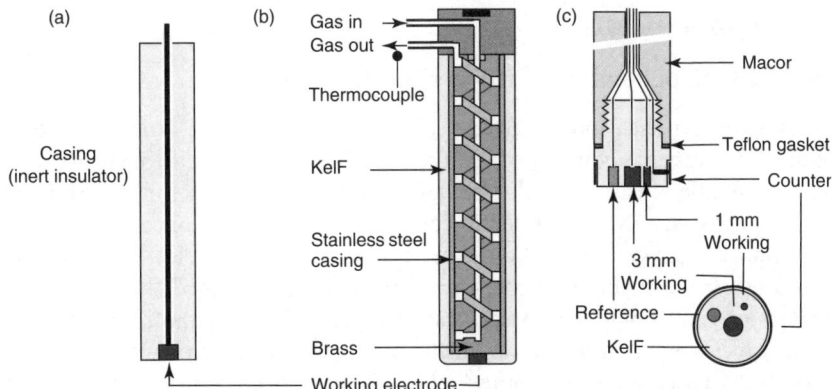

Figure 1.3 Working electrode designs for external reflectance SEC cells. (a) Single element, (b) temperature controlled,[14] and (c) multielectrode assembly.[16]

refocused using a spherical mirror to give a beam focus and path as close as possible to that of the undiverted beam. All three mirrors are built into a mount that allows rotation about axes coincident with their reflective surfaces and beam centre, in addition, the two plane mirrors have tilt axes normal to the rotation axis.

A schematic diagram of an electrode suitable for low-temperature use is shown in Figure 1.3(b).[14] The temperature of the brass pin attached to the working electrode is controlled by the flow rate and temperature of dinitrogen gas that passes through its structure. Using this approach it is difficult to monitor directly the temperature of the thin layer of solution, however, our experience (based on the freezing of solvent/supporting electrolyte) suggests only a small temperature difference ($<ca.$ 10 °C) between that of the outgoing gas. An approach that minimises the distance between the electrodes and also facilitates the construction of small solution volume cells involves incorporation of the working, counter and reference electrodes into a single KelF pin.[16] An additional small (1 mm diameter) working electrode is also incorporated into the assembly so as to provide a means of obtaining better electrochemical characterisation of the solution. In more recent designs (Figure 1.3(c)) the KelF pin is attached to a machineable ceramic cylinder (Macor). The greater dimensional stability provided by this material provides advantages in establishing an airtight seal between the electrode assembly and the cell body.

Relative to the transmission cells described in Section 1.3, external reflectance cells are less simple to construct and require additional optics for incorporation into the optical path of the spectrometer. The advantages associated with the approach relate to the well-defined nature of the working electrode, the wider range of materials suitable for this purpose, and the greater control over the thickness of the thin layer of solution. These factors contribute to a much faster ($>10\times$) rate of electrosynthesis for external reflectance compared to transmission cells.

1.5   Applications

1.5.1   Electrochemically Reversible Reactions

The simplest application of SEC techniques involves the study of electro-chemically reversible couples where the oxidised and reduced forms are stable on the timescale of electrosynthesis. Depending on the species this might involve careful control of solvent, supporting electrolyte and solute concentration, strict exclusion of dioxygen, and choice and preparation of the electrodes. Even in cases where a reversible electrochemical response is obtained the SEC experiments may yield complicated spectral changes indicating the involvement of several species with little, or no, recovery of the starting complex following application of an appropriate potential. In these cases it is important to consider the different timescales and concentration profiles that apply in cyclic voltametric and SEC experiments. In those cases where the electrochemistry is well behaved, quantitative and reversible electrosynthesis may easily be performed using transmittance or external reflectance IR-SEC cells. The potential dependence of the concentration of the oxidised and reduced species follows the expected Nernstian form. In cases where the heterogeneous electron-transfer reaction is slow the half-wave potential ($E_{1/2}$) may be obtained reliably using SEC techniques. This approach has been particularly important for studies of redox proteins.

An example of a system showing quasireversible electrochemistry and reversible chemistry is provided by $Fe_3S_2(CO)_9$ (Figure 1.4). The markedly different pattern of $\nu(CO)$ bands evident for the starting material and the one-electron reduced product is indicative of structural rearrangement, however, on the timescale of the IR-SEC experiment reduction and reoxidation proceed with comparable rates and with near-complete recovery of the starting material following redox cycling (Figure 1.4). The presence of well-defined isosbestic points for both oxidation and reduction (Figures 1.4(a) and (b)) indicate that there is no significant involvement of additional iron-carbonyl species in the reaction.

1.5.2   Redox-Activated Chemical (EC) Reactions

Whereas electrochemical techniques can be effective in terms of quantifying different thermodynamic and kinetic parameters in cases where chemical reaction follows redox activation (EC reactions) the identification of the intermediates formed during the reaction is problematic. In cases where the electrochemical response matches that of a known compound then this may provide a satisfactory basis for identification, but usually it is necessary to resort to other approaches. It is in relation to problems of this sort that SEC techniques can be of most value.

The group 6 complexes, $M(L-L)_2(CO)_2$ (where L-L designates various bidentate ligands with P/As donor atoms) undergo well-defined redox reactions that feature stereochemical change linked to the formal oxidation state of the

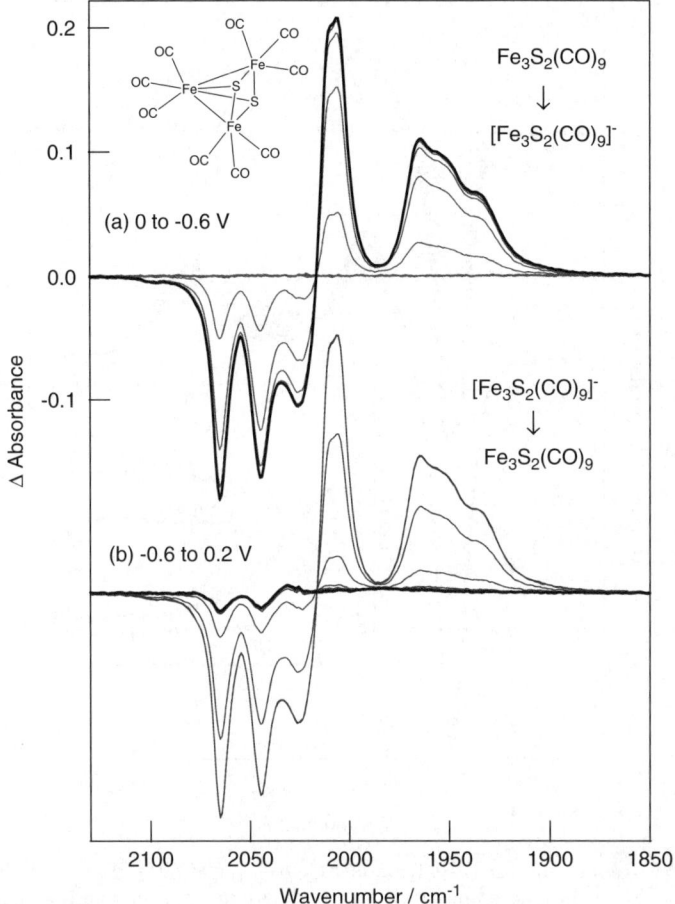

Figure 1.4 IR-SEC of the reduction (a) and reoxidation (b) of a CH$_3$CN solution of Fe$_3$S$_2$(CO)$_9$. Spectra were recorded at 1-s intervals. In this, and the other IR-SEC spectra, a reference spectrum is collected immediately before the application of a reducing potential and this is used to calculate the absorbance or, more properly, differential absorbance spectra. The depletion of species initially present in the thin layer of solution presents as negative bands, whereas electrogenerated species give bands with positive absorbance. The last spectrum of each set is emphasised.

metal. The more stable forms are the *cis* M^0 and *trans* M^I species.[17,18] The v(CO) bands are a particularly useful probe of the chemistry since a shift in wave number is indicative of a change in the electron richness of the metal and the stereochemistry is flagged by the presence of one (*trans*) or two (*cis*) IR-active bands. The reduction of *trans*-[Cr(dppm)$_2$(CO)$_2$]$^+$ (dppm = Ph$_2$PCH$_2$PPh$_2$) in an external reflectance SEC experiment is shown in Figure 1.5. As the reduction progresses the initial product band at 1810 cm^{-1} due to metastable *trans*-Cr(dppm)$_2$(CO)$_2$ is replaced by two bands (1780 and 1843 cm^{-1}) consistent with

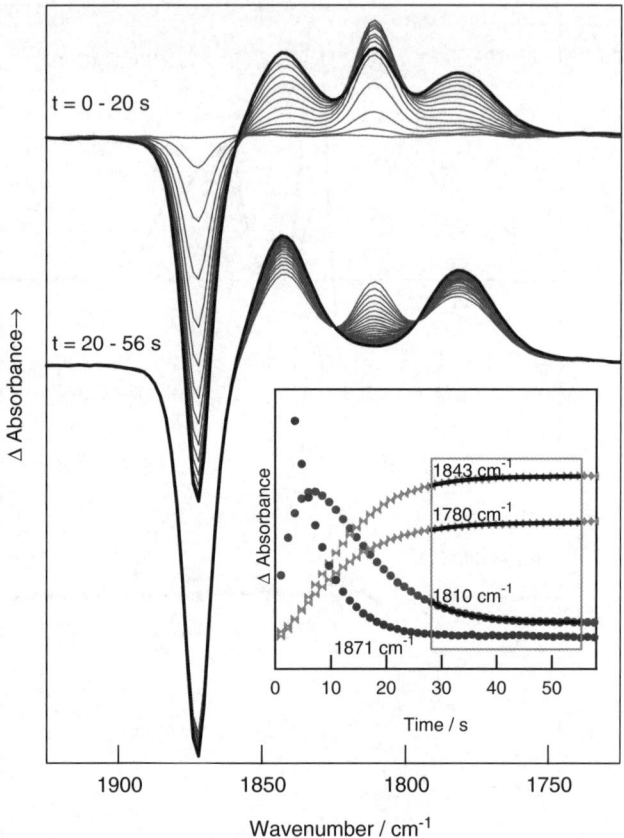

Figure 1.5   Reduction of *trans*-[Cr(dppm)$_2$(CO)$_2$]$^+$ in CH$_2$Cl$_2$ (60-μm thick layer). The inset shows the time traces of the absorbance for the bands associated with the starting material, the *trans*-neutral intermediate and the *cis*-neutral final product. The last spectrum of each set is emphasised. The fit of the absorbance changes at 1780, 1810 and 1843 cm^{-1} over the time range 28–53 s to a single exponential function ($k = 0.144$ s^{-1}) is shown.

rearrangement to give the more stable *cis*-neutral form. As the concentration of the starting material in the thin layer of solution reaches a low, steady-state, value the spectral changes are dominated by the isomerisation reaction. In this case the nonredox character of the second step of the reaction is easily established by the identification of the products, but this can also be confirmed by monitoring the current response during the SEC reaction as discussed in the following sections.

The satisfactory isosbestic points obtained for the isomerisation reaction suggest that the system is amenable to kinetic analysis. The absorbance changes associated with the decay of the *trans*-neutral and appearance of the *cis*-neutral species fit to a single exponential with a rate constant of 0.14 (2) s^{-1} (Figure 1.5 inset). This value is in good agreement with the reported value of 0.17 (3) s^{-1} at

22 °C obtained by double-potential-step chronoamperometry in acetone.[17] It is clear that the rate of electrosynthesis is of vital importance when examining reactions of this sort. Clearly, the half-life of the reaction should be at least comparable to the time required for electrosynthesis and the data shown in Figure 1.5 is close to that limit. If spectra of adequate quality can be obtained using thinner films (10–15 μm) then the time required for quantitative electrosynthesis can be reduced to several seconds.

1.5.3 Electrocatalytic Reactions

Two different types of catalytic reaction will be considered, electron-transfer catalysis (ETC) and the more usual case in which a kinetically slow electrochemical reaction is accelerated by the presence of a second redox couple (EC′ reaction). The distinction between the two types of catalytic reaction is based on whether the net reaction involves a change in redox state (Scheme 1.1).

For ETC reactions a small injection of charge into the thin layer of solution should be sufficient to achieve full conversion of the initial compound. The SEC response for such reactions is illustrated by PR_3/CO substitution in $Fe_2(\mu\text{-pdt})(CO)_6$, pdt $= S(CH_2)_3S$, (3S) in tetrahydrofuran (THF). A solution of 3S with 10 equivalents of $P(OPh)_3$ is initially subject to a potential of 0 V and spectra are collected at a rate of 1 Hz. A reducing potential of –1.2 V is applied for 2 s and the potentiostat is switched to open circuit. The differential absorbance spectra are calculated using the spectrum recorded immediately before switching the potentiostat to open circuit as the reference spectrum. In the following spectra the remaining 3S is quickly depleted from the thin layer and this is replaced by the spectrum of the neutral phosphite product, $Fe_2(\mu\text{-pdt})(CO)_5P(O^iPr)_3$ (Figure 1.6). It is important to note that the reaction must be conducted under conditions where the lifetime of the reactive reduced species is sufficient for the substitution reaction to be competitive with other modes of deactivation. In this case it is essential that the THF be dry and carefully deoxygenated. In solvents such as CH_3CN, where the one-electron reduced complex has a shorter lifetime, the ETC reaction is generally quenched well before complete depletion of the 3S reactant. In cases where the ETC reaction can be followed using SEC techniques spectral subtraction can be used to identify the presence of low concentrations of species formed following redox

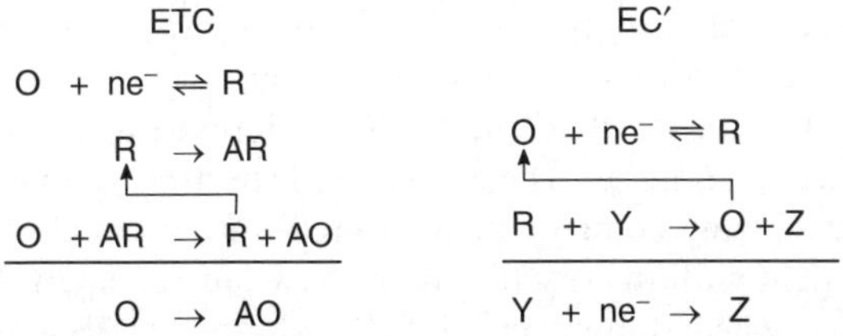

ETC

$$O + ne^- \rightleftharpoons R$$
$$R \rightarrow AR$$
$$O + AR \rightarrow R + AO$$

$$O \rightarrow AO$$

EC′

$$O + ne^- \rightleftharpoons R$$
$$R + Y \rightarrow O + Z$$

$$Y + ne^- \rightarrow Z$$

Scheme 1.1 Distinction between electron-transfer catalytic (ETC) and electrocatalytic (EC′) reactions.

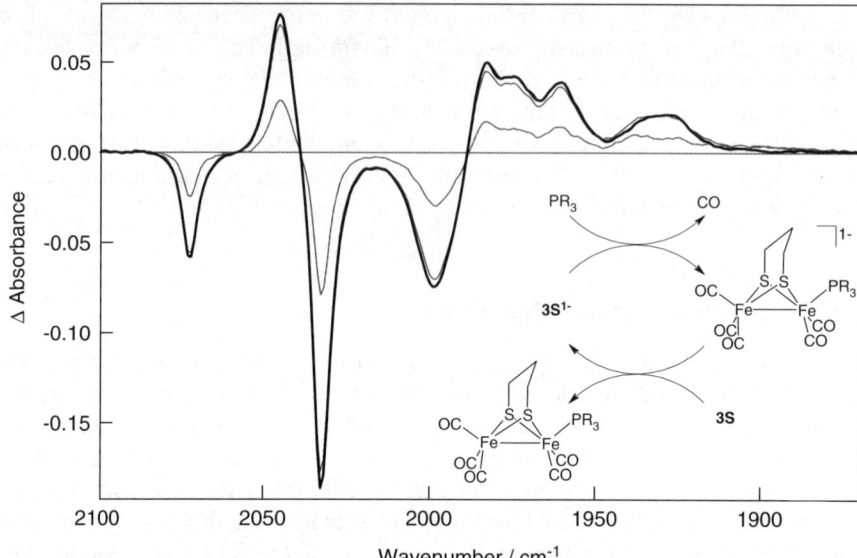

Figure 1.6 Spectra recorded from a solution of Fe$_2$(μ-pdt)(CO)$_6$ (**3S**) with 10 equivalents of P(O^{i}Pr)$_3$ following reduction for 2 s followed by switching the potentiostat to open circuit. The reference spectrum was recorded immediately prior to switching the potentiostat to open circuit and the last spectrum of the series is highlighted.

initiation. The concentration profiles of the transiently stable species are of key mechanistic importance to such reactions.

In contrast to ETC reactions, where the injection of small amounts of charge into the thin layer can result in dramatic spectral change, for EC' reactions the injection of substantial charge into the thin layer may result in minimal spectral change. This is well illustrated by electrocatalytic proton reduction by **3S**. Reduction of **3S** in the absence of protons is complete within *ca.* 30 s (Figure 1.7(a)). After this time there is a reduced current flow due to diffusion of the neutral complex into the thin-layer region. Stepping the potential to slightly more positive values leads to oxidation of one of the minor reduction products and application of much more positive potentials leads to *ca.* 85% recovery of the starting material. In the presence of 7 equivalents of acid, reduction of a layer of solution of similar thickness (estimated using the absorbance of a solvent band) results in a substantial current flow, but with minimal spectral change. As the reaction proceeds the starting complex is removed from the thin-layer region at an increasing rate until it is nearly depleted. Reoxidation leads to only *ca.* 50% recovery of the starting complex. The current and spectral response (Figure 1.7(b)) requires that the starting complex be recovered during the electrocatalytic cycle and this is consistent with the reaction path shown in Scheme 1.2(a). Satisfactory simulation of the cyclic voltammetry of THF solutions of **3S** with between 0 and 10 equivalents of *p*-toluene sulfonic acid (HOTs) has been obtained for a reaction involving consecutive electron–proton additions (Scheme 1.2(a)) together with a

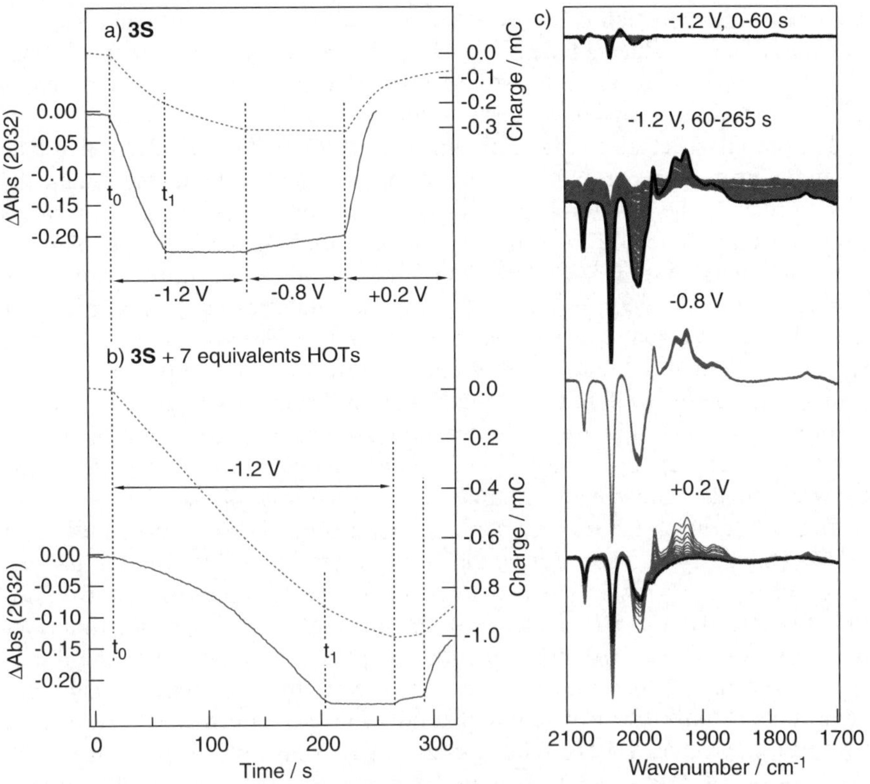

Figure 1.7 Electrocatalytic proton reduction by **3S** in THF. The current response (dashed line) and absorbance change (solid line) showing depletion of the starting material for (a) the reaction in the absence of acid and (b) after addition of 7 equivalents of *p*-toluene sulfonic acid (HOTs). (c) Spectral changes recorded corresponding to the reaction shown in (b). The last spectrum of the series is highlighted.

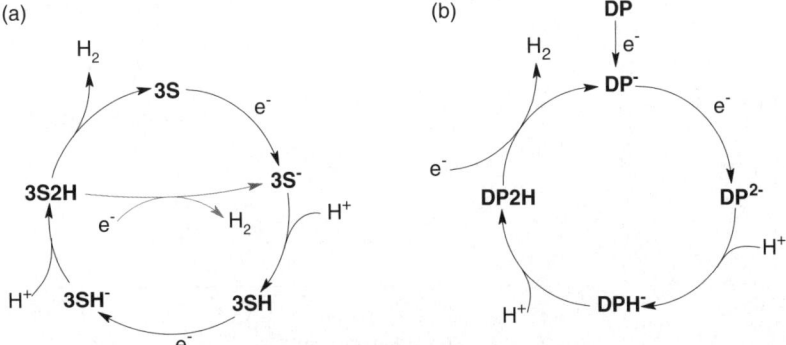

Scheme 1.2 Proposed EC′ reaction paths for electrocatalytic proton reduction by (a) **3S**[19] and (b) Fe$_2$(μ-PPh$_2$)$_2$(CO)$_6$ (**3P**).[20]

path involving reduction of the two-electron, two-proton product that becomes available at more reducing potentials.[19] It is important to note that in the absence of **3S** reduction of HOTs at the vitreous-carbon working electrode at a potential of –1.2 V yields an insignificant current flow.

The initial spectral changes obtained in SEC studies of electrocatalytic reactions provide a particularly effective means of identifying the kinetically inert species of the reaction cycle. For electrocatalytic proton reduction by **3S** the small magnitude of the spectral changes obtained during the initial phase of the reaction (Figure 1.7) requires that the reaction cycle must recover the resting state of the complex. This requires that dihydrogen elimination occur from the two-electron reduced form of the complex. Clearly, in cases where the reaction cycle involves more highly reduced levels of the complex the resting state of the complex will not be recovered and substantial spectral changes will accompany the initial phase of the reaction. Accordingly, SEC studies allow a distinction between EC' reactions that must include the resting state of the complex (small or no spectral changes during the initial phase of the reaction) and those for which the starting complex is a precursor to the species involved in the catalytic cycle. The differing SEC signatures obtained for these situations is provided by electrocatalytic proton reduction by **3S** (Figure 1.7) and $Fe_2(\mu\text{-}PPh_2)_2(CO)_6$, **DP**. For **DP**, depletion of the starting material from the thin layer proceeds as soon as a reducing potential is applied, although there are differences in the initial products for reactions conducted in the presence and absence of protons (Figure 1.8).[20] The initial product in this case is the two-electron, two-proton product **DP2H**. This species can be generated independently by protonation of **DP**$^{2-}$ and shown to be unreactive in terms of dihydrogen

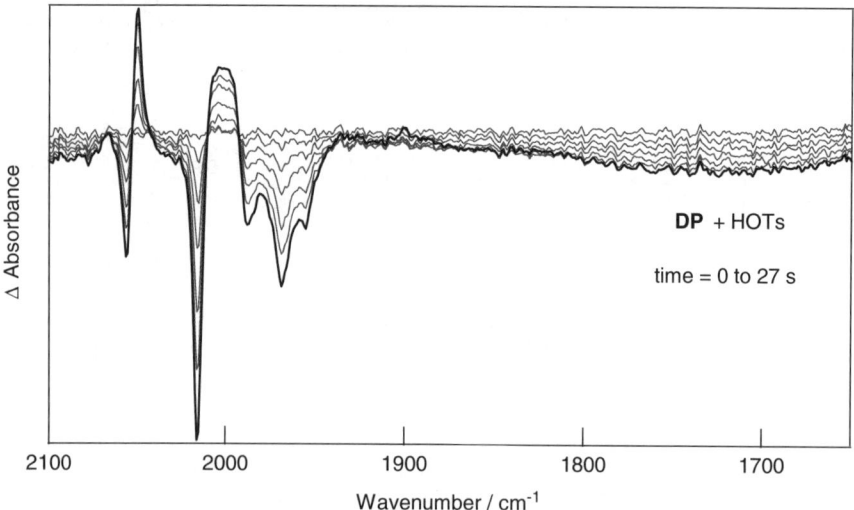

Figure 1.8 Initial stages of the reduction of **DP** with 9 equivalents of HOTs. The time between spectra is *ca.* 4 s. The last spectrum of the set is highlighted.

elimination. Further reduction is needed to give a significant rate of dihydrogen elimination (Scheme 1.2(b)).[20]

When carefully examined, and with the aid of spectral subtraction, more subtle spectral changes recorded during electrocatalysis provide a potential source of the spectral signatures of longer-lived species that are formed during the reaction cycle. During the early stages of electrocatalysis by **3S** (Figure 1.7(c), 100–150 s) the spectrum of a previously unidentified species was detected and the spectrum extracted.[19] A similar spectrum is also obtained from corresponding experiments from the related compound, $Fe_2(\mu\text{-}S(CH_2)_2S)(CO)_6$.[21] While having a spectrum seemingly related to that of the CO-inhibited oxidised form of the enzyme, more recent studies suggest that this species is a rearranged side product.[22]

Electrochemical simulations of the concentration and scan-rate dependence of the voltammetry potentially provide the composition of the intermediates formed during the reaction cycle together with estimates of the rate and equilibrium constants. As shown in the preceding section spectroscopic information can greatly assist the elucidation of the molecular details of these reactions, however, reliable deduction of the structure is greatly enhanced by the incorporation of structural and computational information (Section 1.6). The rapid advance in computer power and implementation of density-functional theory allows a more quantitative approach for evaluation of proposed structures based on spectroscopic information and estimation of the relative energies of the proposed species.[22–24] The recent computational study of the electrocatalytic reaction cycle proposed for **3S**[22] illustrates the opportunities presented by the approach.

1.5.4 Chemically Reactive Species

A range of different strategies need to be considered in cases where the lifetime of the initial product of the redox reaction is short relative to the rate of electrosynthesis. The relatively slow rate of electrosynthesis of spectroscopically significant quantities of product relative to chemical reaction may make the application of stopped- or continuous-flow techniques more suitable than SEC approaches. The advantage to be gained by study of such systems using SEC approaches derives from the control of the oxidising/reducing potential that, in turn, allows the chemical reversibility of the redox reaction to be examined. In most cases quantitative electrosynthesis of a 10-μm layer of solution contained between the working electrode and the IR-transmitting window will be of the order of 10 s. While this places a limitation on the range of species that may successfully be studied, there is the advantage that standard FTIR spectrometers with photovoltaic (*e.g.* MCT) detectors allow spectra to be collected at good resolution ($2\,cm^{-1}$) in *ca.* 1 s and are ideally suited for this work.

The second oxidation of the $Cr(dppm)_2(CO)_2$ complex considered earlier in the section provides a good illustration of the advantage provided by the rapid rate of electrosynthesis when the redox activated products are reactive. In this case one-electron oxidation of the neutral complex is accompanied by *cis/trans* isomerisation. A further oxidation of the *trans*-cation occurs at +1.5 V where the initial reduction product has a $v(CO)$ band at 2024 cm^{-1} but is relatively unstable and over a longer timescale reduction does not result in the recovery of a significant fraction of the starting material. The assignment of the 2024-cm^{-1} band to *trans*-[Cr(dppm)$_2$(CO)$_2$]$^{2+}$ requires rapid reduction at moderate reducing potentials to give the more stable *trans*-[Cr(dppm)$_2$(CO)$_2$]$^+$. For the experiment shown in Figure 1.9 only *ca.* 60% of the trans-cation is recovered following redox cycling to the dication. Despite this, there is a clear connection established between the species responsible for the 2024- and 1872-cm^{-1} bands with the rate of reaction and potential for the interconversion consistent with assignment of the 2024-cm^{-1} band to *trans*-[Cr(dppm)$_2$(CO)$_2$]$^{2+}$.

While the 3D plots give an overview of the spectral changes associated with a reaction, it is more often important to discern whether the interconversion reactions proceed with formation of significant concentrations of intermediate species. The presence of isosbestic points is of key importance in this regard and such spectral features are more easily examined in overlay plots.

Clearly, the examination of intermediates is much more straightforward if their lifetime can be extended, and in many cases this may be achieved by careful attention to the chemistry of the system (*e.g.* choice of solvent and/or

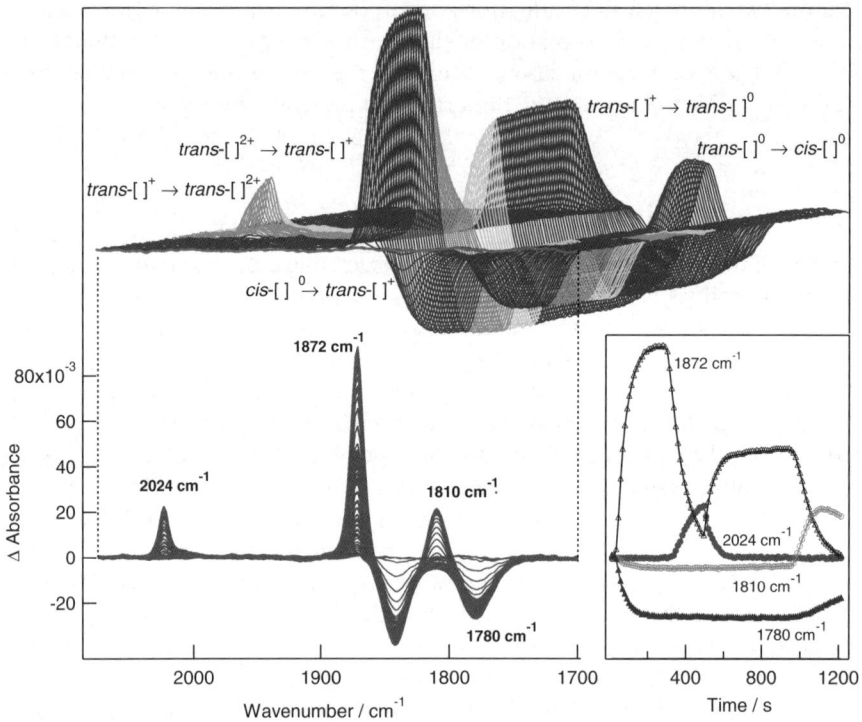

Figure 1.9   Oxidation and reduction of Cr(dppm)$_2$(CO)$_2$ in CH$_2$Cl$_2$. The time between spectra is *ca.* 1 s. The top series of spectra show the time evolution of the spectra in a 3D format and replotted in overlay mode below. The time traces are shown in the lower right panel.

supporting electrolyte) or by reducing the temperature. For studies conducted in organic solvents such as THF, butyronitrile, or dichloromethane electrochemical reactions may be carried out at temperatures as low as $-50\,°C$. Both transmission and external reflectance SEC cells may be modified for low/controlled-temperature operation and this may simply involve thermostatting a block in contact with the cell body. In this case it is necessary to take precautions against condensation on the outer surface of the IR-transmitting window. An alternative strategy is to use the working electrode to provide temperature control and this may be achieved in the manner outlined in Section 1.4. In this case the working electrode also serves as reflecting and cooling element for the thin layer of solution trapped between it and the window. Over short times the extent of cooling of the external surface of the window is low and condensation of water vapour rarely presents a problem.

Application of low-temperature SEC approaches has permitted identification of the spectrum of an intermediate species formed during the reduction of the [FeFe]-hydrogenase H-cluster structural analogue, $Fe_2((\mu-SCH_2)_2C(Me)CH_2SPh)(CO)_5$. At $10\,°C$ the reduction proceeds to give a product with a significantly altered $v(CO)$ spectrum (Figures 1.10(a) and (b)), suggesting substantial structural change. In the spectrum collected immediately after application of a reducing potential there is a subtly different spectral profile suggestive of the involvement of an intermediate species. Analogous experiments conducted at *ca.* $-40\,°C$ (Figure 1.10(c)) give similar results but with much clearer definition of the spectrum of the short-lived species. At lower temperature the spectral changes are sufficiently developed for the spectrum of the intermediate species to be revealed (Figure 1.10(d)).

While a lowering of the temperature will lead to a lowering of the rate of following reactions it is important to note that this will also have an impact on the rate of heterogeneous electron transfer. The improved spectroscopic detection of intermediate species depends on their relative rates of formation and decay. While in the previous example cooling leads to improved detection of the intermediate species there are a number of instances where poorer SEC responses are obtained at lower temperature.

The interconversion between electrogenerated species may be more finely examined using potential modulation techniques[25] analogous to the SNIFTIRS (surface-normalised interfacial FTIR spectroscopy)[26] approach developed for the study of surface species. In this case the oxidising/reducing potential is applied to the working electrode for the time required for collection of a single spectrum (usually *ca.* 1 s) and this leads to partial (*ca.* 10%) electrosynthesis of the solute in the thin-layer region. At the end of the sweep of the moving mirror of the FTIR the potential is switched back to its initial value and maintained in that state until the film is returned to its initial state. The cycling of the potential is repeated in order to establish the reproducibility of the potential dependent spectral changes. The chemical reversibility of the reaction can be assessed by the spectral changes that accompany repeated spectral cycling. In cases where the spectral changes are reversible the spectra recorded at the two potentials can be co-added in order to improve the signal-to-noise ratio. The synchronisation

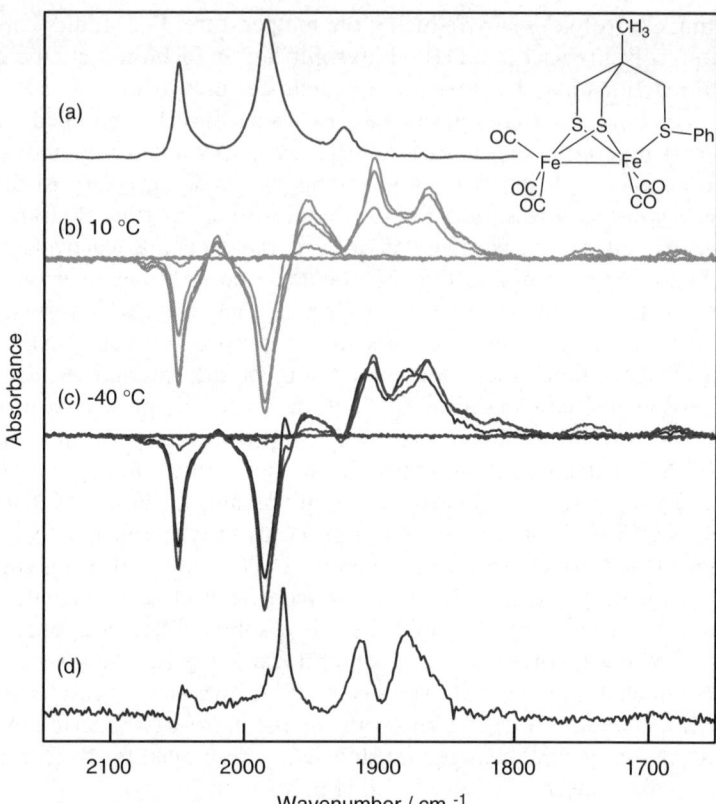

Figure 1.10   (a) Fe$_2$((μ-SCH$_2$)$_2$C(Me)CH$_2$SPh)(CO)$_5$ in CH$_3$CN, (b) SEC reduction
(−1.7 V) at 10 °C, (c) analogous experiment conducted at *ca.* −40 °C,
(d) spectrum of the short-lived product obtained from the lower-
temperature SEC experiment.

between the change of applied potential and the collection of data from the
FTIR (during the forward scans of the moving mirror) is shown in Figure 1.11.
   An example of the application of potential modulation techniques is provided by reduction of Fe$_4$S$_4$(NO)$_4$ between its dianion and trianion forms.[27]
At room temperature the dianion has limited stability in dichloromethane and
experiments were conducted at lower temperatures in order to diminish the
effects of decomposition. Reversible conversion between the dianion and trianion forms is clearly evidenced by the spectra where the rates of reduction and
reoxidation are similar and the waveform for the experiment was symmetric
(*i.e.* in Figure 1.12, $n = 0$).

1.5.5   Determination of the Charge State

Whereas thin-layer SEC approaches provide an effective means of identifying
stable and transiently stable electrogenerated species, in cases where these species

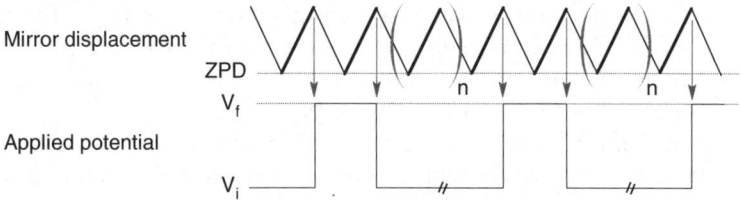

Figure 1.11 Synchronisation between the collection of FTIR spectra and the potential applied to the working electrode during potential modulation techniques. It is assumed that single-sided interferograms are collected during the forward sweep of the moving mirror (ZPD = zero path difference for the two paths of the interferometer).

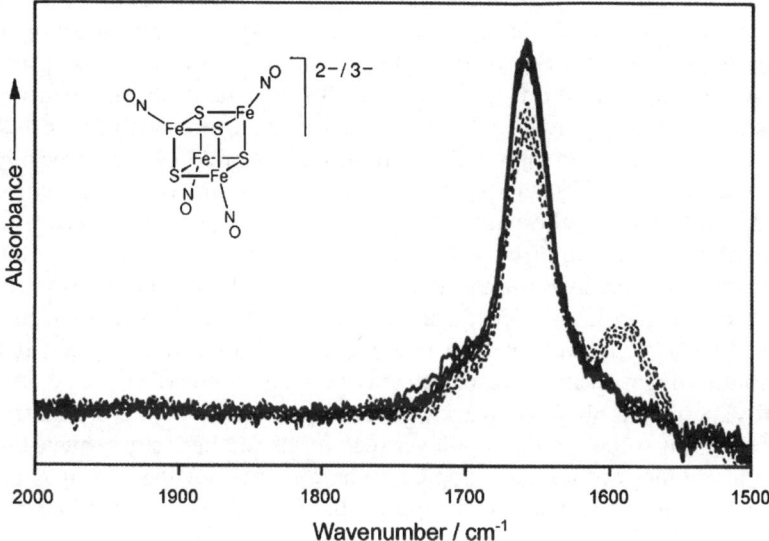

Figure 1.12 Sequential reduction for 4 s at –2.0 V (dashed line) and oxidation for 10 s at –1.0 V (solid line) cycles of $[Fe_4S_4(NO)_4]^{2-}$ in CH_2Cl_2 at –3 °C.

cannot be isolated it is often difficult to assign a charge state. For longer-lived species the current response obtained following a potential step may be correlated with the spectroscopic results and this may provide a sufficiently good estimate of the charge transferred per molecule. This approach relies on the distinction between the charge transferred into the thin-layer region relative to the "bulk" solution. Clearly this distinction is greater for thinner layers of solution ($< 20 \, \mu m$) where the rate of electrosynthesis is fast. An alternate strategy, that may be used even when there is a mixture of species present in solution, is to identify whether the intermediate species undergoes disproportionation reactions.

Reduction of **3S** proceeds through a short-lived intermediate, **3S$_A$**, *en route* to formation of three distinct products (**3S$_B$**, **3S$_C$**, and **3S$_D$**) where the relative concentrations of the latter products are sensitive to the identity of the solvent and the presence of free CO in solution. Conditions can be identified where the reaction proceeds quantitatively to $3S_B$ and in this case the coulometric results indicate a two-electron reduction of 3S to $3S_B$.

The determination of the charge state of $3S_A$ presents more serious difficulties. Whereas the electrochemical (cyclic voltammetry, rotating-disc electrode) response of the system indicates that the reduction proceeds in one-electron steps, this is not of itself sufficient to associate $3S_A$ with the one-electron reduced product. Indeed the average shifts of the terminal $v(CO)$ bands of $3S_A$ and $3S_B$ are similar and this suggests a similar electron richness at the iron centres. The proposition that the redox level of $3S_A$ is intermediate between those of 3S and $3S_B$ can be confirmed by its disproportionation. This may be established by generating $3S_A$ in a SEC experiment then switching the potentiostat to open circuit where the decay of the concentration of $3S_A$ with concomitant growth in the concentration of 3S flags disproportionation, whereas conversion of $3S_A$ solely to $3S_B$ would suggest a nonredox transformation. The concentration dependence of 3S, $3S_A$, and $3S_B$ in such an experiment is shown in Figure 1.13. The first spectrum recorded after switching the potentiostat to open circuit shows that the loss of 3S ceases. Crucially, as the concentration of $3S_A$ decays there is an increase in the concentration of both 3S and $3S_B$. An alternative explanation for the recovery of the concentration of 3S in the thin-layer region involves diffusion from the bulk solution. The contribution made to the concentration change of 3S from diffusion can be estimated from its (time)$^{1/2}$ dependence and provided that there is sufficient $3S_A$ in the thin layer at the time the potentiostat is switched to open circuit a contribution from chemical reactions may be readily discerned.[19]

For experiments that require a quantitative estimate of the concentration of the different species in the thin-layer region this can be easily achieved when the band profiles are not overlapping. Where this is not the case it is necessary to use more sophisticated approaches (*i.e.* chemometrics[28]). In this example, well-defined spectra of 3S, $3S_A$, and $3S_B$ can be obtained either directly or indirectly by the use of spectral subtraction. These spectra may be used as vectors in a multicomponent analysis and this may be performed using specialist or general plotting and statistical analysis programs such as Igor Pro(Wavemetrics). When approaches such as these are used it is important to check that the mass balances remain consistent through the experiment.

1.5.6   Reactions of Redox-Activated Complexes with Gaseous Substrates

The activation of gaseous molecules such as H_2, N_2, O_2, CO, and CO_2 is important both for environmental and economic reasons and the SEC study of catalysts for activation of these species is facilitated by cells capable of operation at elevated pressures. This feature may be incorporated into external reflection SEC designs and a schematic diagram of such a cell is shown in

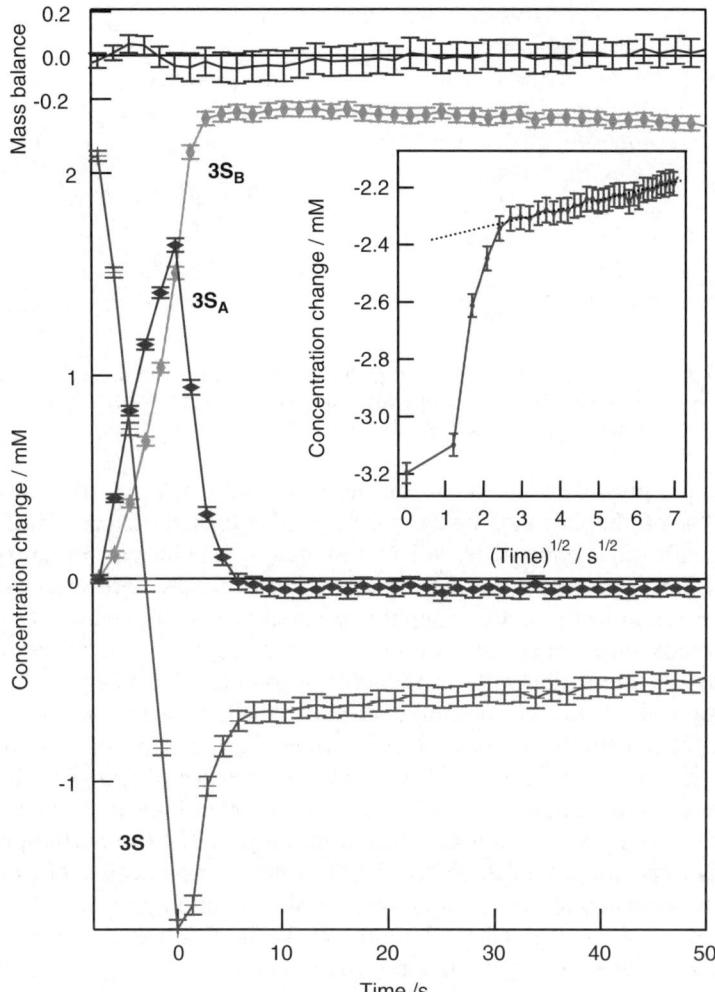

Figure 1.13 Time evolution of the concentration of **3S** and products formed following reduction for *ca.* 6 s, at this time ($t = 0$ s) the potentiostat was switched to open circuit. The concentration change is obtained by multicomponent analysis of the differential absorption IR spectra. The values of the concentration change for **3S** are offset by 2.1 mM. The error bars are drawn at the 3 e.s.d. level. The mass balance corresponds to the sum of the concentration changes. The inset shows the change in concentration of **3S** on a root timescale. Modified from ref. 19.

Figure 1.14.[16] The outer casing of the cell is constructed from stainless steel and the IR beam enters and departs through a 7-mm thick CaF_2 window. The solution is contained by a Teflon insert and this is fashioned so as to minimise the volume required for operation. The connections to the electrodes are most easily managed if a multielectrode assembly is used.

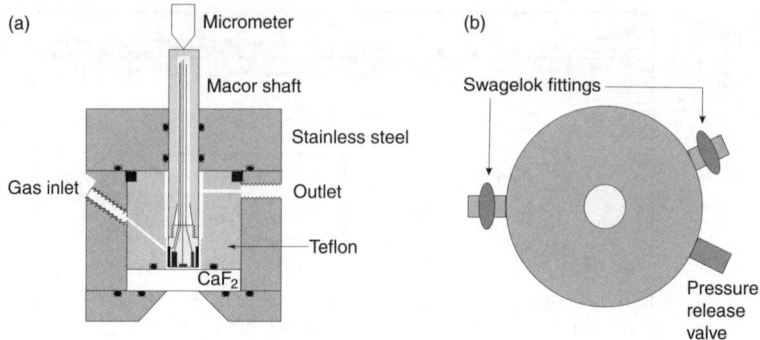

Figure 1.14   Cross-sectional (a) and top (b) schematic views of an external reflection
SEC cell suitable for operation at gas pressures to 1 MPa.[16] The details of
the multielectrode assembly are shown in Figure 1.3(c).

An upper pressure limit for operation of the cell is set at 1 MPa, this is less
than 10% of the calculated safe pressure. The change in concentration of
gaseous substrates effected by a 1-MPa pressure is sufficient for satisfactory
study of most systems. In addition to operation at elevated pressures the cell is
very well suited to the study of highly air-sensitive compounds.

The redox-state dependence of substrate binding to metal clusters is illustrated by CO interaction with the reduced forms of $[Fe_4S_4(SPh)_4]^{2-}$.[29] Under an
inert atmosphere the cluster undergoes two well-behaved one-electron reductions at potentials of –0.9 and –1.7 V (inset Figure 1.15), of these only the
second process is significantly affected by the presence of CO. The increased
cathodic current response and a lowered reversibility indicate significant CO
interaction with the tetra-anionic form of the cluster. The CO-partial-pressure-
dependent spectral changes obtained following the application of potentials
sufficient to generate the tetra-anion are shown in Figure 1.15. Since the
starting complex does not have bound CO the spectral changes in the $v(CO)$
region are dominated by growth bands (free CO absorbs weakly in the IR). The
absence of depletion bands in the differential absorption spectra confirms the
low reactivity of the complex in its resting state to CO. Reduction to the tetra-
anion is marked by the rapid growth of intense $v(CO)$ bands reflecting the
formation of iron carbonyl species, where the time and concentration dependence of the spectra indicate a complex reaction mixture. At high CO
partial pressures there is an increased prominence of the band at 1750 cm^{-1}
attributable to $[Fe(CO)_4]^{2-}$ and this signals extensive cluster decomposition and
accounts for the increased cathodic current at high CO partial pressures.

At elevated CO partial pressures the reaction between the cluster and CO
occurs following reduction to the trianion state. Two distinct products are
formed during the reaction where the initial stage of the reaction is dominated
by a product featuring a single $v(CO)$ band at 1812 cm^{-1}, at longer times a less
distinct set of bands centred on 1930 cm^{-1} grow slowly into the spectrum, sig-
nalling the appearance of a product, or products, with terminally bound CO
groups. Reoxidation of the thin layer gives a rapid set of spectral changes where

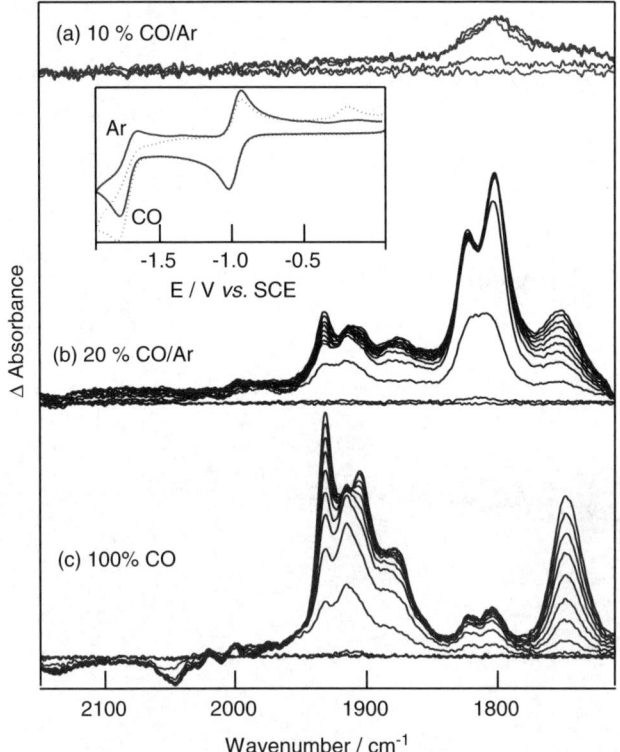

Figure 1.15 $Fe_4S_4(SPh)_4$ reduced at $-1.8\,V$ in CH_3CN in (a) 10, (b) 20, and (c) 100% CO/Ar gas mixtures at 0.36 MPa. The inset shows the cyclic voltammetry recorded under an inert atmosphere and for CO-saturated solvent.

oxidation of the species having the terminally bound CO groups leads to a product having bands at 1945, 2011 and $1966\,cm^{-1}$. Reoxidation of the species responsible for the $1812\text{-}cm^{-1}$ band, most likely having a bridging CO group, appears to result in dissociation of the CO group from the (di)iron centre. This may have relevance for understanding biological systems, such as the 4Fe-3S half of FeMoco, the N_2-fixing centre in the nitrogenase enzyme, also binds CO in a bridging mode at low CO partial pressures with reversible dissociation of CO on reoxidation (Figure 1.16).[30]

1.5.7   Biomolecules

Application of IR-SEC to biomolecules brings additional challenges and these are mostly related to the availability of material, its maximum concentration, interference from other vibrational modes of the macromolecule, and the need to work in highly absorbing solvents such as water. The water solubility of most salt window materials means that studies must be conducted using insoluble

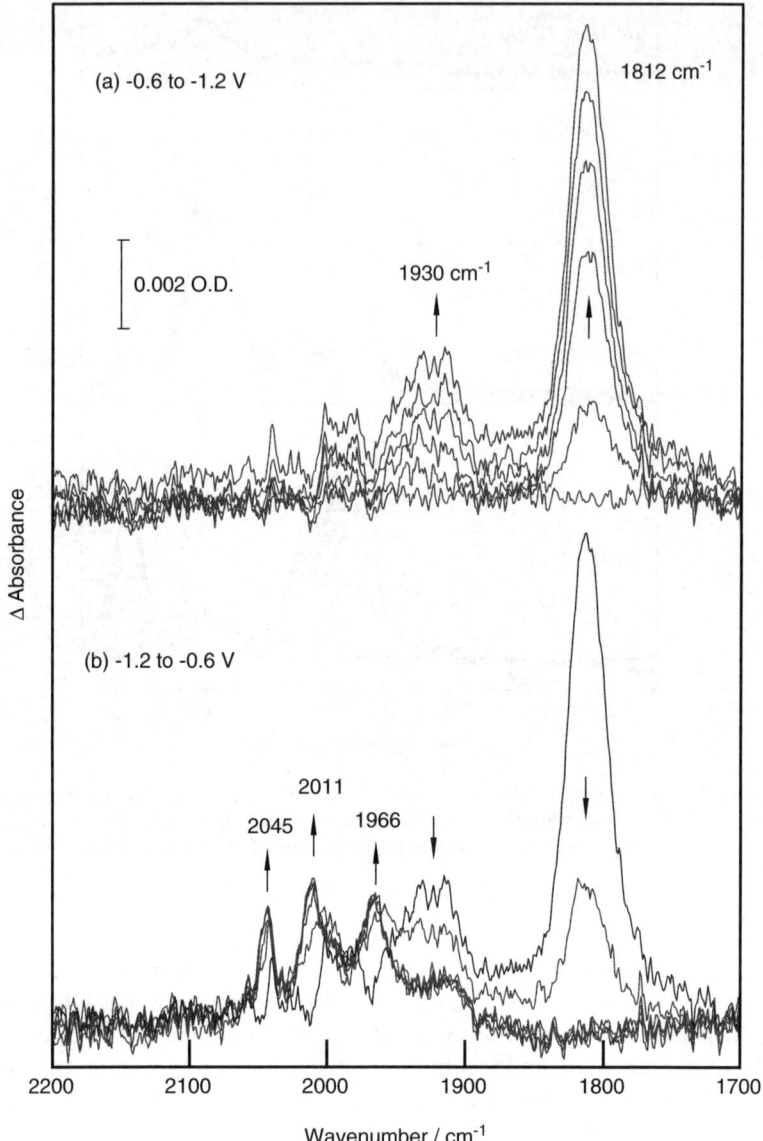

Figure 1.16 SEC of $[Fe_4S_4(SPh)_4]^{2-}$ in CH_3CN at 0.36 MPa of CO. (a) Reduction of the trianion to the tetra-anion and (b) reoxidation to the trianion, 1.2 s between spectra.

materials and these tend either to have a high wave number cutoff (*e.g.* CaF$_2$) or to give low transmission due to high reflection losses (*e.g.* ZnSe). Despite having intense, broad absorptions at *ca.* 1600 and 3400 cm^{-1} water does present a useful spectral window, particularly in experiments with a short optical pathlength, and this may be extended by the use of D_2O or H_2O/D_2O mixtures.

Although characteristic IR bands or regions have been identified for amino acid side chains or protein folding motifs (α-helix *vs.* β-sheet for example), interpretation of the spectra for whole proteins is extremely complex. OTTLE cells of the type reported by Moss *et al.*[10] (Figure 1.1(b)) have been used in both UV-Vis and FTIR studies of redox-linked conformational changes in a number of heme proteins including myoglobin and hemoglobin,[31] cytochrome c_3,[32] cytochrome c oxidase[33] and the cytochrome bc_1 complex.[34] Difference spectra show redox-dependent changes in vibrational modes arising not only from the peptide backbone and side chains, but also from the porphyrin rings, providing information on local and more distant structural changes resulting from redox transitions at the heme cofactor centres. For the small, soluble, electron-transfer protein cytochrome c_3, diffusion is sufficiently efficient that direct, rather than mediated, electrochemistry is possible, but for work with larger proteins, a cocktail of redox mediators is added to the protein solution to improve electron-transfer rates. The gold minigrid working electrode is generally coated with a thiolate surface modifier such as cysteamine to protect the proteins from denaturation at the bare metal.

As with investigations of small molecules, the IR-SEC technique applied to proteins is most powerful when addressing redox changes at metal centres bound by ligands having vibrational modes that have a large IR cross section and are well separated in energy from other vibrational modes of the system. By far the most extensive application has been in studies of hydrogenase enzymes that incorporate CO and CN⁻ ligands at their di-iron or nickel-iron active sites and undergo a complex series of redox-potential-induced interconversions between active and inactive states. These studies, also using cells of the Moss design, have permitted identification of the conditions that favour formation of each state and the potentials at which redox interconversion occur. From the pH dependence of the Nernst plots it is also possible to identify the extent to which the redox steps are coupled to proton transfer. In this case a cocktail of redox mediators are added to the protein solution to improve electron transfer to the bulky protein molecules. Critically, control of the potential allows generation of well-defined oxidation states of the enzyme that are difficult to achieve by other means. In conjunction with IR spectroscopic studies in conventional solution cells (in which the pH and partial pressure of H_2 and sample history can be readily varied) the IR-SEC studies have filled in gaps in the catalytic schemes deduced from low-temperature EPR studies where EPR-silent states cannot be characterised. For example, it was assumed that reduction of the paramagnetic inactive, oxidised states of [NiFe]-hydrogenases, known as ready and unready, respectively, led to a single species until the reduced products were found to have subtly different IR signatures associated with the active site CO and CN⁻ ligands.[35] More than ten states of the [NiFe]-hydrogenases have now been identified by IR-SEC, and the kinetics for interconversions between many of these states revealed at physiologically relevant temperatures.[36] For the [FeFe]-hydrogenases, IR-SEC confirmed evidence gleaned from the crystallographic characterisation of distinct samples that a CO ligand that bridges the two metals in the catalytically active oxidised state [v(CO) 1802 cm^{-1}] moves to a terminal position in the reduced enzyme.[37] In analysis of the spectra of hydrogenases, a multipoint "baseline" is typically subtracted to distinguish the tiny ligand vibrations from the large solvent and protein background and this manipulation leaves the v(CO) and v(CN) bands with an artificially "triangular" shape. The v(CO) and v(CN) bands are then sufficiently intense (for millimolar hydrogenase samples) that absolute rather than difference spectra are presented. A review of the SEC studies of hydrogenase enzymes has recently been published.[38]

1.6 Integration with other Electrochemical and Spectroscopic Techniques

The delineation of the chemistry following redox activation is dependent on the collection of as detailed information as possible on all the species formed during the reaction. While the previous section has highlighted the information able to be extracted from IR-SEC studies, it is important to recognise that in most cases this represents a small fraction of the detail needed to characterise the chemistry of the system. In addition to the application of a broad range of complementary spectroscopic techniques, described in other chapters of the book, electrochemical simulations can assist in the identification of mechanistically important species. An example of the complementary nature of the electrochemical and SEC approaches is evident in the recent study of electrocatalytic proton reduction in the presence of $Fe_2(\mu\text{-pdt})(CO)_6$.[19] It is the molecular structure, however, that is the key to the development of an *in-silico* description of the reaction chemistry. A useful strategy for obtaining this information from transiently stable species involves the use of XAFS (X-ray absorption fine structure) techniques, however, in such investigations it is critical to be able to characterise the sample under investigation. In this regard SEC techniques can play a central role in the identification of target species and the validation of the samples subject to XAFS analysis. While a number of *in-situ* designs of SEC XAFS cells have been reported,[39] the duration of XAFS data collection is generally long (>1 h) it is difficult to maintain the reactive samples in the required state. In this case it can be more effective to generate the desired species and freeze-quench the sample for low-temperature XAFS data collection. Continuous-flow electrosynthesis approaches may be used for this purpose, where the form of the sample is verified spectroscopically, in line, prior to freezing.[19,20,40] Experiments of this sort can provide reasonably accurate (typically to 0.02 Å) core geometry and this can be used to build the starting geometry for DFT structure modelling. Good agreement between the final DFT calculated structure and spectroscopy with the XAFS-determined bond lengths and the SEC determined spectroscopy can then provide a strong basis for assignment of the structure.[23,41]

1.7 Summary

IR-SEC techniques are extremely effective for the study of the redox-activated chemistry of a broad range of chemical, and biochemical, systems that display a broad range of electrochemical behaviour. The structural detail added to the electrochemical response facilitates the building and testing of reaction schemes and mechanisms. While further advances in cell design are to be expected, the field is in a relatively mature state with well-developed designs for cells operating under a range of physical conditions of temperature and pressure and able to accommodate sample volumes of the order of 1 μL. Given access to workshop facilities these cells can be fabricated at modest cost and excellent results can be obtained from most commercial FTIR spectrometers. Improvements in computational methods, electrochemical simulation and *ab-initio* calculations, when integrated with the spectroscopy of the intermediate species offer the promise of a far more detailed analysis of the complex reaction chemistry that follows redox activation. These details underpin advances in our understanding of the catalytic reactions that economically and biologically sustain life.

Acknowledgements

SPB gratefully acknowledges the contribution to his work in this area by the workshop staff at University College London and the University of Melbourne, most particularly Frank Ambrose and Marino Artuso, and to the contributions of a number of undergraduate and postgraduate students, particularly Slava Ciniawskii, David Truccolo, Chris Shaw, Karolina Padusinski, Michael Cheah, and Mark Bondin. Thanks are extended to Professor Ralph Cooney, who encouraged us to develop external reflectance SEC cells, and to Professor Chris Pickett for his brilliant insights into the chemistry, electrochemistry and spectroscopy of many of these systems. The SERC (UK) and ARC (Australia) are thanked for financial support.

References

1. A. J. Bard and L. R. Faulkner, '*Electrochemical Methods: Fundamentals and Applications*', 2nd ed., Wiley, New York, 2001; '*Spectroelectrochemistry: Theory and Practice*', R. J. Gale, ed., Plenum Press, New York, 1988; '*Electrochemical Interfaces: Modern Techniques for In-Situ Interface Characterisation*', H. D. Abruña, ed., VCH, New York, 1991; S. Pons, J. K. Foley, J. Russell and M. Seversen, *Mod. Aspects Electrochem.*, 1986, **17**, 223.
2. R. L. Birke and J. R. Lombardi, in '*Spectroelectrochemistry: Theory and Practice*', R. J. Gale, ed., Plenum Press, New York, 1988; M. I. Boyer, S. Quillard, M. Cochet, G. Louarn and S. Lefrant, *Electrochim. Acta,* 1999, **44**, 1981; Q. Hu and A. S. Hinman, *Anal. Chem.*, 2000, **72**, 3233; P. Damlin, C. Kvarnstrom, A. Petr, P. Ek, L. Dunsch and A. Ivaska, *J. Solid State Electrochem.*, 2002, **6**, 291; T. Itoh and R. L. McCreery, *J. Am. Chem. Soc.*,

2002, **124**, 10894; K. Crowley and J. Cassidy, *J. Electroanal. Chem.*, 2003, **547**, 75; L. Kavan, M. Kalbac, M. Zukalova, M. Krause and L. Dunsch, *ChemPhysChem*, 2004, **5**, 274; E. Alessio, S. Daff, M. Elliot, E. Iengo, L. A. Jack, K. G. Macnamara, J. M. Pratt and L. J. Yellowlees, *Trends Mol. Electrochem.*, 2004, 339.

3. M. Iwaki, G. Yakovlev, J. Hirst, A. Osyczka, P. L. Dutton, D. Marshall and P. R. Rich, *Biochemistry*, 2005, **44**, 4230.
4. W. N. Hansen, R. A. Osteryoung and T. Kuwana, *J. Am. Chem. Soc.*, 1966, **88**, 1062.
5. M. Witek, J. Wang, J. Stotter, M. Hupert, S. Haymond, P. Sonthalia, G. M. Swain, J. K. Zak, Q. Chen, D. M. Gruen, J. E. Butler, K. Kobashi and T. Tachibana, *J. Wide Bandgap Mater.*, 2002, **8**, 171.
6. J. Stotter, J. Zak, Z. Behler, Y. Show and G. M. Swain, *Anal. Chem.*, 2002, **74**, 5924.
7. J. McEvoy and J. S. Foord, *Electrochim. Acta*, 2005, **50**, 2933.
8. R. W. Murray, W. R. Heineman and G. W. O'Dom, *Anal. Chem.*, 1967, **39**, 1666.
9. M. Krejcik, M. Danek and F. Hartl, *J. Electroanal. Chem.*, 1991, **317**, 179.
10. D. Moss, E. Nabedryk, J. Breton and W. Mäntele, *Eur. J. Biochem.*, 1990, **187**, 565.
11. K.-S. Yun, S. Joo, H.-J. Kim, J. Kwak and E. Yoon, *Electroanalysis*, 2005, **17**, 959.
12. B. Beden and C. Lamy, in *'Spectroelectrochemistry: Theory and Practice'*, R. J. Gale, ed., Plenum Press, New York, 1988.
13. M. E. Rosa-Montanez, H. D. Jesús-Cardona and C. R. Cabrera-Martínez, *Anal. Chem.*, 1998, **70**, 1007; C. Geskes and J. Heinze, *J. Electroanal. Chem.*, 1996, **418**, 167.
14. S. P. Best, R. J. H. Clark, R. C. S. McQueen and R. P. Cooney, *Rev. Sci. Instrum.*, 1987, **58**, 2071.
15. I. S. Zavarine and C. P. Kubiak, *J. Electroanal. Chem.*, 2001, **495**, 106; D. Kardash, J. Huang and Korzeniewski, *J. Electroanal. Chem.*, 1999, **476**, 95.
16. S. J. Borg and S. P. Best, *J. Electroanal. Chem.*, 2002, **535**, 57.
17. A. M. Bond, B. S. Grabaric and J. J. Jackowski, *Inorg. Chem.*, 1978, **17**, 2153.
18. A. Vallat, M. Person, L. Roullier and E. Laviron, *Inorg. Chem.*, 1987, **26**, 332.
19. S. J. Borg, T. Behrsing, S. P. Best, M. Razavet, X. Liu and C. J. Pickett, *J. Am. Chem. Soc.*, 2004, **126**, 16988.
20. M. H. Cheah, S. J. Borg, M. I. Bondin and S. P. Best, *Inorg. Chem.*, 2004, **43**, 5635.
21. S. J. Borg, M. I. Bondin, S. P. Best, M. Razavet, X. Liu and C. J. Pickett, *Biochem. Soc. Trans.*, 2005, **33**, 3.
22. C. Greco, G. Zampella, L. Bertini, M. Bruschi, P. Fantucci and L. D. Gioia, *Inorg. Chem.*, 2007, **46**, 108.

23. S. J. Borg, J. W. Tye, M. B. Hall and S. P. Best, *Inorg. Chem.*, 2007, **46**, 384.
24. S. Zilberman, E. I. Stiefel, M. H. Cohen and R. Car, *J. Phys. Chem. B*, 2006, **110**, 7049; J. W. Tye, M. Y. Darensbourg and M. B. Hall, *J. Comput. Chem.*, 2006, **27**, 1454; S. Zilberman, E. I. Stiefel, M. H. Cohen and R. Car, *Inorg. Chem.*, 2006, **45**, 5715.
25. S. P. Best, S. A. Ciniawsky and D. G. Humphrey, *J. Chem. Soc., Dalton Trans.*, 1996, 2945.
26. A. Bewick and S. Pons, in *"Advances in Infrared and Raman Spectroscopy"*, R. J. H. Clark and R. E. Hester, eds., Wiley, New York, Vol. 12, 1985, p. 1.
27. S. A. Ciniawsky, 'Infrared Spectroelectrochemistry of Coordination Compounds', PhD, University College London, London, 1992.
28. J. C. Miller and J. N. Miller, *"Statistics and Chemometrics for Analytical Chemistry"*, Prentice Hall, Harlow, 4th edn, 2000; M. J. Adams, *'Chemometrics in Analytical Spectroscopy'*, Royal Society of Chemistry, Cambridge, 2nd edn, 2004.
29. K. A. Vincent, 'Elucidation of the Chemistry of FeMoCo', PhD, The University of Melbourne, Melbourne, 2002.
30. C. J. Pickett, K. A. Vincent, S. K. Ibrahim, C. A. Gormal, B. E. Smith and S. P. Best, *Chem. – Eur. J.*, 2003, **9**, 76.
31. D. D. Schlereth and W. Mäntele, *Biochemistry*, 1992, **31**, 7494.
32. D. D. Schlereth, V. M. Fernández and W. Mäntele, *Biochemistry*, 1993, **32**, 9199.
33. P. Hellwig, J. Behr, C. Ostermeier, O.-M. H. Richter, U. Pfitzner, A. Odenwald, B. Ludwig, H. Michel and W. Mäntele, *Biochemistry*, 1998, **37**, 7390.
34. F. Baymann, D. E. Robertson, P. L. Dutton and W. Mäntele, *Biochemistry*, 1999, **38**, 13188.
35. A. L. de Lacey, E. C. Hatchikian, A. Volbeda, M. Frey, J. C. Fontecilla-Camps and V. M. Fernandez, *J. Am. Chem. Soc.*, 1997, **119**, 7181.
36. B. Bleijlevens, F. A. Broekhuizen, A. L. Lacey, W. Roseboom, V. M. Fernandez and S. P. J. Albracht, *J. Biol. Inorg. Chem.*, 2004, **9**, 743; A. L. de Lacey, V. M. Fernandez, M. Rousset, C. Cavazza and E. C. Hatchikian, *J. Biol. Inorg. Chem.*, 2003, **8**, 129.
37. Y. Nicolet, A. L. de Lacey, X. Vernède, V. M. Fernandez, E. C. Hatchikian and J. C. Fontecilla-Camps, *J. Am. Chem. Soc.*, 2001, **123**, 1596.
38. S. P. Best, *Coord. Chem. Rev.*, 2005, **249**, 1536.
39. C. Hennig, J. Tutschku, A. Rossberg, G. Bernhard and A. C. Scheinost, *Inorg. Chem.*, 2005, **44**, 6655; D. A. Smith, R. C. Elder and W. R. Heineman, *Anal. Chem.*, 1985, **57**, 2361; H. D. Dewald, J. W. Watkins, II, R. C. Elder and W. R. Heineman, *Anal. Chem.*, 1986, **58**, 2968; D. H. Igo, R. C. Elder and W. R. Heineman, *J. Electroanal. Chem.*, 1991, **314**, 45; N. R. S. Farley, S. J. Gurman and A. R. Hillman, *Electrochem. Commun.*, 1999, **1**, 449; M. Giorgetti, I. Ascone, M. Berrettoni, P. Conti, S. Zamponi

and R. Marassi, *J. Biol. Inorg. Chem.*, 2000, **5**, 156; M. R. Antonio, L. Soderholm, C. W. Williams, J.-P. Blaudeau and B. E. Bursten, *Radiochim. Acta*, 2001, **89**, 17; A. Cognigni, I. Ascone, S. Zamponi and R. Marassi, *J. Synch. Rad.*, 2001, **8**, 987.

40. M. I. Bondin, G. Foran and S. P. Best, *Aust. J. Chem.*, 2001, **54**, 705.
41. M. I. Bondin, S. J. Borg, M.-H. Cheah and S. P. Best, *Radiat. Phys. Chem.*, 2006, **75**, 1878; M. I. Bondin, S. J. Borg, M. H. Cheah, G. Foran and S. P. Best, *Aust. J. Chem.*, 2006, **59**, 263.

CHAPTER 2

UV-Vis Spectroelectrochemistry of Selected Iron-Containing Proteins

SURAJ DHUNGANA[a] AND ALVIN L. CRUMBLISS[b]

[a] Laboratory of Respiratory Biology, National Institute of Environmental Health Sciences, NIH, DHHS, 111 T.W. Alexander Drive, PO Box 12233, Research Triangle Park, NC 27709, USA; [b] Department of Chemistry, Duke University, Durham, NC 27708, USA

2.1 Introduction and Chapter Scope

In this chapter we will explore the application of spectroelectrochemistry to selected iron-containing proteins. Our choice of chemical systems is dictated by a number of factors and we will include a discussion of both nonheme- and heme-containing proteins. We emphasise iron due to its central importance in biology. Iron is an essential nutrient for virtually every living cell, yet the iron paradox is that this essential element is also toxic.[1] Consequently, the location, concentration, and chemical environment of iron in a biological system must be carefully controlled. Proteins play an integral part in iron transport, storage, homeostasis and biological function, and consequently investigations of iron proteins are of significance. At the centre of iron's essential functions in biology, and its toxic effects, is the ease and propensity with which it undergoes redox reactions. Hence, a detailed investigation of the redox reactions of iron is of major significance.

As a transition metal with a partially filled d-subshell of electrons, the UV-Vis spectroscopy of iron is a sensitive measure of both the first coordination shell environment (ligand field), and oxidation and spin state of the metal.[2] Ligand field bands are an excellent experimental handle for spectroelectrochemistry, and in the case of heme proteins the Soret band provides an additional probe for

Spectroelectrochemistry
Edited by Wolfgang Kaim and Axel Klein
© Royal Society of Chemistry, 2008

oxidation state change. Spectroelectrochemistry is an ideal tool to explore the bioinorganic redox chemistry of iron as it relates to iron proteins.

Aspects of the aqueous and coordination chemistry of iron provide specific opportunities and challenges associated with spectroelectrochemical investigations of iron proteins. Depending on the coordination environment of iron, an oxidation state change from +3 to +2 can significantly change the binding-site affinity (thermodynamics of binding) as well as first coordination shell turnover kinetics (kinetics of dissociation).[2–4] This may be an important feature of the mechanism for iron transport and release, but it also adds a level of complexity to spectroelectrochemical data workup and interpretation, as discussed in Section 2.3. Oxidation-state change may also effect spin-state changes between high- and low-spin electron configurations. Further complications can arise from acid-base chemistry associated with iron in an aqueous environment, where iron behaves as a Bronsted acid, leading to hydrolysis and the formation of multiple-hydroxyl containing species that may precipitate.[4] In certain instances, these hydrolysis reactions must be considered in spectroelectrochemical data analysis.

The most common oxidation states of iron in biology are +2 and +3, although the +4 state is also found, usually as an oxo species when iron is in a porphyrin environment (*e.g.* in cytochrome P_{450}). Here, we will focus on the spectroelectrochemistry studies that investigate the driving force and mechanism for redox changes between the +2 and +3 oxidation state. The driving force for this change ($E_{1/2}$) is strongly dependent on the groups directly bound to iron (first coordination shell), as illustrated in Scheme 2.1. In addition to this first coordination shell effect on $E_{1/2}$, there is a second coordination shell and a medium effect on $E_{1/2}$. This is particularly important in metalloproteins, as the protein and protein structure controls the second coordination shell through variations in residues present in the binding pocket. Protein folding also controls the exposure of the iron to the external environment. These influences on $E_{1/2}$ are also illustrated in Scheme 2.1 and discussed in Section 2.4.

Finally, in proteins with multiple binding sites (*e.g.* transferrin), or in complex protein assemblies with multiple subunits (*e.g.* hemoglobin) there is an opportunity for allosteric interactions between redox-active iron sites. These allosteric interactions are readily probed by the spectroelectrochemical technique, as illustrated in Sections 2.3.2.2 and 2.4.2.3.

The protein environment also provides a challenge in investigating redox processes since it acts as a barrier to electron flux in and out of the iron centre. This is due to the steric bulk of the folded polypeptide chain that can also act as an electronic insulator. As there is distance dependence in the outer-sphere electron transfer at the protein/solid electrode interface, these heterogeneous electron-transfer rates tend to be slow. This problem is overcome by the use of a small-molecule mediator as discussed in Section 2.2.3.

In this chapter we describe the spectroelectrochemical technique as it applies to nonheme and heme iron proteins containing single or multiple iron-binding sites, and as single protein or multiple subunit assemblies. In this discussion, we will address the influence of the following on the spectroelectrochemical

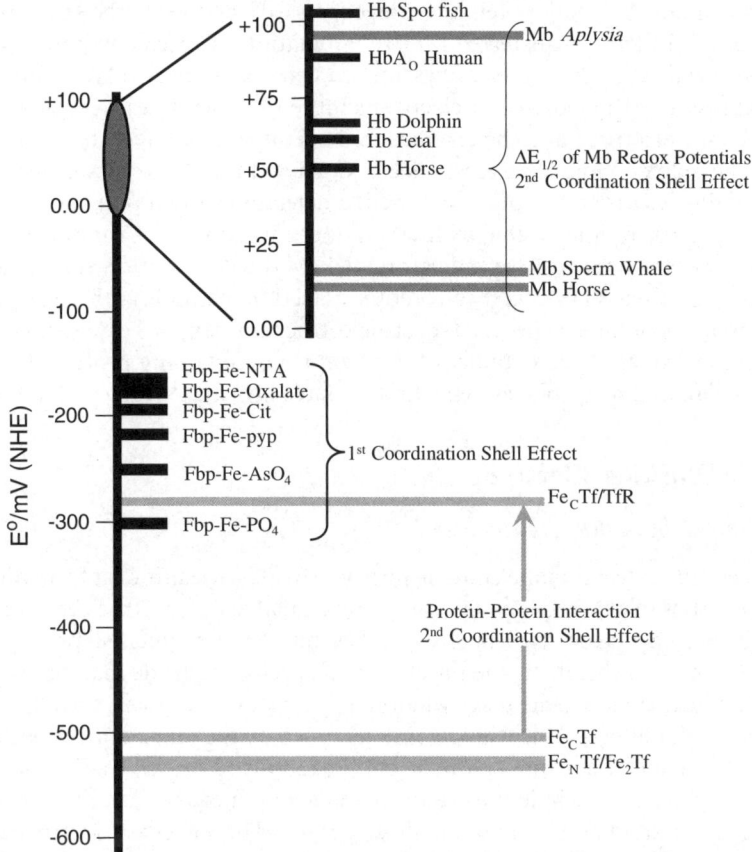

Scheme 2.1   Summary of redox potentials for various heme and non heme iron proteins, which illustrates first and second coordination shell effects.

experiment: first and second coordination shell and medium effects on $E_{1/2}$, iron dissociation and nonideality/reversibility, cooperativity, and receptor binding. While there is a wealth of excellent data available in the literature to illustrate these points, for reasons of space, we have limited our choice of examples to those from our own laboratory.

2.2   Experimental Design

Spectroelectrochemistry utilises the difference in the spectroscopic signature between the oxidised and the reduced form of a system to probe its redox properties. An incremental application of potential gradually changes the spectral profile of a system corresponding to the changes in the population of oxidised and reduced species. These spectral changes as a function of applied potential allow for the determination of various redox properties including, the half-potential, $E_{1/2}$ (all potentials are expressed relative to NHE), and the number of electrons transferred, n. A combination of three components must be specifically chosen for each system studied: i) an electrode material and configuration; ii) a mediator or electron-shuttle that facilitates electron transfer between the electrode and the chemical redox couple of interest (necessary only when the redox couple of interest is slow to react at the electrode surface, which is generally the case for proteins); and iii) a detection system that can probe the relative concentrations of the oxidised and/or reduced state of the system. Since UV-Vis spectra are sensitive indicators of Fe^{3+}/Fe^{2+} oxidation state changes, we will only discuss UV-Vis spectroscopic detection methods in this chapter. As we present these three aspects of spectroelectrochemistry, our discussion will be geared towards the redox studies of different iron-containing proteins found in mammalian and microbial systems that we have investigated in our laboratory.

2.2.1 Working Electrode

2.2.1.1 Electrode Materials

Advances in conventional electrochemistry provide a wealth of information on the properties of electrode materials that are available for spectroelectrochemical experiments. Spectroelectrochemistry relies on the fundamental principles of electrochemistry; therefore, the selection of a specific electrode material is based on the effective electrochemical window it provides in a given solvent system (solvent with buffers). This window is determined by the range of over potentials the electrode material exhibits for solvent breakdown. For example, a gold mesh electrode provides a window to study systems with more negative redox potentials compared to a platinum mesh electrode. The effective electrochemical window of a material can often be extended to access the redox potential of a biological system by modifying its surface.[5] Our studies on human transferrin required us to extend the limits of a gold mesh electrode to more negative potentials and this was achieved by forming a mercury/gold amalgam.[6] Further discussion of the electrochemical window exhibited by different electrode materials in different solvent systems can be found in various electrochemistry monographs.[7,8] Additionally, electrode materials for UV-Vis spectroelectrochemistry are required to be optically transparent since monochromatic light is used to probe the oxidation states of the proteins. Gold, gold/mercury amalgam and platinum mesh electrodes all allow the transmission of monochromatic light and are ideal for the spectroelectrochemistry experiments discussed in this chapter.

2.2.1.2 Preparation and Cleaning of the Working Electrode

A general idea of the anticipated redox range of the system under investigation is usually adequate to determine the ideal optically transparent electrode material; however, the success of the spectroelectrochemical experiment may be determined by the pristine condition of the electrode. The working electrode, whether platinum or gold, must be carefully cleaned between each experiment to obtain reproducible results with high accuracy. Methods to clean various electrode materials are well described in the literature and can be directly applied to the cleaning of the optically transparent, Pt and Au mesh electrodes.[9,10] The redox potential of a given protein system can often be very negative and this requires a special electrode that permits the application of potentials below -500 mV without the evolution of H_2 gas, resulting from the H_2O solvent breakdown. Amalgamation of a Au electrode with Hg extends the operational electrochemical window and allows for the application of more negative potentials to the system. The amalgamation procedure consists of electrochemical deposition of Hg ($E_{dep} = 544$ mV) from 0.05 M $Hg(NO_3)_2$ in 2.0 M HNO_3 for 10–15 min.[6]

2.2.2 Optically Transparent Thin-Layer Cell (OTTLE)

Various spectrophotometric methods are suitable for monitoring the spectral changes associated with the changes in the redox state of the system brought about by the applied potentials. The electronic properties of iron-proteins are well defined in the UV-Vis spectral region, allowing for the absorbance bands to be used as spectrophotometric handles during the reduction/oxidation process. The use of an optically transparent thin-layer electrode (OTTLE) is ideal for UV-Vis spectral monitoring of redox profiles as it requires only a very small sample volume and allows for maximum exposure of the sample to the electrode surface. The electrochemistry component of spectroelectrochemistry is in fact bulk electrolysis with a volume of 30–60 μL when confined within an OTTLE cell (0.02–0.04 cm pathlength) as described here. This results in a very thin diffusion layer, which is ideal for rapid electrolysis. This setup reduces the time required for the system to achieve equilibrium after each applied potential and significantly cuts down the overall time required to complete the experiment, thus preventing the protein from being exposed to an electrical potential for extended periods, which can cause denaturation.

A schematic of an OTTLE cell designed for spectroelectrochemistry experiments using UV-Vis spectra as a detection mode is illustrated in Figure 2.1. These experiments can be carried out using a two- or three-electrode system (working, reference, and auxiliary); the OTTLE cell construction and the electrode arrangement discussed here are for a three-electrode system. The cell construction involves the use of a conventional acrylamide disposable or quartz spectrophotometer cuvette with 1 cm pathlength. The spectral region of interest determines the choice of the cuvette material. A 52 mesh platinum or a gold gauze working electrode (~ 0.5 cm $\times$ 2 cm) is sandwiched between the inside of the cuvette optical face and a piece of silica or quartz glass aligned parallel to the optical face and perpendicular to the light path. The sandwiched electrode setup is held in place by a small Tygon® spacer, such that its location at the bottom of the cuvette does not interfere with the light path. The optical

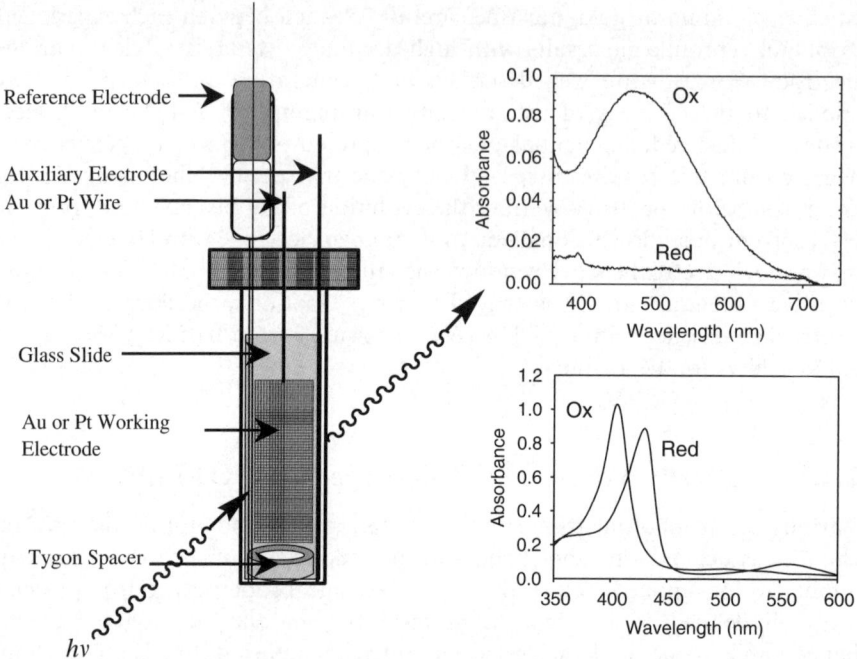

Figure 2.1 Schematic representation of the OTTLE cell and representative spectra obtained in a spectroelectrochemistry experiment involving ferric-binding protein (top insert) and myoglobin (bottom insert).

pathlength of the resulting OTTLE cell assembly is in the range of 0.02–0.04 cm and is determined by the electrode material and mesh size, and the tightness with which the electrode is sandwiched against the wall of the cuvette. Once the working electrode is fixed, the cuvette is made airtight by securely placing a rubber septum over the top. A series of vacuum pump/inert gas (N_2 or Ar) flush cycles renders the cell free from O_2 and ready for electrode and sample loading.

The reference and auxiliary electrodes, required to complete the electrochemical cell, need special preparation prior to being fitted into the OTTLE cell. A Ag/AgCl reference microelectrode is fixed into the broad end of a Pasteur pipette and a salt bridge within the pipette is created by plugging the bottom with agar gel prepared in 0.2 M KCl/buffer (the same buffer as is used to prepare the sample protein solution). An auxiliary electrode, 2×50 mm platinum or 1×50 mm gold wire, is inserted in one corner of the OTTLE cell touching the bottom of the cell. Although not required, often it is a good idea to compartmentalise the auxiliary electrode inside a capillary tube salt bridge similar to the reference electrode. The salt-bridge solution used to establish the connection in both reference and auxiliary electrodes contains the buffer and at least 0.2 M background electrolyte. After an inert atmosphere is established inside the cell, the reference and the auxiliary electrodes are inserted into the two opposite corners of the OTTLE cell by making a small puncture in the septum covering the cuvette. The electrodes stand against the glass or quartz piece (holding the working electrode against the wall) and they are additionally held in place by the Tygon® spacer at the bottom and the rubber septum at the top. They should be fixed tightly so that they sandwich the working electrode firmly against the wall of the cuvette. A tightly held working electrode arrangement narrows the distance between the cuvette wall and the glass/quartz piece, so when the sample is loaded, the liquid rises up on to the working electrode easily *via* capillary action. The circuit between the three-electrode system is completed *via* the sample solution pool at the bottom of the cuvette. Once the three-electrode setup is established under N_2 or Ar positive pressure the sample needs to be carefully placed at the bottom of the cuvette using a syringe.

Although this design of the OTTLE cell may appear cumbersome, it has several advantages:

- The ability to easily construct and dismantle the OTTLE cell allows one to tailor the cell to the needs of a specific experiment and thoroughly clean the system (cell and electrodes) once the experiment is complete.
- Electrolysis can be carried out in very small sample volumes (350–500 μL total volume), which allow for rapid equilibration at each applied potential and makes the time-dependent denaturation of the protein less of a concern.
- A very thin layer of working solution and the minigrid electrode significantly reduces the diffusion layer and decreases the ohmic-level drop.
- Anaerobic conditions can be easily maintained inside the cell.

2.2.3 Mediator

A primary challenge in protein electrochemistry is to facilitate efficient heterogeneous electron transfer between the protein and the electrode. Heterogeneous electron transfer in most proteins is sluggish compared to small molecules and thus requires longer application of a specific potential, and may result in requiring a significant over potential. Long exposure times can often cause the protein to denature and aggregate at the surface of the electrode. Such an aggregation disrupts the heterogeneous electron transfer and gives an inaccurate redox profile. To alleviate these problems, an electrochemical mediator is used to "shuttle" electrons between the electrode and the redox-active site within the protein. To serve its function, a mediator needs to be relatively small, able to undergo rapid reversible electron transfer, and have a redox potential ($E_{1/2}$) in the general region of the expected redox potential of the protein.

Mediator-facilitated reduction/oxidation of a protein requires a rapid turnover of the reduced and the oxidised form of the mediator for both the heterogeneous electron transfer at the electrode and homogeneous electron transfer with the redox-active protein. Additional factors need to be considered while selecting a mediator for a given protein; a few important characteristics of good mediators for spectroelectrochemistry experiments are as follows:

- The mediator must be spectroscopically silent in the region of interest (both in oxidised and reduced state) so that it does not interfere with the UV-Vis absorbance profile of the protein in the region used to monitor the reaction.
- The mediator should not interact with the protein and alter the redox profile. Charged mediators can bind to oppositely charged patches on the protein and can hinder efficient electron transfer, as well as change the redox behaviour of the protein. For example, an allosteric site with positively charged residues in hemoglobin can bind negatively charged mediators and alter any cooperative interactions associated with electron transfer.[11]
- The redox potential of the mediator must be close to the expected redox potential of the protein under investigation. This reduction half-potential requirement for a mediator can be estimated by the following relationship:

$$\left| E_m^0 - E_p^0 \right| \leq \frac{2RT}{n_m F}, \tag{2.1}$$

where E_m^0 is the redox potential of mediator, E_p^0 is the redox potential of the protein, and n_m is the number of electrons transferred for the mediator.
- Kinetically, the mediator should be able to undergo rapid reversible electron transfer (heterogeneous and homogeneous) at the electrode surface and the redox-active site in the protein. At any given time a sufficient amount of the oxidised and reduced forms of mediator should be present in solution with the concentration of the mediator in excess over the concentration of the protein to facilitate favourable electron-transfer kinetics.

Typical mediators used for iron proteins include ferricyanide,[5,12–14] methylviologen,[15,16] substituted quinones,[17] and hexammineruthenium(III).[11,18]

2.3   Data Analysis

Spectroscopic data obtained from spectroelectrochemical experiments require careful and case-specific analysis. The Fe^{3+}/Fe^{2+} redox couple has a unique role in different iron-containing proteins. It is hypothesised that the mammalian iron-transport protein transferrin uses the Fe^{3+}/Fe^{2+} redox couple as a switch that controls the time and site-specific release of iron,[19] while other iron-containing proteins, such as myoglobin, are able to hold on to iron in both oxidation states. Therefore, it is very important to evaluate the protein and its interaction with both the oxidised and reduced states of iron and accordingly develop a data-analysis model. The spectroelectrochemical response of an iron binding protein can be ideal Nernstian, non-Nernstian resulting from coupled chemical equilibria (*e.g.* Fe^{2+} dissociation), or non-Nernstian because of allosteric interactions between multi-iron sties. The data analysis involved in these characteristic responses is illustrated below with specific examples.

2.3.1   Nernstian Response

An ideal Nernstian response during an electrochemical experiment may be observed for a single Fe^{3+}/Fe^{2+} redox centre that exhibits a fully reversible electron transfer. These electron-transfer processes should not be limited by the kinetics of heterogeneous electron transfer or be coupled with a chemical re-action (such as dissociation of Fe^{2+}) following reduction at the metal centre. The UV-Vis spectral changes observed during the reduction/oxidation of such ideal systems are a true response of the applied electrode potential. The changes in absorbance, therefore, represent the changes in the absolute concentration of the oxidised and reduced species at various applied electrode potentials and can be converted into the concentration ratio of oxidised to reduced species using Beer's law, eqn (2.2), where A_E is the absorbance of the solution at equilibrium

$$\frac{[Ox]}{[Red]} = \frac{A_E - A_R}{A_O - A_E} \tag{2.2}$$

for any applied potential E_{app}, A_O is the absorbance of the fully oxidised protein, and A_R is the absorbance of the fully reduced protein, all at a fixed wavelength (λ). Once the equilibrium concentration ratio of oxidised to reduced species is determined at different applied potentials, the $\log \frac{[Ox]}{[Red]}$ ratio is plotted as a function of the applied potential E_{app} following a rearranged form of the Nernst equation, eqn (2.3), where E_{app} is the applied potential, $E_{1/2}$ is the midpoint potential where

$$\log \frac{[Ox]}{[Red]} = \left(\frac{nF}{RT}\right) E_{app} - \left(\frac{nF}{RT}\right) E_{1/2} \tag{2.3}$$

$\frac{[Ox]}{[Red]} = 1$, R is the gas constant in $J\,K^{-1}\,mol^{-1}$, T is temperature in Kelvin, n is the number of electrons transferred, F is the Faraday constant, and $[Ox]$ and $[Red]$ are the concentrations of oxidised and reduced species, respectively. This analysis allows for a direct determination of the midpoint potential, $E_{1/2}$; when $\log \frac{[Ox]}{[Red]} = 0$, then $E_{app} = E_{1/2}$. Additionally, the slope of the plot can be used to calculate n, the number of electrons transferred during the redox process. An ideal Nernstian system will give whole number n values corresponding to the number of electrons transferred in the system. A single iron centre that undergoes a one-electron reduction will give $n = 1$.

Myoglobin is a classic example of a protein with a single Fe^{3+}/Fe^{2+} redox centre that exhibits a reversible Nernstian response. The kinetics of homo-geneous electron transfer are reasonably rapid in a myoglobin system despite the tertiary globin structure surrounding the heme iron. Additionally, the porphyrin

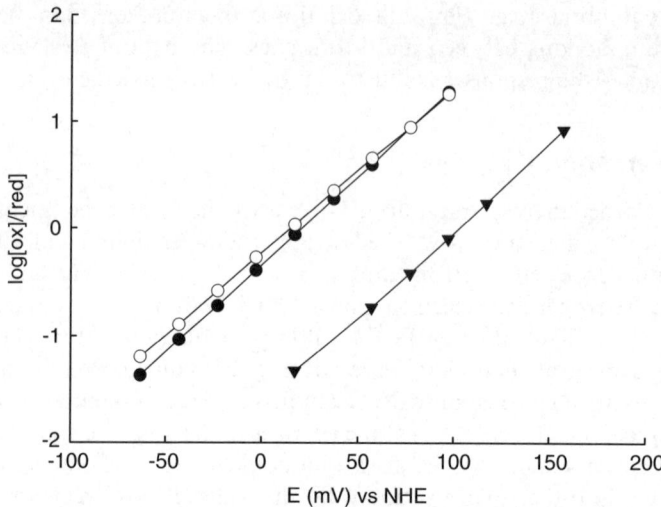

Figure 2.2   Nernst plots from spectroelectrochemistry of myglobins (Mb) from different species: ○ sperm whale (swMb, $E_{1/2} = 29$ mV, $n = 1.0$), ● horse (hMb, $E_{1/2} = 29$ mV, $n = 1.0$), and ▼ *Aplysia* (aMb, $E_{1/2} = 103$ mV, $n = 1.0$). Conditions: [heme] = 0.06–0.08 mM; [MOPS] = 50 mM at pH 7.1; [Ru(NH$_3$)$_6$Cl$_3$] = 1 mM; 20 °C. Figure adapted from ref. 20 and used with permission.

macrocycle binds iron tightly as both Fe^{3+} and Fe^{2+}, and prevents the dissociation of iron. The ideal Nernstian behaviour of myoglobin during spectroelectrochemistry experiments is often used to calibrate a newly constructed experimental setup and to check the proper functioning of electrodes and other associated components. An illustrative example of a Nernst plot obtained for a spectroelectrochemical investigation of myoglobin is given in Figure 2.2.[20]

2.3.2   Non-Nernstian Response

2.3.2.1   Coupled Chemical Equilibria

High-spin iron in a nonheme environment exhibits a significant change in the thermodynamics and kinetics of protein binding on reduction from Fe^{3+} to Fe^{2+}. This is illustrated by the mammalian serum iron-transport protein, transferrin. The thermodynamic affinity for Fe^{3+} is ~10^{20} in the presence of carbonate as a synergistic anion, and is reduced to ~10^3 on reduction to Fe^{2+}.[21,22] Iron-ligand turnover is also enhanced upon reduction.[2,3] The net result is a non-Nernstian spectroelectrochemical response because of an electrochemically driven reduction followed by a coupled equilibrium dissociation of Fe^{2+} as illustrated in eqns (2.4) and (2.5):

$$hTfFe^{3+} + e^- \rightleftharpoons hTfFe^{2+} \tag{2.4}$$

$$hTfFe^{2+} \rightleftharpoons Fe^{2+}_{aq} + hTf \qquad (2.5)$$

The Nernst plot slope in this case doesn't yield an n value that corresponds to the true number of electrons transferred and the observed apparent midpoint potential ($E^{obs}_{1/2}$) is also shifted positive relative to the "true" Nernstian value. Consequently, the observed midpoint potential, $E^{obs}_{1/2}$, associated with the coupled equilibrium eqns (2.4) and (2.5) must be corrected for the equilibrium dissociation reaction (eqn (2.5)) to obtain the true or Nernstian midpoint potential ($E_{1/2}$) associated with redox eqn (2.4). This is illustrated in the following paragraphs using the spectroelectrochemistry of human transferrin as an example.[6]

The reduction of Fe^{3+} to Fe^{2+} within the C-lobe of human transferrin, Tf_C, is followed by the dissociation of Fe^{2+}, due to a low affinity of Tf_C for Fe^{2+}. The degree of Fe^{2+} dissociation depends upon the dissociation constant of the reduced form of the protein (K_d) and its concentration. The K_d for the C-lobe of human transferrin has been estimated to be $10^{-2.7}\,M^{-1}$.[22] Here, we present a sample calculation that addresses the dissociation of Fe^{2+} from the C-terminal lobe of the transferrin. The coupled equilibria are shown in eqns (2.4) and (2.5),

$$K_d = \frac{[Tf_C][Fe^{2+}]}{[Tf_C Fe^{2+}_C]} \qquad (2.6)$$

The K_d expression for eqn (2.5) is shown in eqn (2.6), and this equation plus the mass-balance equations described below (eqns (2.7)–(2.9)) are used to correct the observed redox potential ($E^{obs}_{1/2}$) to give the Nernstian $E_{1/2}$ for eqn (2.4). For the fully oxidised system:

$$C_O = [Fe_{tot}] = [Tf_C Fe^{3+}_C] = A_0/\varepsilon l \qquad (2.7)$$

where $[Fe_{tot}]$ is the total concentration of iron in the system, $Tf_C Fe^{3+}$ is the iron in the C-lobe in the fully oxidised form, A_0 is the absorbance of the fully oxidised form, ε is the molar absorptivity of $Tf_C Fe^{3+}$ for the iron in the C-lobe, and l is the pathlength. For the fully reduced system:

$$C_O = [Fe_{tot}] = [Tf_C Fe^{2+}_C] + [Fe^{2+}_{aq}] \qquad (2.8)$$

where $[Fe_{tot}]$ is the total concentration of iron in the system, $Tf_C Fe^{2+}_C$ is the iron bound in the C-lobe in the reduced form, and Fe^{2+}_{aq} is the free aquated Fe^{2+} resulting from dissociation. For systems at intermediate potentials all three forms of iron will be present in solution, $Tf_C Fe^{23+}_C$, $Tf_C Fe^{2+}_C$, and Fe^{2+}_{aq}. The total Fe concentration for these intermediate potentials is described by:

$$C_O = [Fe_{tot}] = [Tf_C Fe^{3+}_C] + [Tf_C Fe^{2+}_C] + [Fe^{2+}_{aq}] \qquad (2.9)$$

The exact amount of Fe^{2+} that dissociates from the reduced $Tf_CFe_C^{2+}$ can be accounted for by the use of K_d, the dissociation constant, and the concentrations of intermediate Fe species involved in the reduction process can be properly determined using the matrix described below.

A typical experiment starts with the initial concentration of $Tf_CFe_C^{3+} = C_O$, prior to any reduction. Referring to the matrix shown below, when a potential is applied, some $Tf_CFe_C^{3+}$ is reduced to $Tf_CFe_C^{2+}$. Here, we are assuming the system has not reached equilibrium, thus there is no dissociation of reduced Fe^{2+} (second row of the matrix). At this point, if the concentration of $Tf_CFe_C^{3+}$ is C_1, then the concentration of $Tf_CFe_C^{2+}$ will be C_O-C_1. However, when the system reaches equilibrium, dissociation of Fe^{2+} follows the reduction. If the experimentally observed concentration of $Tf_CFe_C^{3+}$ at equilibrium is C_1' (this concentration can be calculated from the experimental absorbance reading at 465 nm using Beer's Law), then the concentration of $Tf_CFe_C^{2+}$ adjusted for the dissociation of Fe^{2+} is $C_O-C_1'-x$ and the concentration of Fe^{2+} is x. This is shown in the third row of the matrix. The matrix below shows the calculation required to correct the concentration of various iron species at different equilibrium potentials after accounting for the dissociation:

	$Tf_CFe_C^{3+}$	$Tf_CFe_C^{2+}$	Tf_C	Fe_{aq}^{2+}
No applied potential	C_o	0	0	0
Initially at a given applied potential	C_1	C_O-C_1	0	0
At equilibrium[a]	C_1'	$(C_O-C_1')-x$	x	x

Note: [a]"Equilibrium" is defined as the point at which the concentration reading, i.e. the absorbance, is constant at a given applied potential.

Rewriting the concentration using the above-defined K_d (eqn (2.6)), we obtain:

$$K_d = \frac{[(x)(x)]}{[(C_O - C_1') - x]} \tag{2.10}$$

The above equation can be rearranged to obtain:

$$x^2 + xK_d - (C_0 - C_1')K_d = 0 \tag{2.11}$$

For each equilibrium position during the redox experiment, the concentration C_1' is calculated from the absorbance reading at 465 nm (the λ_{max} for Fe^{3+}-Tf) using Beer's law (absorbance = (concentration) × (extinction coefficient) × (pathlength)). Equation (2.11) is then solved to obtain x, the free Fe_{aq}^{2+} concentration. The corrected value for $[Tf_CFe_C^{2+}]$ is then derived using $(C_0-C_1'-x)$ and a new corrected Nernst plot is generated using $[Red] = [Tf_CFe_C^{2+}]$ as determined from the above calculations. Such an analysis changes the $\frac{[Ox]}{[Red]}$ ratio and rotates the Nernst plot, such that the resulting plot will have a greater slope. This method of correction for a coupled dissociation that follows the Fe^{3+}/Fe^{2+}

reduction is identical for different iron-proteins. The K_d is the only parameter that will differ from system to system and will determine the magnitude of the correction. For example, Figure 2.3 shows the correction made for mammalian transferrin following the above-described method.[6]

2.3.2.2 Allosteric Behaviour due to Protein Subunit–Subunit Interactions

Proteins with more than one iron-binding site often display interactions between iron centres that influence their redox activity. For example, subunit–subunit interactions or allosteric interactions are clearly seen in the $\alpha_2\beta_2$ hemoglobin (Hb) tetramers. The allosteric effects responsible for the differential O_2 binding between the four subunits of Hb also gives rise to a non-Nernstian spectroelectrochemical response. These allosteric interactions in Hb-mediated O_2 transport can be modeled using a two-state model involving a "relaxed" or R-state with high O_2 affinity and a "tense" or T-state of low O_2 affinity.[23–25] This same R $\rightleftharpoons$ T state equilibrium model can also be used to interpret the non-Nernstian behaviour of Hb redox monitored by spectroelectrochemistry. Reduction or oxidation of interacting electroactive centres in Hb gives rise to a nonlinear Nernst plot. The changing slope of the Nernst plot, n, can no longer be simply interpreted as the number of electrons transferred. Nonlinear Nernstian plots obtained from such spectroelectrochemical data require a special treatment.

The absorbance change obtained from a typical spectroelectrochemical experiment of Hb can be plotted similarly as a simple Nernstian system using eqn (2.3). However, the plot of log $\frac{[Ox]}{[Red]}$ as a function of the applied potential, E_{app}, results in a curve rather than the straight line seen for the Nernstian system. These data can be fitted using a curve-fitting program (*e.g.* "Prostat" Ward and Reeves, IBM Version or ORIGIN Scientific Graphing and Analysis Software) that allows for the generation of a plot of the first derivative of the Nernst plot. The reduction half-potential for the protein, $E_{1/2}$, can be directly read from the Nernst plot as it corresponds to the x-axis value when the y-axis, log $\frac{[Ox]}{[Red]}$, is equal to zero. The derivative plot allows for the determination of an n value at various points along the Nernst plot and the n value corresponding to $E_{1/2}$ is often reported as $n_{1/2}$ or n_{50}. A subscript corresponding to the ratio of $\frac{[Ox]}{[Red]}$ is used to designate the n values along the curve (*e.g.* n_{10} for 10% oxidised, *etc.*). Often the highest n value is found at potentials greater than the $E_{1/2}$ and is reported as n_{max} and the corresponding E value is reported as E_{max}. These parameters, along with the sigmoidal shape of the Nernst plot, are illustrated in Figure 2.4.[26]

The n value and the constant change in slope ($= (RT/F)n$) along the curve cannot be interpreted as the number of electrons transferred during the oxidation/reduction process; however, the n value at the midpoint potential, $n_{1/2}$, is indicative of the level of cooperativity between different subunits. The $E_{1/2}$ value allows the comparison of different Nernst plots as a function of the

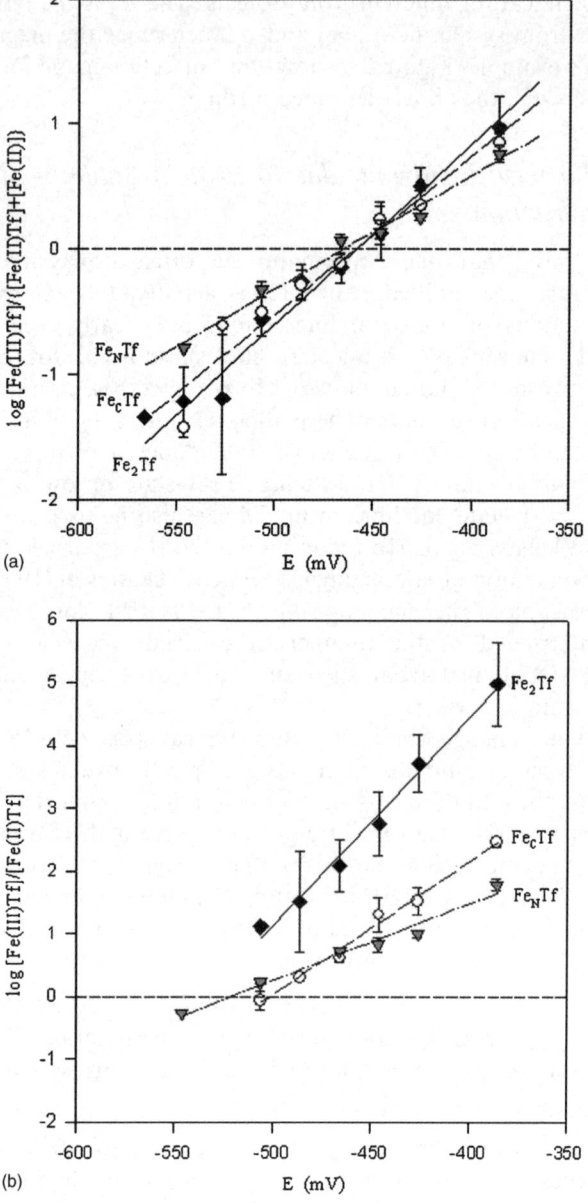

Figure 2.3 Uncorrected (a) and corrected (b) Nernst plots for ◆ diferric (Fe$_2$Tf), ○ C-terminal (Fe$_C$Tf) and ▼ N-terminal (Fe$_N$Tf) monoferric transferrin. Conditions: $[MV^{2+}] = 0.2–0.4$ mM; $[KCl] = 500$ mM; $[MES] = 50$ mM at pH = 5.8; 20 °C; Fe$_2$Tf (0.11–0.19 mM in Fe); Fe$_C$Tf (0.18–0.22 mM in Fe); Fe$_N$Tf (0.68 mM in Fe). Error bars represent standard deviations for the average of 2–3 independent experiments. Data obtained below –530 mV for Fe$_2$Tf and Fe$_C$Tf in panel (a) were not used for the corrected Nernst plots in panel (b) due to the low absorbance changes measured at these low potentials. Figure adapted from ref. 6 and used with permission.

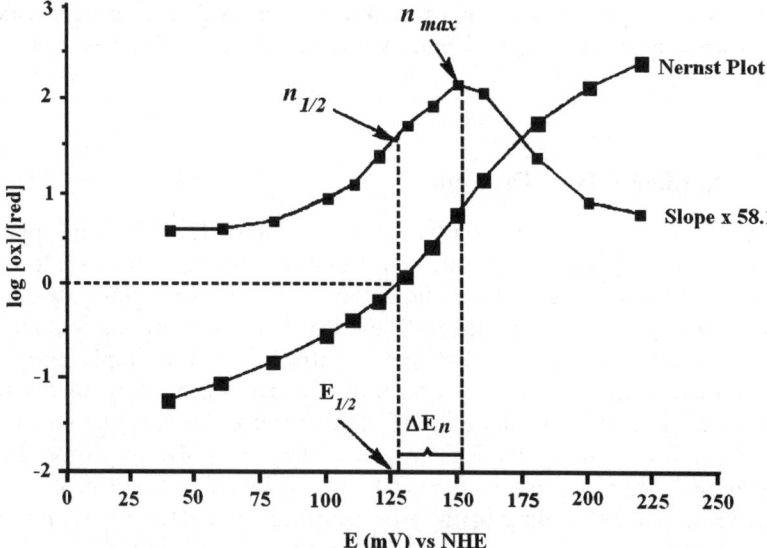

Figure 2.4 Plot illustrating various parameters that may be obtained from a Nernst plot of Hb that shows cooperativity. The lower line is a plot of eqn (2.3) for human hemoglobin A (Hb A_0). The upper line is a plot of the changing slope of the Nernst plot multiplied by 58.1 (F/RT). These data serve to illustrate the parameters $E_{1/2}$, E_{max}, n_{50} and n_{max}. Figure adapted from ref. 26 and used with permission.

midpoint potential (ease of oxidation or reduction), while the absolute value of the difference between the midpoint potential and the potential at the maximum slope ($|E_{1/2} - E_{max}|$) gives a general sense of the asymmetry of the curve.

2.4 Redox Behaviour of Selected Iron-Containing Proteins

The first and the second coordination shell of iron together influence the redox chemistry of iron proteins. The ligands directly coordinated to iron, whether they are amino-acid residues of the protein, or a heme group, or exogenous ligands, define the first coordination shell and strongly influence the redox properties of the metal centre. Nonheme and heme iron proteins have distinct redox properties determined by the first coordination shell of iron. Often, the first coordination shell includes a labile synergistic anion that can easily be replaced; such a replacement alters the redox properties at the iron centre. The second coordination shell of iron also has significant influence on the redox properties at the iron centre; however, its magnitude of influence vastly differs from system to system.[27] In the following sections, the first coordination shell and the second coordination shell effects on nonheme and heme iron proteins will be systematically illustrated along with other important aspects of iron-protein spectroelectrochemistry, using selected examples from our laboratory.

2.4.1 Nonheme Iron Proteins

Bacterial and mammalian iron-transport proteins, ferric-binding protein (FbpA)[28] and transferrin,[21,29] are nonheme-containing proteins requiring thorough control of the iron redox potential.[6,16,19,30] Undesired release of iron and its participation in the Haber–Wiess cycle result from the inability to control the redox chemistry at the metal centre.[1,31,32] More importantly, the time- and site-specific delivery of iron in biological systems is regulated by a redox switch. Reduction from Fe^{3+} to Fe^{2+} thermodynamically and kinetically favours iron release while making the iron centre susceptible to proton-driven dissociation.[4] Thus, the electrochemical investigations of these proteins are crucial in our understanding of the iron transport and release mechanism in biological systems.

2.4.1.1 *Ferric-binding Protein (Bacterial Transferrin)*

The ferric-binding proteins (FbpA) have been identified in several species of Gram-negative bacteria including pathogenic *Haemophilus influenzae* (*h*FbpA), *Neisseria gonorrhoeae*, and *Neisseria meningitidis* (*n*FbpA).[28] FbpA resides within the periplasmic space between the outer and the cytoplasmic membranes in these Gram-negative bacteria and functions as the primary periplasmic iron transport agent by tightly sequestering iron as Fe^{3+} ($K_d \approx 10^{-18}$).[16,30] Iron binding by *n*FbpA is sensitive to Fe^{3+}/Fe^{2+} reduction that is in the redox potential range of NADH-driven reductases, and the reduced Fe^{2+} is bound by *n*FbpA at significantly lowered affinity relative to Fe^{3+}.[16] This provides a compelling argument that a reductase may be an integral component of FbpA-mediated iron transport and release at the cytoplasmic membrane.

The first coordination shell of iron in *n*FbpA is occupied by four amino-acid residues and an exogenous anion. The exogenous or synergistic anion is essential for iron binding and a PO_4^{3-} anion is found to serve this purpose in the wild-type protein, $Fe^{3+}n$FbpA-PO_4.[33] The reduction potential of $Fe^{3+}n$FbpA-PO_4 was determined by applying an increasingly negative (reducing) potential to an anaerobic OTTLE cell containing an optically transparent electrode, the oxidised protein $Fe^{3+}n$FbpA-PO_4, and methylviologen mediator at the periplasmic pH of 6.5. The concentrations of the protein samples for these analyses were in the range of 0.6–1.2 mM and were thoroughly degassed prior to transfer into the anaerobic OTTLE cell. The characteristic broad absorption band associated with the broad LMCT band in $Fe^{3+}n$FbpA-PO_4, centred at 481 nm, decreased in intensity with the application of increasingly negative potentials (E_{app}) over the range -100 to -400 mV (NHE). A gold mesh electrode is ideal for these spectroelectrochemical experiments as its redox window in an aqueous

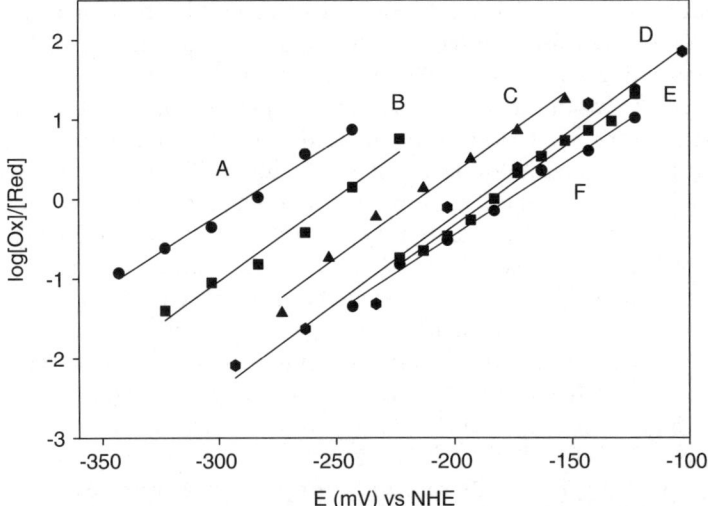

Figure 2.5 Nernst plots obtained by spectroelectrochemistry illustrating the influence of the exogenous anion, X, in recombinant ferric-binding protein, $Fe^{3+}nFbpA-X$. (A) X = phosphate, $E_{1/2} = -300$ mV; (B) X = arsenate, $E_{1/2} = -251$ mV; (C) X = pyrophosphate, $E_{1/2} = -212$ mV; (D) X = citrate, $E_{1/2} = -191$ mV; (E) X = oxalate, $E_{1/2} = -186$ mV; (F) X = NTA, $E_{1/2} = -184$ mV. Conditions: Au gauze (52 mesh) working electrode; $[Fe^{3+}nFbpA-X] = 0.63–1.2$ mM; [methylviologen] = 4.4–8.4 mM; [MES] = 50 mM at pH 6.5; 20 °C. Figure adapted from ref. 30 and used with permission.

system encompasses the range of −100 to −400 mV (NHE). A representative spectral profile is shown in the inset of Figure 2.1.[16,30]

The relationship between the concentration ratio of the oxidised to reduced form of the protein and the absorbance readings has been demonstrated in Section 2.3.1 and can be used to create a Nernst plot using eqn (2.3) (Figure 2.5).[30] The result is a well-behaved Nernstian plot as evidenced by its slope $(nF/RT\,\text{mV}^{-1})$ corresponding to a single electron-transfer process $(n = 1)$. The reduction is expected to facilitate the dissociation of the reduced Fe^{2+} because of its increased lability and lower affinity for the protein. In the presence of a mediator, the equilibria involved in the OTTLE cell can be as follows,

$$Fe^{3+}nFbp-X + e^- \rightleftarrows Fe^{2+}nFbp-X \qquad (2.12)$$

$$Fe^{2+}nFbp-X \rightleftarrows Fe^{2+}_{aq} + apo-nFbp-X \qquad (2.13)$$

However, the ideality of the Nernstian plot (Figure 2.5) indicates that there is little or no Fe^{2+} dissociation occurring in the electrochemical cell during the spectroelectrochemical experiment. The occurrence of significant dissociation would result in a Nernstian plot whose slope $(nF/RT\,\text{mV}^{-1})$ would not yield a single electron-transfer process ($n = 1$). Experimentally, upon air oxidation the LMCT bands are fully restored, which is independent evidence of little or no Fe^{2+} dissociation and fully reversible redox behaviour.

The Nernst plots in Figure 2.5 are a series of parallel lines (with slope, nF/RT, corresponding to $n = 1$) that are displaced along the applied potential axis, depending on the identity of the exogenous anion X. The intercepts correspond to $E_{1/2}$ values for ferric-binding proteins containing different exogenous anions. These $E_{1/2}$ values, which are extremely sensitive to the nature of the synergistic anion, are listed in the legend for Figure 2.5. The synergistic anion and the coordinating amino-acid residues from the protein control the first coordination shell environment of the iron and the $E_{1/2}$ values.

The periplasmic space where ferric-binding proteins reside is rich in anion composition.[28,34] It has been shown that the exogenous anion can undergo facile exchange with other anions to generate a series of iron-nFbpA protein assemblies, $Fe^{3+}n$FbpA-X.[30,35] The large difference in the $E_{1/2}$ ($> 100\,mV$; Figure 2.5) observed for different $Fe^{3+}n$FbpA-X is a prime example of first coordination shell influence on the redox potential of a metal ion in a protein. Within the periplasmic space, this difference in $E_{1/2}$ may have a profound effect on the *in-vivo* iron-release process, if redox is involved. This is a reasonable hypothesis given the high Fe^{3+} affinity exhibited by FbpA ($K_d \approx 10^{-18}$) and the approximately 12 orders of magnitude loss in affinity upon reduction to Fe^{2+}.[16,30] All of the $E_{1/2}$ values for the $Fe^{3+}n$FbpA-X studied lie above $-300\,mV$ (Figure 2.5) and fall therefore within a range where reduction by NADH or NADPH cofactors may readily occur; however, a facile anion exchange can apparently modulate the redox potential of $Fe^{3+}n$Fbp assembly. For example, the replacement of PO_4^{3-} by citrate would shift the $E_{1/2}$ positive by *ca* $110\,mV$ (Figure 2.5), making the Fe^{3+} reduction more energy efficient ($\sim 11\,kJ$). Consequently, we propose that the anion composition of the periplasm can influence the identity of the synergistic anion bound to the periplasmic iron-transport protein $Fe^{3+}n$FbpA-X, and consequently can influence the ease of delivery of iron to the cytoplasm *via* a reductive mechanism.[30]

2.4.1.2   Mammalian Transferrin

Human serum transferrin (*h*Tf) tightly sequesters Fe^{3+} ($K_d = 10^{-20}$), which enables it to function in blood as an iron-transport and bacteriostatic agent.[36] Each *h*Tf consists of two homologous lobes, the N-lobe and C-lobe, which reversibly bind an Fe^{3+} ion to give the diferric form of the protein, Fe_2Tf.[37-39] The Fe^{3+} binding site in each lobe is defined by two dissimilar domains, which together form a deep binding cleft. The Fe^{3+} in each lobe is bound octahedrally to four amino-acid ligands (two tyrosine, one aspartate and one histidine) and two oxygen atoms of the exogenous anion, carbonate. The absence of the exogenous carbonate anion abolishes the Fe^{3+}-binding properties of *h*Tf. Although the first coordination shells of Fe^{3+} in the two lobes are identical, there are differences in the second coordination shell. More importantly, the reduction potential at the Fe^{3+} centre of this very mobile protein is thoroughly controlled to prevent undesired Fe^{2+} release and to avoid catalytic generation of reactive oxygen species *via* the Haber–Weiss cycle.[1,32]

Time- and site-specific delivery of iron by transferrin requires recognition by iron-demanding cells, as well as compartmentalisation and a well-mediated iron-release process. These stringent requirements of iron delivery are fulfilled *via* a cellular process known as receptor-mediated endocytosis.[21,40–42] During receptor-mediated endocytosis, iron-demanding cells express the transferrin receptor, TfR.[43] At extracellular serum pH = 7.4, TfR discriminately binds iron-bound holo-*h*Tf with greater affinity over iron-free apo-*h*Tf. The newly formed transferrin–transferrin receptor complex, Fe_2Tf/TfR, is then encapsulated and internalised by the cell through the formation of an ATP-driven proton-pumping endosome. Within the endosome Fe_2Tf/TfR is thoroughly compartmentalised and is subjected to a drop in pH.[44,45] A lower pH of 5.5 inside the endosome facilitates iron release from transferrin in a controlled manner. Subsequent to release from *h*Tf, iron exits the endosome *via* the membrane divalent metal ion transporter now known as DMT1, which accepts iron only as Fe^{2+}.[46–48] After completing its iron-donating function the endosome returns to the cell surface and releases metal-free apo-*h*Tf to the blood for another cycle of iron transport, thus completing the receptor-mediated endocytosis cycle.

A fundamental and still unanswered question in iron metabolism is how iron is released from *h*Tf and when and where is it reduced to the ferrous state before being transported across the endosomal membrane to the cytoplasm by divalent metal transporter DMT1? Experiments carried out at endosomal conditions indicate that iron released from Tf to most physiological iron binders is easily reducible.[49] However, the time required for the release of iron from transferrin as Fe^{3+} to physiological chelators, even at endosomal pH, is much longer (>6 min) than that compared to the cell-cycling time of transferrin, which may be as little as 1–2 min.[21] Transferrin completes some 100–200 cycles of iron uptake, transport and delivery to cells during its lifetime in the circulation, thus demanding a sophisticated and efficient iron-release process.[50,51]

The reduction of Fe^{3+} to Fe^{2+} is an attractive hypothesis for initiating iron release from transferrin in the endosome. This is because reduction thermodynamically and kinetically facilitates iron dissociation, and additionally favours the iron-release process by making the reduced Fe^{2+}-complex susceptible to protonation at the ligand donor sites.[4] Reduction of $Fe^{3+}_2 h$Tf results in the loss of ~ 17 orders of magnitude in iron-binding affinity, and the ligand-exchange kinetics of high-spin Fe^{2+} are known to be several orders of magnitude higher than for Fe^{3+}.[2,3,21,22] Use of traditional electrochemical techniques, such as bulk electrolysis and electrochemical titration, to investigate redox properties of transferrin is confounded by the inefficient heterogeneous electrons transfer between the electrode and the metal centre, as well as the dissociation of Fe^{2+} from the binding site upon reduction.[22,52,53] The spectroelectrochemical technique described here allows for a careful investigation of the redox properties of transferrin while specifically addressing the above-mentioned confounding factors. The use of a redox mediator enables efficient heterogeneous electron transfer, while the data analysis can accurately account for the dissociation of the reduced Fe^{2+}. This allows for a more accurate determination of the reduction half-potential of the iron in transferrin and the true redox properties associated with the iron-loaded protein.

Studies investigating the redox potential of transferrin have focused on data collected at the serum pH of 7.4.[22,52,53] The redox potential for Fe_2Tf was determined to be $-400\,mV^{52}$ and $-520\,mV^{53}$ (*vs*. NHE) in two different investigations. The $-400\,mV$ value is not corrected for Fe^{2+} dissociation (see Section 2.3.2.1), while the $-520\,mV$ is a corrected potential, but was carried out in high background electrolyte concentration (2.0 M KCl). Spectroelectrochemical experiments carried out to investigate the redox potential of Fe-saturated diferric transferrin, Fe_2Tf, as well as monoferric C-terminal lobe (Fe_CTf) and monoferric N-terminal lobe (Fe_NTf) of full-length transferrin under endosomal conditions (pH = 5.5 and $[Cl^-]$ = 500 mM) requires careful experimental design and data interpretation. The redox potentials are very negative and beyond the electrochemical window available to a Pt or Au electrode.[6] This required the use of an amalgamated Au electrode, as described in Section 2.2.1.2. Data analysis also requires correction for the dissociation of Fe^{2+}, as described in Section 2.3.2.1. Given the low affinity of apo-transferrin for Fe^{2+} and the relatively high lability of Fe^{2+}, the overall process taking place in the OTTLE cell is in the combination of eqns (2.4) and (2.5), where $n = 1$ for the monoferric system and 2 for the diferric system. Mammalian transferrin has two Fe binding sites and Fe^{2+} dissociation from both sites must be included in the correction. The dissociation constant of Fe in the two lobes, C-terminal lobe and the N-terminal lobe, are different and require use of corresponding K_d values to correct for the dissociation.[22] The correction, described in Section 2.3.2.1, allows for the construction of the corrected Nernst plots and the determination of the corrected reduction half-potentials.

The redox potentials of diferric transferrin (Fe_2Tf), C-terminal lobe (Fe_CTf) and N-terminal lobe (Fe_NTf) in full-length transferrin exhibit very little variation with pH.[6,53] A typical spectral profile of a spectroelectrochemical experiment is shown in Figure 2.6[6] and the corresponding observed and corrected Nernst plots are shown in Figure 2.3. The necessary correction to the experimental data due to Fe^{2+} dissociation was carried out as described in Section 2.3.2.1. The corrected redox potential for Fe_2Tf at endosomal conditions is $-526\,mV$ (NHE),[6] which is identical to that determined at pH 7.4[53] and in a high electrolyte concentration (2 M KCl). The number of electrons transferred, n, calculated from the slope of the Nernst plot indicate two electrons are transferred from the electrode to the protein, which is consistent with the two reducible Fe sites in Fe_2Tf. The reduction half-potential for monoferric C-terminal lobe (Fe_CTf) of full-length transferrin was determined to be $-501\,mV$ (NHE) and $n = 1.2$. The monoferric N-terminal lobe (Fe_NTf) of full-length transferrin under endosomal pH showed a slightly more negative reduction potential of $-520\,mV$ (NHE) ($n = 0.7$) compared to the C-terminal lobe (Fe_CTf).[6] The amino-acid residues and the synergistic anion (CO_3^{2-}) coordinated to Fe are identical in both C- and N-terminal lobes; however, the amino-acid sequence and the secondary/tertiary protein structure around the

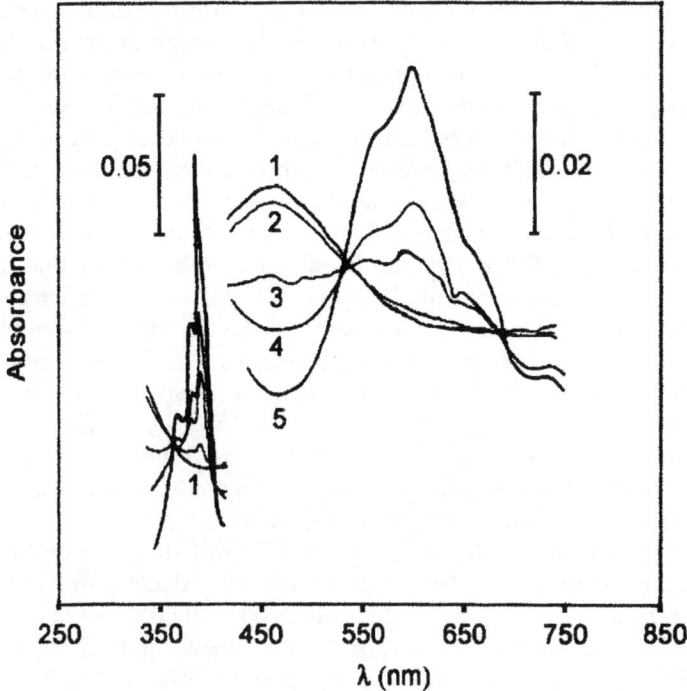

Figure 2.6   A representative visible spectral profile of iron-loaded mammalian transferrin during the course of a typical spectroelectrochemical reduction using methylviologen mediator. The absorbance maximum due to Fe^{3+} bound to transferrin is at 465 nm and the methylviologen peaks are observed at 395 and 602 nm. The numbered spectra correspond to applied potentials of −206 (1), −386 (2), −426 (3), −486 (4), and −506 (5) mV. Conditions: [Fe_CTf = 0.20 mM, [methylviologen] = 0.4 mM, [KCl] = 500 mM, [MES] = 50 mM at pH 5.8 and 20 °C. Figure adapted from ref. 6 and used with permission.

two Fe-binding pockets are different.[37] Thus, the small difference in the redox potential observed for the two lobes of transferrin may be a second coordination shell effect controlled by the differences in the second coordination shell of Fe. Regardless of the differences, these $E_{1/2}$ values are very negative even in endosomal conditions and are inaccessible to physiological reductants. Therefore, these data are not consistent with a NADH- or NADPH-driven reduction of transferrin within an endosome without the participation of a Fe^{2+} scavenging ligand or some other coupled reaction.[6]

2.4.1.3   Mammalian Transferrin and Transferrin Receptor Complex

A salient feature of the receptor-mediated transferrin-to-cell endocytic cycle in human iron metabolism is the persistence of the transferrin–transferrin receptor assembly (hTf/TfR) throughout the cycle: transferrin free of its receptor is not known to exist within the cell.[21,54] The crystal structure of the ectodomain of the TfR shows that it is a homodimeric transmembrane protein (80 KDa) with each monomer independently capable of binding one hTf molecule.[55] During the recognition/binding process, the C-lobe of hTf appears to have most of the interaction with the TfR. In an attempt to mimic the conditions of the endosome as closely as possible, we have used the spectroelectrochemistry technique to investigate the redox properties of the complex between iron-loaded C-lobe half-transferrin (Fe_CTf) and the water-soluble ectodomain of the transferrin receptor (TfR) at endosomal pH (5.8) and 500 mM Cl^- anion concentration. Our purpose was to determine if formation of the Fe_CTf/TfR assembly has an influence on the ease of reduction of the sequestered Fe^{3+}. This is of particular significance with respect to the reductive iron-release hypothesis, since as noted above the redox potential of receptor free Fe_2Tf is very negative ($E_{1/2} < -500$ mV $vs.$ NHE);[6,53] and therefore Fe^{3+}/Fe^{2+} reduction is not attainable under physiological conditions using common available oxidoreductase cofactors such as NADH or NADPH.

Optical spectra of transferrin C-lobe docked with the transferrin receptor showed a characteristic broad absorption band centred at 465 nm, just as in the receptor-free *holo*-protein (Figure 2.1 inset).[19] The intensity of this absorbance band declined as more negative potentials were applied in a spectroelectrochemistry experiment, but did not qualitatively change in its overall features. An EPR spectrum of the Fe_C/TfR complex at pH 5.8, recovered from the OTTLE cell after completion of spectroelectrochemical studies allowed us to conclude that the first coordination shell of Fe^{3+} in transferrin is intact and unperturbed when C-lobe is complexed with TfR.[19] Consequently, we assume that C-lobe and Fe_C/TfR complex have similar if not identical Fe^{3+} and Fe^{2+} binding constants, and so we take K_d for binding of Fe^{2+} in the protein–receptor complex to be 10^{-3} M as calculated for free Tf.[22] This value was used to correct the observed Nernst plot data by accounting for the dissociation of Fe^{2+} that occurs upon reduction. Nernst plots for the observed spectroelectrochemical data for Fe_CTf/TfR, and data corrected for Fe^{2+} dissociation, are presented in Figure 2.7.[19] The corrected plot exhibits typical Nernstian behaviour for a one-electron transfer and a $E_{1/2}$ value of -285 mV.

Figure 2.8 compares corrected Nernst plots for C-lobe half-transferrin free in solution and bound to the transferrin receptor, at endosomal pH.[19] These data clearly demonstrate that docking iron-loaded C-lobe transferrin at the transferrin receptor at pH 5.8 makes it energetically more favourable to reduce Fe^{3+} to Fe^{2+} by ~ 200 mV. Furthermore, receptor-docking places Fe^{3+} reduction in a range accessible to NADH or NADPH cofactors, consistent with the hypothesis that reduction is the initial event in iron release from transferrin in the endosome. Fe^{2+} is bound by hTf at least 14 orders of magnitude more weakly than Fe^{3+}, so that reductive release of iron bound to hTf in the transferrin–transferrin receptor complex is then physiologically and thermodynamically feasible, and the barrier to transport across the endosomal membrane is lifted. The transferrin receptor, therefore, is more than a simple conveyor of

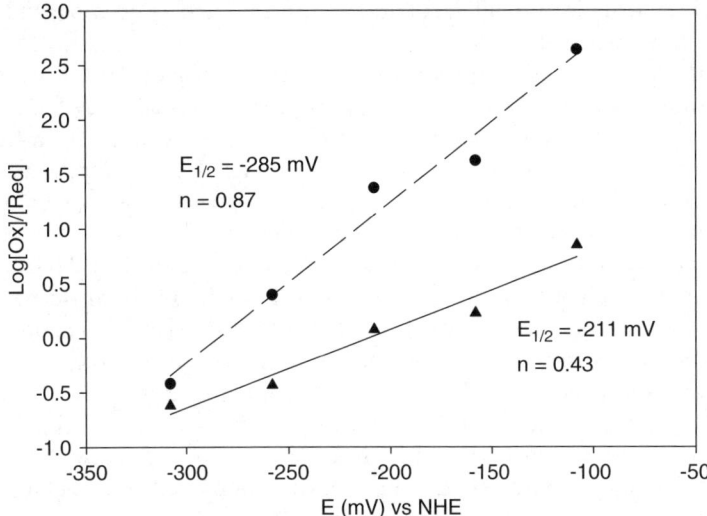

Figure 2.7   Representative Nernst plots corresponding to the reduction of Fe^{3+} in the Fe_C/TfR assembly. (▲) Observed data; (●) Data corrected for Fe^{2+} dissociation from Fe_C/TfR. Conditions: Au mesh electrode; [Fe_CTf/TfR] = 0.19 mM; [MES] = 50 mM at pH 5.8; [methyl viologen] = 1.4 mM; [KCl] = 500 mM. Figure adapted from ref. 19 and used with permission.

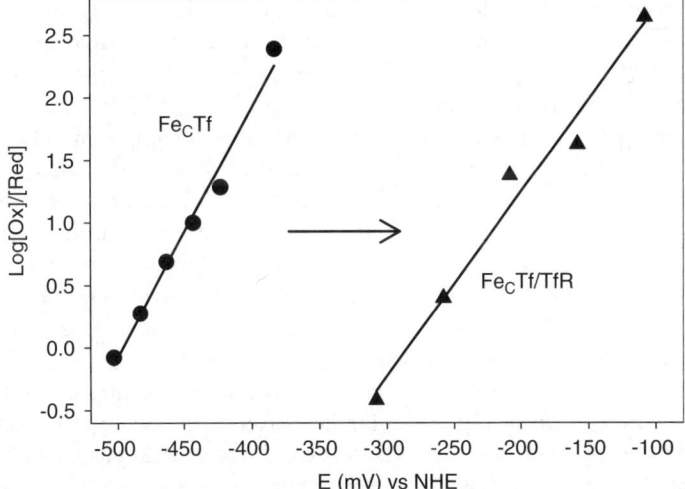

Figure 2.8   Nernst plot for reduction of Fe^{3+} in the C lobe of human transferrin (●, Fe_CTf) and C-lobe transferrin complexed to the transferrin receptor (▲, Fe_CTf/TfR) at endosomal pH 5.8. Data are corrected for the dissociation of Fe^{2+}. These data illustrate that redox potential of Fe^{3+} in the C lobe of human transferrin is shifted positive by ∼200 mV when it is complexed to the transferrin receptor. Conditions: Au mesh electrode; [Fe_CTf/TfR] = 0.19 mM; [Fe_CTf] = 0.20 mM; [MES] = 50 mM at pH 5.8; [methyl viologen] = 1.4 mM; [KCl] = 500 mM. Figure adapted from ref. 19 and used with permission.

transferrin and its iron, while protein–protein interaction goes beyond simple association of two molecules.[19]

Now the question becomes why is the energy cost of Fe^{3+} reduction in C-lobe transferrin complexed to its receptor greatly decreased by the receptor-induced rise in reduction potential to $-285\,mV$ (Figure 2.8)? The iron-free apo-*h*Tf is bound weakly, if at all, to TfR at extracellular pH, but the relative affinity of the receptor is documented to be greater for apo-*h*Tf than for holo-*h*Tf at endosomal pH.[56] This differential binding energy of TfR for apo- and holo-*h*Tf may provide the thermodynamic driving force required to shift the transferrin redox potential higher to enable the *in-vivo* reduction of transferrin bound Fe^{3+}.[19] Any changes in the reduction half-potential arising from the Tf/TfR interaction may be considered a second coordination shell influence, since the first coordination shell of Fe^{3+} remains unperturbed, as indicated by UV-Vis and EPR spectra of the Fe_CTf/TfR assembly. The second coordination shell interaction between a protein and its receptor and the influence of this inter-action on redox properties at the first coordination shell of a metal is still a novel concept.

2.4.2   Heme Iron Proteins

Heme (iron-protoporphyrin) proteins are a class of metalloproteins that con-tain one or more heme prosthetic groups. Heme proteins serve a number of biological functions including electron transport, catalysis, signaling, and O_2 storage and transport.[57] Fe^{2+} present in each heme will readily undergo a re-versible one-electron oxidation to Fe^{3+}. Here we will present the results of representative spectroelectrochemical investigations of the heme-containing proteins hemopexin (Hpx), myoglobin (Mb), and hemoglobin (Hb). Hemo-pexin and Mb each contain a single heme moiety and Hb consists of four subunits, each of which contains a heme moiety, to form an $\alpha_2\beta_2$ tetramer. The biological functions of these three heme proteins are diverse and the results of their UV-Vis spectroelectrochemical investigations provide an excellent illus-tration of the variety of information that can be obtained from this technique.

Hemopexin is a glycoprotein that noncovalently binds heme, a lipophilic moiety, and maintains its solubility in monomeric form.[58] Hemopexin is a heme scavenger and is the second line of defence against the hemoglobin-mediated oxidative damage during intravascular hemolysis. The heme-Hpx complex is taken in by hepatic cells and heme is degraded within the cytoplasm by heme-oxygenase. Therefore, during the heme transport process it is critical for Hpx to control the redox potential of the heme iron and/or to prevent the access of redox-reactive molecules to the metal centre.

Myoglobin, a single-heme system, and hemoglobin, a multiheme system, both contain an identical heme prosthetic group; however, their O_2 binding properties are significantly different, as is the second coordination shell of iron.[25] Consequently, this leads these two important proteins to have different Fe^{3+}/Fe^{2+} redox properties. The redox and the O_2 binding properties of Mb parallel each other, and yet there is a significant difference in the O_2 affinity and redox potential of Mbs from different organisms. The same is true for Hbs.[12] Spectroelectrochemical investigations of Mbs and Hbs provide insight into the electron-transfer processes at the heme centre. Additional critical factors include the formal redox potential, the number of electrons transferred in the redox process, cooperative effects between the redox centres (in multiheme systems), allosteric sites and their influence on the redox potentials, and the effects of diverse homotropic and heterotropic effectors on $E_{1/2}$ and the level of cooperativity.

In the sections that follow we will describe representative spectroelectrochemical investigations of different Hpxs, Mbs and Hbs and we will discuss the observed differences in the various parameters obtained from a redox analysis of these heme proteins.

2.4.2.1 Hemopexin

Heme cofactors in biological systems are essential to carry out a cascade of reactions, and yet the presence of a free heme group can be toxic and lead to oxidative damage. Hemopexin (Hpx) scavenges the heme groups that are liberated during hemolysis by encapsulation and controls heme-facilitated cellular oxidation, while maintaining the solubility of lipophilic heme in its monomeric form. During the heme transport/recycling process Hpx controls the unnecessary heme-catalysed oxidation reactions by both controlling the redox potential at the Fe centre and by making the heme physically inaccessible for catalysis.

The chemistry and biochemistry of Hpx has been reviewed[58] and a crystal structure is available.[59,60] Hemopexin is present in serum at about 10 μM and its primary function is to transport released heme to its degradation site in the parenchymal cells of the liver *via* receptor-mediated endocytosis. Encapsulation of a single heme by Hpx occurs *via* bis-histidyl protein side-chain coordination of the Fe. Spectroelectrochemical investigation of the heme-Hpx assembly gives insight into the role of Hpx in controlling the reduction potential of the heme Fe, the efficiency of electron transfer at the metal centre, the influence of bishistidyl coordination at the Fe centre, and the possible role of Fe redox in the Hpx-mediated transport and recycling of heme.

We have reconstituted heme-Hpx in the form of meso porphyrin–hemopexin (MHpx) and protoporphyrin IX–hemopexin (PPHpx). Spectroelectrochemical measurements were made using the Soret bands for both MHpx (405 nm and 412 nm for the oxidised and reduced forms) and PPHpx (414 nm and 428 nm for the oxidised and reduced forms) at pH 7.2 in the presence of various background electrolytes.[61] Well-behaved Nernst plots were obtained with slopes consistent with a one-electron transfer as expected. Our $E_{1/2}$ results are summarised in Table 2.1 and are in the range reported for other heme proteins. The reduction potentials of meso heme and protoporphyrin IX encapsulated in Hpx are shifted ~100 mV positive relative to the free heme groups. The $E_{1/2}$ values for MHpx and PPHpx are sensitive to heme structure and the trend in Table 2.1 suggests a more stable Fe^{3+} state for MHpx. This is consistent

Table 2.1   Redox potentials ($E_{1/2}$) for the hemopexins MHpx and PPHpx at
pH 7.2 in the presence of various background electrolytes.[a,b]

Hemopexin	Background electrolyte			
	KCl	*NaCl*	*K₂HPO₄*	*NaNO₃*
MHpx	6	52	23	48
PPHpx	45	88	62	–

[a]Conditions: 25 °C; 0.03–0.1 mM MHpx; 0.5–2.5 mM $K_3Fe(CN)_6$; 0.5 M HEPES and 0.2 M background electrolyte at pH 7.2. $E_{1/2}$ values are in mV relative to NHE and represent the average of
two or more independent measurements.
[b]Ref. 61.

with reports in the literature where replacement of the heme vinyl groups
with alkyl groups resulted in a stabilisation of the Fe^{3+} form of the protein.[62,63]
The salt effects illustrated in Table 2.1 are consistent with significant heme-
moiety solvent exposure. Salt effects on Hpx $E_{1/2}$ values are also consistent with
the reported cation and anion influence on the folding and stability of the
protein assembly as determined by melting experiments.[58,64] The crystal
structure[59,60] of Hpx shows that Na^+, Cl^-, and PO_4^{3-} interact with various sites
on the surface of the protein, both near and far from the heme site, which is
also consistent with the significant anion and cation effects on the observed
$E_{1/2}$ values.

The influence of pH on the formal reduction potential was also examined
using the spectroelectrochemical technique.[61] The $E_{1/2}$ value for MHpx at pH
7.2 in the presence of 0.2 M KCl was shifted from 6 mV to 105 mV on dropping
the pH to 5.5. Therefore, as the pH is lowered the ease of Fe^{3+} reduction is
increased. The Nernst equation predicts a 59 mV shift in $E_{1/2}$ for each pH unit
change for a redox equilibrium that involves a single proton. Consequently, the
observed 100 mV positive shift in $E_{1/2}$ that occurs with a decrease in pH from
7.2 to 5.5 is consistent with an equilibrium reaction involving a single H^+ that
stabilises the reduced form of the protein.

The redox sensitivity of hemopexin-encapsulated heme to electrolyte composition and pH illustrate the importance of first coordination shell (bis-
histidine ligation and heme structure) and second coordination shell (protein
structure/folding and environment) effects in these heme proteins. These ob-
servations also suggest a possible role for Fe^{3+}/Fe^{2+} redox in hemopexin-
mediated heme transport/recycling, as high chloride anion concentration and
low pH are known conditions for the endosome where the heme is released.

2.4.2.2   Myoglobins

Myoglobin is a heme protein with a single heme prosthetic group. The biological function of Mb is to accept O_2 from the transport protein Hb and store
it until needed. The resting state for O_2 storage is Fe in the +2 oxidation
state (deoxy-Mb), which can readily undergo a reversible one-electron loss
to the oxidised Fe^{3+} form (met-Mb). The Soret bands may be used for

spectroelectrochemical experiments with Mb; large absorbance changes are seen at ~ 410 nm (the absorption maximum corresponding to Fe(III)-Mb or met-Mb) and at ~ 435 nm (the absorption maximum corresponding to Fe(II)-Mb, deoxy-Mb). Mbs from different species may have small shifts in absorption maxima and molar absorptivities for the Soret band corresponding to the Fe(III)/Fe(II) oxidation state; however, these bands are always distinct for Fe(III) and Fe(II) oxidation states and can be used as spectroelectrochemical handles. Monitoring the spectroscopic profile at the intense Soret bands will suffice to obtain the data required for the Nernst plot; however, it is a good practice to monitor the UV-Vis spectral profile from 340 nm to 700 nm as a function of applied potential. Over this spectral region, one can observe five isosbestic points (at *ca.* 420, 462, 522, 606, and 662 nm) for a successful oxidation/reduction experiment. This complete spectral analysis is often used to detect any problems associated with the condition or concentration of the protein (*e.g.*, denaturation).

Spectroelectrochemical experiments with Mb can be executed starting with met-Hb and sweeping the potential in the negative (reducing) direction, or the reverse. For heme proteins it is prudent to do a few experiments in both directions to ensure that the redox process is completely reversible. This is possible for heme proteins since Fe^{2+} dissociation does not occur. In the case of Mb (and Hb) it is procedurally easier to start with the oxidised protein, as it will not bind O_2 and therefore trace amounts of the oxy-form of the protein are not present. The absorbance of the fully oxidised Fe(III)-Mb (A_O) can be generated by applying $+400$ mV (NHE) to the OTTLE cell until equilibrium is reached (time required can be from 15–45 min). Similarly, the absorbance of the fully reduced (A_R) Fe(II)-Mb can be obtained by applying a potential of -250 mV (NHE). The spectroelectrochemical profile of oxidation or reduction can be obtained by incremental application of potential (20 mV increments) to the OTTLE cell. The time required for the system to attain equilibrium at each applied potential can vary significantly (15–45 min) and will depend upon the protein, mediator, and the background electrolyte concentrations.

Nernst plots corresponding to the spectroelectrochemical investigation of sperm whale, horse, and *Aplysia* Mbs are shown in Figure 2.2.[20] All display ideal Nernstian behaviour ($n = 1$), with the major difference being the reduction half-potential. The first coordination shell of Fe^{2+} in Mb is the same in all forms of Mb; however, the differences in the second coordination shell of iron, overall amino-acid sequence, and the composition of amino acids around the heme pocket can be significantly different from species to species. In the case of *Aplysia* Mb a valine (Val63 (E7)) is present in the binding pocket in place of the distal histidine in sperm whale and horse Mb. Furthermore, Aplysia remains five-coordinate in the met form, while sperm whale and horse Mb incorporate a H_2O ligand in the met form.[65] Thus, the differences in reduction half-potentials observed for different Mbs are influenced by the second coordination shell of iron, which includes amino-acid residues around the heme pocket, and the difference in the Mb secondary and tertiary structure resulting from the differences in the amino-acid sequence. This is illustrated by the differences observed in $E_{1/2}$ for sperm whale ($E_{1/2} = 29\,\mathrm{mV}$), horse ($E_{1/2} = 29\,\mathrm{mV}$) and *Aplysia* ($E_{1/2} = 103\,\mathrm{mV}$) Mbs (Figure 2.2).[20,66] This difference in the redox potential is also graphically illustrated in Scheme 2.1, which further illustrates that the first and second coordination shells influence the metal redox properties.

2.4.2.3 Hemoglobins

The first coordination shell of Fe in Hb is identical to that of Mb, but there are other major differences that influence the spectroelectrochemical/Nernst plot profile of Hb and make it distinct from that of Mb. Hb is a tetrameric protein with four heme-containing subunits ($\alpha_2\beta_2$), each of which is redox-active. Differences between Mb and Hb include amino-acid sequence, which results in different redox potentials for the α and β chains, plus subunit–subunit interactions that lead to allostery in Hb. This allosteric interaction leads to cooperative electron transfer that gives rise to a non-Nernstian redox profile that requires special consideration for data analysis and interpretation of results.[66,67]

The experimental setup for spectroelectrochemical analysis of Hb redox is identical to that of Mb. Intense Soret bands in the UV region are excellent spectroscopic handles for monitoring the progress of the reduction or the oxidation of Fe as a function of the applied potential. The absorption maxima for the oxidised form of human Hb A_o (Fe(III)-Hb or met-Hb) and its reduced form (Fe(II)-Hb or deoxy-Hb) are shifted relative to that of Mb, with maxima at 406 nm for Fe(III)-Hb A_o and at 430 nm for Fe(II)-Hb A_o, and isosbestic points at 415, 455, 524, and 598 nm. Similar to Mb, the spectroscopic changes during the reduction of Fe can most easily be monitored by following the changes in the Soret bands, but it is recommended that the entire UV-Vis spectral range be investigated, as noted above for Mb, in order to check for hemichrome formation and other undesirable side effects. We have shown that the positively charged mediator $\mathrm{Ru(NH_3)_6^{3+}}$ works well in Hb spectro-electrochemistry as it does not bind at the allosteric binding site and is optically transparent.[26]

When the Hb spectroelectrochemical data are plotted according to the re-arranged form of the Nernst equation (eqn (2.3)), the resulting plot is a sig-moidal curve (Figure 2.9)[67] with a maximum slope (corresponding to n_{max}) greater than that seen for Mb. The increased slope of the Hb Nernst plot corresponds to n values greater than 2.0 in some cases.[11,26,66,67] Section 2.3.2.2 and Figure 2.4 describe the treatment of this non-Nernstian behaviour resulting from subunit–subunit interaction and the subsequent extraction of reduction potential and n values along the Nernst plot.[66,67]

A general trend is evident that ease of Hb oxidation (lower $E_{1/2}$) generally correlates with increased O_2 affinity (low $P_{1/2}$) for various Mbs, Hbs and their mutants.[26] This is illustrated in Figure 2.9 where the Nernst plot for horse Hb is displaced towards lower potentials relative to that of human Hb A_0.[67] One of the structural differences between human and horse Hb is the substitution of a

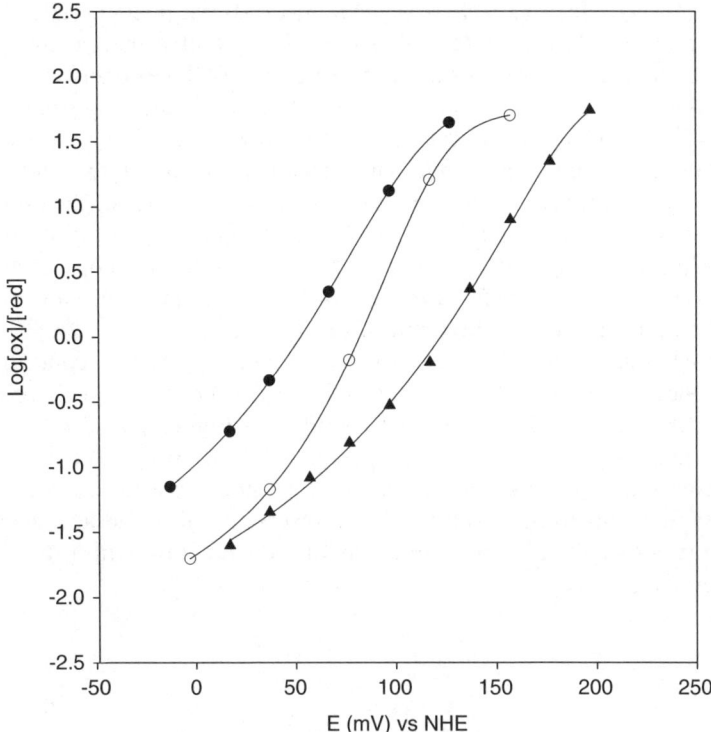

Figure 2.9 Nernst plots from spectroelectrochemistry of Hbs showing the influence of Hb structure and allosteric effector. ($\bigcirc$) Hb A_0 ($E_{1/2} = 85$ mV, $n_{1/2} = 2.0$) ($\bullet$) Horse Hb ($E_{1/2} = 55$ mV, $n_{1/2} = 1.7$) and ($\blacktriangle$) Hb A_0 in the presence of 200 mM KCl ($E_{1/2} = 125$, $n_{1/2} = 1.2$). Conditions: Pt mesh electrode; [Ru(NH$_3$)$_6$Cl$_3$] = 1 mM; [heme] = 0.06 mM; [MOPS] = 50 mM at pH 7.1 ([HEPES] = 50 mM for horse Hb at pH 7.5) and 20 °C. Figure adapted from ref. 67 and used with permission.

serine in horse Hb for alanine at position E14β. This change is beyond the first coordination shell, but creates a more hydrophilic distal heme pocket in horse Hb, increasing the ease of oxidation and oxygenation.[12,20,68,69]

This effect of changing the globin chain and/or the heme binding pocket on $E_{1/2}$ is distinct from that caused by heterotropic effectors (*e.g.*, protons, anions, *etc.*) bound at the allosteric binding site, or other sites spatially removed from the heme prosthetic group. Heterotropic effectors influence the oxidation process in a manner similar to the oxygenation process. As discussed earlier, this influence is reflected in the $E_{1/2}$ value and the midpoint, $n_{1/2}$, and maximum slope, $n_{\max}$, of the Nernst plot. During the oxygenation process, the influence of the heterotropic effector is reflected in the oxygen affinity expressed as $P_{1/2}$, and the slope of the Hill plot (commonly referred to as the Hill coefficient).[12] This behaviour may be interpreted in the context of a two state allosteric model involving a T-state of lower O_2 affinity and an R-state of higher O_2 affinity.[23,24,70] The model invokes a T $\rightleftharpoons$ R equilibrium that shifts between the

T [deoxy, Fe(II)-Hb] to the R (oxygenated) or R-like [met, Fe(III)-Hb] conformation of the Hb. Extensive studies have shown that structural changes that stabilise either the T- or the R-state conformation of Hb tetramer typically have comparable influence on both the oxidation and the oxygenation process.[11,26,67,71] (When these effects are not comparable, this also provides information concerning the underlying mechanism of redox and oxygen transport cooperativity in those specific cases.) The influence of the presence of Cl^-, a heterotropic ligand and allosteric effector, on the Nernst plot for Hb A_0 is illustrated in Figure 2.9.[67] The Nernst plot is shifted to more positive potentials, consistent with a decreased ease of Hb A_0 oxidation and a shift in the T $\rightleftharpoons$ R equilibrium in the direction of the T-state.

Although anions can act as heterotropic effectors, high anion concentrations can produce a complex effect on the redox parameters obtained from spectroelectrochemistry of Hbs. This is illustrated in Figure 2.10 for Hb A_0, Hb Presbyterian and spot fish Hb in the presence of Cl^- and P_i.[20] Increases in $E_{1/2}$ with increasing anion concentrations are consistent with heterotropic ligand binding and a shift to the T-state, while a reverse in this trend at higher anion concentrations suggests an environmental or second coordination shell effect is also operative.

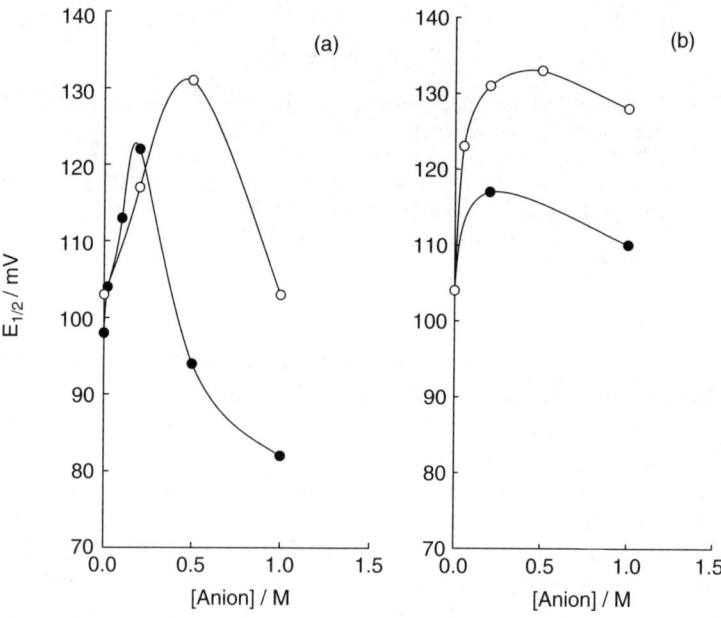

Figure 2.10   Anion effects on Hb $E_{1/2}$ values. (a) $E_{1/2}$ of Hb A_0 ($\bullet$) and Hb Presbyterian ($\bigcirc$) plotted as a function Cl^- concentration in 0.05 M MOPS pH 7.1 and (b) $E_{1/2}$ of spot fish Hb as a function of ($\bullet$) Cl^- concentration and ($\bigcirc$) inorganic phosphate concentration. Conditions: Pt mesh electrode: [heme] = 0.06–0.08 mM; [Ru(NH$_3$)$_6$Cl$_3$] = 1 mM; [MOPS] = 50 mM at pH 7.1; 20 °C. Figure adapted from ref. 20 and used with permission.

Unlike the midpoint slope ($n_{1/2}$) of an ideal Nernstian plot, the slope of a non-Nernstian response cannot be interpreted as the number of electrons involved in the oxidation/reduction process. For the Hbs, the n parameter is influenced by site–site heterogeneity and allosteric effects.[66,67] The n parameter is an indicator of the level of cooperativity that is operative; high n values indicate a high level of cooperativity, while low n values indicate reduced cooperativity. The sensitivity of the n parameter to heterotropic effectors may be seen in Figure 2.11.[26] The trend illustrated is consistent with the two-state (R and T) model for Hb.[23,24,70] Maximum cooperativity is indicated by the highest values for n_{max} (defined in Figure 2.4) as illustrated for HbA_0 in the absence of a heterotropic effector. The T-state is stabilised by heterotropic effectors (data points 1–4), which results in an increase in ease of reduction (increase in $E_{1/2}$) and a decrease in cooperativity (decrease in n_{max}) due to a diminished ease of T $\rightleftharpoons$ R shift as a result of T-state stabilisation. R-state stabilisation occurs in HbCPA and horse Hb (data points 6–9), which is characterised by an increase in ease of oxidation (lower $E_{1/2}$) and reduced cooperativity as illustrated by diminished n_{max} values.

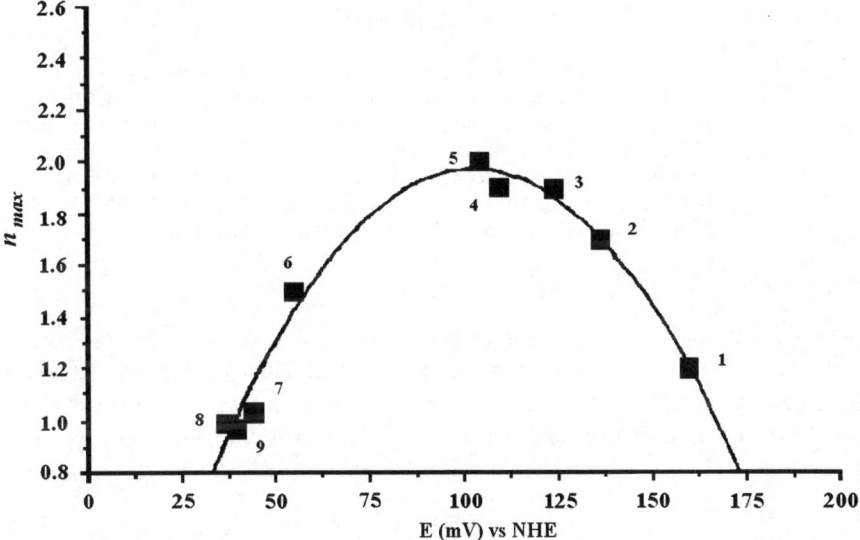

Figure 2.11 Plot showing the influence of the presence and absence of heterotropic ligands for various hemoglobins. n_{max}, as defined in Figure 2.4, plotted as a function of the midpoint potential for: *1*, Hb A_0 in 0.05 M MOPS, 0.2 M NaNO₃, and 0.25–0.6 mM IHP (in excess over [heme] which varied from 0.1 to 0.23 mM); *2*, Hb A_0 in 0.05 M MOPS, 0.2 M NaNO₃; *3*, Trout I Hb in 0.05 M MOPS, 0.2 M NaNO₃; *4*, Hb A_0 in 0.2 M MOPS; *5*, Hb A_0 in 0.05 M MOPS; *6*, HbCPA in 0.05 M MOPS, 0.2 M NaNO₃, and 0.25–0.6 mM IHP (in excess over [heme]); *7*, HbCPA in 0.05 M MOPS, 0.2 M NaNO₃; *8*, HbCPA in 0.2 M MOPS; *9*, hMb in 0.05 M MOPS, 0.2 M NaNO₃. Additional conditions: Pt mesh electrode; [Ru(NH₃)₆Cl₃] = 0.30–1.1 mM; [heme] = 0.1–0.23 mM; pH 7; 20 °C. HbCPA is carboxypeptidase digested Hb. Figure adapted from ref. 26 and used with permission.

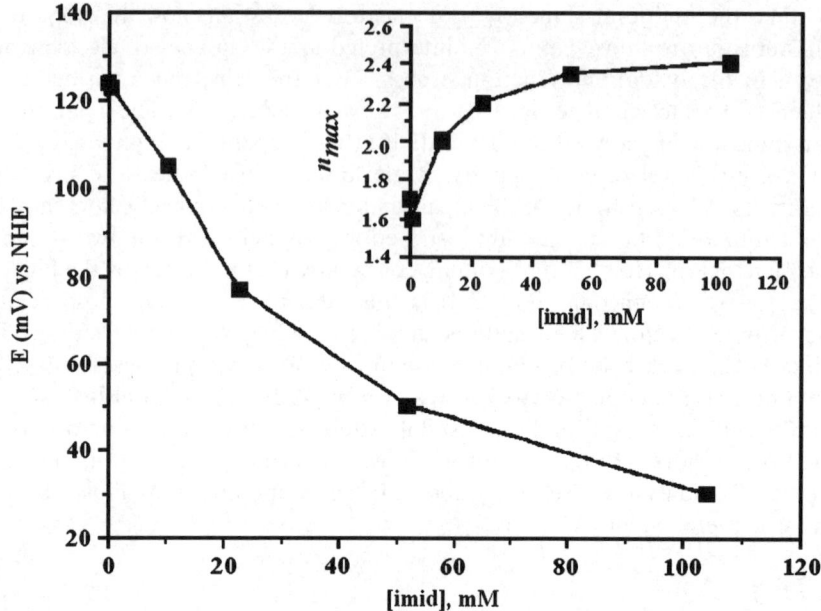

Figure 2.12 Plot of $E_{1/2}$ for Hb A_0 as a function of imidazole concentration that illustrates the influence of homotropic effector equilibrium reaction 14 on the ease of reduction and level of cooperativity (inset). Parameters obtained by spectroelectrochemistry; n_{max} is defined in Figure 2.4. Conditions: Pt mesh electrode; [heme] = 0.1–0.23 mM; [Ru(NH$_3$)$_6$Cl$_3$] = 0.30–1.1 mM; [NaNO$_3$] = 200 mM; [MOPS] = 50 mM at pH 7.1; 20 °C. Figure adapted from ref. 11 and used with permission.

In general, the presence of homotropic effectors will stabilise the R-state and shift the $E_{1/2}$ to lower values (*i.e.* the Hb is more easily oxidised). This is illustrated by the data in Figure 2.12, where the $E_{1/2}$ and n parameter variations are due to the homotropic effector equilibrium shown in reaction (2.14), where L = imidazole.[11]

$$Hb + L \rightleftharpoons Hb(L) \qquad (2.14)$$

Nernst plots and the associated descriptive parameters obtained from spectroelectrochemical data are also sensitive to modifications in the globin chains other than changing the amino-acid sequence. This is illustrated in Figure 2.13.[72] Nitrosylation of human Hb A_0 at the β-93 sulfhydryl group results in a significant shift in $E_{1/2}$ to lower potentials. This may have implications in the function of Hb as a transporter of NO, an *in-vivo* signaling molecule.[73] A similar shift in $E_{1/2}$ is observed for conversion of Hb A_0 to carboxypeptidase-digested hemoglobin, CPA-Hb. The lack of sigmoidal shape in this case suggests that the origin of the $E_{1/2}$ change is a shift in the T $\rightleftharpoons$ R equilibrium through stabilisation of the R-state.[26,72]

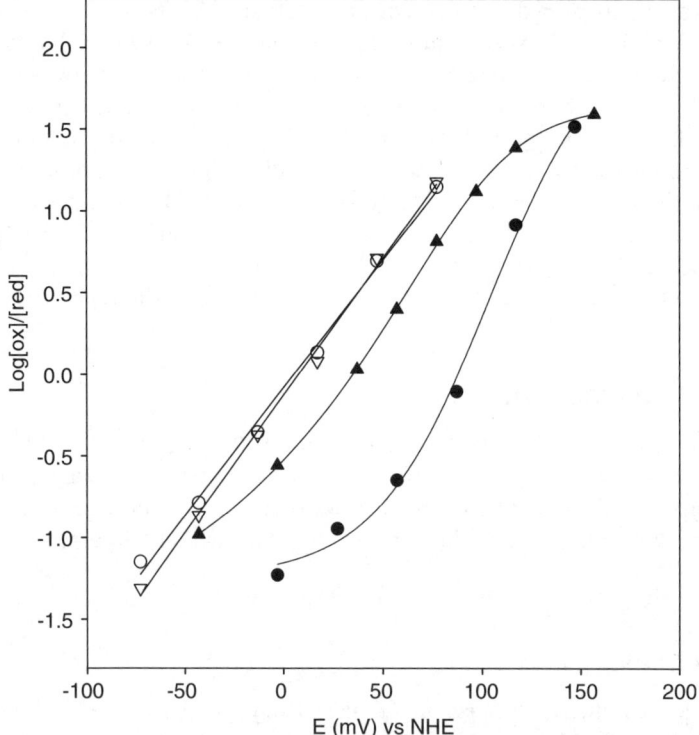

Figure 2.13 Nernst plots illustrating the influence of globin chain modification on $E_{1/2}$ and the level of cooperativity. (●) Hb A_0, (▲) SNO-Hb A_0 (90% β93 nitrosylated), (○) CPA-Hb and (∇) SNO-CPA-Hb (65% β93 nitrosylated). Conditions: [heme] = 0.06 mM to 0.08 mM in 0.05 M HEPES, 0.5 mM EDTA, 1 mM Ru(NH$_3$)$_6$Cl$_3$, 20 °C at pH 7.5. Figure adapted from ref. 72 and used with permission.

The ability of the globin group to control the redox potential of the heme is apparent with the range of redox potentials seen for Hbs and Mbs from various species (Scheme 2.1 and Figures 2.2, 2.9–2.13). This enables globin to protect the heme groups in Hb and Mb from rapid oxidation, which, in turn allows for the reversible binding of dioxygen to the heme prosthetic group. Homotropic and heterotropic effectors, as well as environmental anions, influence the thermodynamic ease of a given Hb oxidation, as well as the level of cooperativity, presumably by influencing the T ⇌ R equilibrium. All of these factors can be readily analysed through spectroelectrochemical investigation.

2.5 Conclusions

Spectroelectrochemistry is a rapid and convenient method to study the redox properties of iron proteins. Nernstian behaviour can be readily identified and used to determine the formal reduction potential ($E_{1/2}$) and number of electrons (n) involved in the oxidation/reduction process. Non-Nernstian behaviour can also be readily identified. Analysis of non-Nernstian plots can provide valuable information about chemical equilibria that may be coupled to the redox process, which by itself may be a reversible Nernstian process. This was illustrated here in the case of human transferrin. The shapes of the Nernst plots for some systems that exhibit non-Nernstian behaviour can provide valuable information about the ease of reduction ($E_{1/2}$) and the possibility of cooperativity and mechanisms for cooperativity. In the case of the Hbs, the parameters obtained through spectroelectrochemistry can be compared with the Hill parameters that characterise dioxygen affinity to aid in our understanding of the origin of cooperative effects in Hb-mediated redox and dioxygen transport.

Acknowledgements

The NSF (Grants CHE 0079066 and CHE 0418006) is gratefully acknowledged for financial support of our work in this area. We are indebted to our collaborators who have contributed to the experimental data base and ideas described in this chapter and whose names appear in the bibliography.

References

1. R. R. Crichton, *"Inorganic Biochemistry of Iron Metabolism: From Molecular Mechanism to Clinical Consequences"*, Wiley, New York, 2nd edn., 2001.
2. W. R. Harris, in *"Molecular and Cellular Iron Transport"*, ed. D. M. Templeton, Marcel Dekker, Inc., New York, 2002, 1.
3. R. G. Wilkins, *"Kinetics and Mechanism of Reactions of Transition-metal Complexes,"* VCH, New York, 2nd edn., 1991.
4. S. Dhungana and A. L. Crumbliss, *Geomicrobiol. J.*, 2005, **22**, 87.
5. S. J. Dong, J. J. Niu and T. M. Cotton, *Method. Enzymol.*, 1995, **246**, 701.
6. D. C. Kraiter, O. Zak, P. Aisen and A. L. Crumbliss, *Inorg. Chem.*, 1998, **37**, 964.
7. A. J. Bard and L. R. Faulkner, *"Electrochemical Methods: Fundamentals and Applications,"* John Wiley & Sons, Inc., New York, 2nd edn., 2001.
8. P. T. Kissinger and W. R. Heineman, *"Laboratory Techniques in Electroanalytical Chemistry,"* Marcel Dekker, Inc., New York, 2nd edn., 1996.
9. H. Durliat and M. Comtat, *Anal. Chem.*, 1982, **54**, 856.
10. J. P. Hoare, *J. Electrochem. Soc.*, 1984, **131**, 1808.
11. K. M. Faulkner, C. Bonaventura and A. L. Crumbliss, *Inorg. Chim. Acta*, 1994, **226**, 187.
12. E. Antonini and M. Brunori, *"Hemoglobin and Myoglobin in their Reactions with Ligands,"* North-Holland Publ. Co., Amsterdam, 1971.
13. S. Dong, Y. Zhu and S. Song, *Bioelectrochem. Bioenerg.*, 1989, **21**, 233.
14. S. Song and S. Dong, *Bioelectrochem. Bioenerg.*, 1988, **19**, 337.
15. M. Rivera, M. A. Wells and F. A. Walker, *Biochemistry*, 1994, **33**, 2161.

16. C. H. Taboy, K. G. Vaughan, T. A. Mietzner, P. Aisen and A. L. Crumbliss, *J. Biol. Chem.*, 2001, **276**, 2719.
17. A. L. Raphael and H. B. Gray, *J. Am. Chem. Soc.*, 1991, **113**, 1038.
18. L. S. Reid, V. T. Taniguchi, H. B. Gray and A. G. Mauk, *J. Am. Chem. Soc.*, 1982, **104**, 7516.
19. S. Dhungana, C. H. Taboy, O. Zak, M. Larvie, A. L. Crumbliss and P. Aisen, *Biochemistry*, 2004, **43**, 205.
20. C. H. Taboy, K. M. Faulkner, D. Kraiter, C. Bonaventura and A. L. Crumbliss, *J. Biol. Chem.*, 2000, **275**, 39048.
21. P. Aisen, *Metal Ions Biol. Sys.*, 1998, **35**, 585.
22. W. R. Harris, *J. Inorg. Biochem.*, 1986, **27**, 41.
23. K. Imai, *"Allosteric Effects in Hemoglobin,"* Cambridge University Press, Cambridge, 1982.
24. J. Monod, J. Wyman and J. P. Changeux, *J. Mol. Biol.*, 1965, **12**, 88.
25. M. Perutz, *"Science Is Not a Quiet Life: Unravelling the Atomic Mechanism of Haemoglobin,"* World Scientific Publ. Co., London, 1997.
26. K. M. Faulkner, C. Bonaventura and A. L. Crumbliss, *J. Biol. Chem.*, 1995, **270**, 13604.
27. M. R. Gunner, E. Alexov, E. Torres and S. Lipovaca, *J. Biol. Inorg. Chem.*, 1997, **2**, 126.
28. T. A. Mietzner, S. B. Tencza, P. Adhikari, K. G. Vaughan and A. J. Nowalk, in *"Curr. Top. Microbiol. Immunol,"* ed. P. K. Vogt and M. J. Mahan, Springer, Berlin, 1998, **225**, 114.
29. R. T. A. MacGillivray and A. B. Mason, in *"Molecular and Cellular Iron Transport,"* ed. D. M. Templeton, Marcel Dekker, Inc., New York, 2002, 41.
30. S. Dhungana, C. H. Taboy, D. S. Anderson, K. G. Vaughan, P. Aisen, T. A. Mietzner and A. L. Crumbliss, *Proc. Nat. Acad. Sci. (USA)*, 2003, **100**, 3659.
31. J-L. Pierre, M. Fontecave and R. R. Crichton, *BioMetals*, 2002, **15**, 341.
32. J-L. Pierre and M. Fontecave, *BioMetals*, 1999, **12**, 195.
33. C. M. Bruns, A. J. Nowalk, A. S. Arvail, M. A. McTigue, K. G. Vaughan, T. A. Mietzner and D. E. McRee, *Nature Struct. Biol.*, 1997, **4**, 919.
34. S. J. Ferguson, in *"Prokaryotic Structure and Function : A New Perspective: 47th Symposium of the Society for General Microbiology,"* ed. S. Mohan, M. Dow and J. Coles, Cambridge University Press, Cambridge, 1991, 311.
35. H. Boukhalfa, D. S. Anderson, T. A. Mietzner and A. L. Crumbliss, *J. Biol. Inorg. Chem.*, 2003, **8**, 881.
36. P. Aisen, A. Leibman and J. Zweier, *J. Biol. Chem.*, 1978, **253**, 1930.
37. S. Bailey, R. W. Evans, R. C. Garratt, B. Gorinsky, S. Hasnain, C. Horsburgh, H. Jhoti, P. F. Lindley, A. Mydin, R. Sarra and J. L. Watson, *Biochemistry*, 1988, **27**, 5804.
38. E. N. Baker, *Adv. Inorg. Chem.*, 1994, **41**, 389.
39. E. N. Baker and P. F. Lindley, *J. Inorg. Biochem.*, 1992, **447**, 147.

40. P. Aisen, in "*Iron Metabolism in Health and Disease,*" ed. J. H. Brock, J. W. Halliday, M. J. Pippard, and L. W. Powell, W. B. Saunders, London, 1994, 1.

41. P. Aisen, C. Enns and M. Wessling-Resnick, *Int. J. Biochem. Cell Biol.,* 2001, **33**, 940.

42. N. C. Andrews, *N. Eng. J. Med.,* 1999, **341**, 1986.

43. C. A. Enns, in "*Molecular and Cellular Iron Transport,*" ed. D. M. Templeton, Marcel Dekker, Inc., New York, 2002, 71.

44. A. Dautry-Varsat, A. Ciechanover and H. F. Lodish, *Proc. Natl. Acad. Sci. (USA),* 1983, **80**, 2258.

45. R. D. Klausner, J. V. Ashwell, J. B. VanRenswoude, J. Harford and K. Bridges, *Proc. Natl. Acad. Sci. (USA),* 1983, **80**, 2263.

46. H. Gunshin and M. A. Hediger, in "*Molecular and Cellular Iron Transport,*" ed. D. M. Templeton, Marcel Dekker, Inc., New York, 2002, 155.

47. M.-T. Nunez, V. Gaete, J. A. Watkins and J. Glass, *J. Biol. Chem.,* 1990, **265**, 6688.

48. N. C. Andrews, M. D. Fleming and H. Gunshin, *Nutr. Rev.,* 1999, **57**, 114.

49. G. R. Buettner, *Arch. Biochem. Biophys.,* 1993, **300**, 535.

50. E. C. Theil and P. Aisen, in "*Iron Transport in Microbes, Plants, and Animals,*" ed. G. Winkelmann, D. van der Helm and J. B. Neilands, VCH, Weinheim, 1987, 491.

51. J. H. Katz, *J. Clin. Invest.,* 1961, **40**, 2143.

52. D. C. Harris, J. A. Rinehart, D. Hereld, R. W. Schwartz, F. P. Burke and A. P. Salvador, *Biochem. Biophys. Acta,* 1985, **838**, 295.

53. S. A. Kretchmar, Z. E. Reyes and K. N. Raymond, *Biochem. Biophys. Acta,* 1988, **956**, 85.

54. Y. Cheng, O. Zak, P. Aisen, S. C. Harrison and T. Watz, *Cell,* 2004, **116**, 565.

55. C. M. Lawrence, S. Ray, M. Babyonyshev, R. Galluser, D. W. Borhani and S. C. Harrison, *Science,* 1999, **286**, 779.

56. B. Ecarot-Charrier, V. L. Grey, A. Wilczynska and H. M. Schulman, *Can. J. Biochem.,* 1980, **58**, 418.

57. J. J. R. Frausto da Silva and R. J. P. Williams, "*The Biological Chemsitry of the Elements,*" Oxford University Press, Oxford, 2nd edn., 2001.

58. W. T. Morgan and A. Smith, *Adv. Inorg. Chem.,* 2001, **51**, 205.

59. H. R. Faber, C. R. Groom, H. M. Baker, W. T. Morgan, A. Smith and E. N. Baker, *Structure,* 1995, **3**, 551.

60. M. Paoli, B. F. Anderson, H. M. Baker, W. T. Morgan, A. Smith and E. N. Baker, *Nat. Struct. Biol.,* 1999, **6**, 926.

61. M. M. Flaherty, K. R. Rish, A. Smith and A. L. Crumbliss, *BioMetals,* 2007, Online First 10.1007/s10534-007-9112-9.

62. K.-B. Lee, E. Jun, G. N. LaMar, I. Rezzano, R. K. Pandey, K. M. Smith, F. A. Walker and D. H. Buttlaire, *J. Am. Chem. Soc.,* 1991, **113**, 3576.

63. L. S. Reid, A. R. Lim and A. G. Mauk, *J. Am. Chem. Soc.,* 1986, **108**, 8197.

64. N. V. Shipulina, A. Smith and W. T. Morgan, *J. Protein Chem.,* 2001, **20**, 145.

65. M. Bolognesi, S. Onesti, G. Gatti, A. Coda, P. Ascenzi and M. Brunori, *J. Mol. Biol.*, 1989, **205**, 529.
66. C. H. Taboy, C. Bonaventura and A. L. Crumbliss, *Bioelectrochem. Bioenerg.*, 1999, **48**, 79.
67. C. H. Taboy, C. Bonaventura and A. L. Crumbliss, *Method. Enzymol.*, 2002, **353**, 187.
68. M. Perutz, *Nature*, 1960, **185**, 491.
69. T. Kleinschmidt and G. Braunitzer, *Biomed. Biochim. Acta*, 1983, **42**, 685.
70. M. Perutz, "*Mechanisms of Cooperativity and Allosteric Regulation in Proteins*," Cambridge University Press, Cambridge, 1990.
71. C. Bonaventura, S. Tesh, K. M. Faulkner, D. Kraiter and A. L. Crumbliss, *Biochemistry*, 1998, **37**, 496.
72. C. Bonaventura, C. H. Taboy, P. S. Low, R. D. Stevens, C. Lafon and A. L. Crumbliss, *J. Biol. Chem.*, 2002, **277**, 14557.
73. C. Bonaventura, A. Fago, R. Henkens and A. L. Crumbliss, *Antiox. Redox Signal.*, 2004, **6**, 979.

CHAPTER 3

Mixed-Valence Intermediates as Ideal Targets for Spectroelectrochemistry (SEC)

WOLFGANG KAIM,[a] BIPRAJIT SARKAR[a] AND GOUTAM KUMAR LAHIRI[b]

[a] Institut für Anorganische Chemie, Universität Stuttgart, Pfaffenwaldring 55, D-70550 Stuttgart, Germany; [b] Department of Chemistry, Indian Institute of Technology, Bombay, Powai, Mumbai-400076, India

3.1 Introduction

Mixed-valence compounds continue to attract attention,[1] not least because of their occurrence as intermediates of multistep redox systems.[2–6] Mixed-valence species are found in the geo- and biosphere, as evident from minerals such as Fe_3O_4 and from metalloproteins, where the $Fe^{II}/Fe^{III}/Fe^{IV}$, Cu^{I}/Cu^{II} and $Mn^{II}/Mn^{III}/Mn^{IV}$ combinations are established.[7–10] Man-made mixed-valence compounds, starting from Prussian Blue in the early 18th century,[11] have raised interest in what is now known as materials science because of their often special optical, electrical and magnetic properties.[1,12] These physical properties then prompted attempts at increasingly sophisticated levels to theoretically understand and computationally reproduce the experimental features of mixed-valence compounds.[1,2,5,13] More recent developments involve the application of mixed-valence systems as models and actual components in the areas of molecular electronics[14] and molecular computing.[15]

In addition, mixed-valence intermediates feature prominently in mechanistic approaches aimed at electron-transfer processes,[2] including ultrafast valence exchange[16] and multielectron catalysis (where mixed-valency during stepwise processes is inevitable).[17] The modelling and the classification derived[1] have

Spectroelectrochemistry
Edited by Wolfgang Kaim and Axel Klein

recently been applied not only to transition-metal compounds but also to coupled organic redox systems[18-20] and even to main-group-element-containing molecules.[21] Another extension concerns the investigation of interactions not only between electron-transfer sites[1-5] but between reaction centres, involving electron *and* atom transfer reactivity, as evident from apparently "irreversible" steps in, *e.g.*, cyclic voltammetry.[22-24]

Some mixed-valence systems are sufficiently stable to be isolated, such as Prussian Blue $(Fe^{III})_4[Fe^{II}(CN)_6]_3$ or the much studied Creutz–Taube ion $\{(\mu\text{-}pz)[Ru(NH_3)_5]_2\}^{5+}$ (**1**), pz=pyrazine.[2-4] However, in many (but not all[25]) cases they are less robust than their homovalent congeners, often being formed at inconveniently high or low redox potentials, or existing only in a very limited electrochemical potential range as quantified by a small comproportionation constant K_c.

$$M^n\text{--}L\text{--}M^n \underset{+e^-}{\overset{-e^-}{\rightleftharpoons}} M^n\text{--}L\text{--}M^{n+1} \underset{+e^-}{\overset{-e^-}{\rightleftharpoons}} M^{n+1}\text{--}L\text{--}M^{n+1} \qquad (3.1)$$

$$\text{Red} \qquad (E_1) \qquad \text{Int} \qquad (E_2) \qquad \text{Ox}$$

$$\text{comproportionation constant } K_c = \frac{[\text{Int}]^2}{[\text{Red}][\text{Ox}]} = 10^{\Delta E/59\,\text{mV}} \qquad (3.2)$$

$$\Delta E = E_2 - E_1$$

$$RT \cdot \ln K_c = nF \cdot \Delta E$$

In such instances the spectroelectrochemical methods come in handy, permitting *in situ* identification and characterisation by suitable spectroscopy of intermediates that exist in equilibrium with other redox forms or that have a limited lifetime under the circumstances. While the Creutz–Taube ion **1** is thus easily isolated,[2] its decacyano analogue **2**,[26] the corresponding di-iron(III,II) complex **3**,[27] and the closely related organometallic species **4**[28] have been conveniently generated and investigated by SEC (UV–Vis–NIR, EPR, IR). Whereas IR or Raman vibrational spectroscopy yield typical absorption bands for mixed-valence *and* homovalent states alike, the mixed-valence situation is frequently distinguished from neighbouring redox states through the low-energy intervalence charge-transfer (IVCT) bands, often found in the otherwise less frequented near-infrared (NIR) region,[29] or by EPR signals due to the paramagnetism in the common case of an odd-electron intermediate. While typical mixed-valence intermediates are part of purely electron-transfer series,[2-4] the spectroelectrochemical method is also suitable for studying "irreversible" electrochemical (*i.e.* redox-activated) reactions in which electron transfer (E) is accompanied by chemical bond-breaking or bond-forming steps (C).[22-24] The spectroelectrochemical analysis of such EC-involving reactions (which may include mixed-valence forms) can be useful for understanding electrocatalysis and general redox catalysis.

	k		M	L_n
$ML_n \rceil^k$ N N L_nM	1	5+	Ru	$(NH_3)_5$
	2	5−	Ru	$(CN)_5$
	3	5−	Fe	$(CN)_5$
	4	1+	Mo	$(CO)_3(PR_3)_2$

A very limited selection of examples will serve here to illustrate the power of the spectroelectrochemical approach, including optical "spectro"electrochemistry (Section 3.3), the selective but often most informative EPR spectroelectrochemistry (Section 3.4 and Chapter 7), and the well-established[30] but also increasingly employed[16,31] infrared vibrational spectroelectrochemistry (Section 3.5 and Chapter 1). Although mainly transition-metal coordination compounds will be discussed, the previously mentioned extension of the concept and methodology to nonmetallic systems[18–21] should be kept in mind.

3.2 Requirements for Using SEC in Mixed-Valence Chemistry

To detect mixed-valence intermediates *via* SEC in solution requires the presence of this intermediate in sufficient concentration. It is thus not necessary to have a nearly 100% participation of the intermediate in an equilibrium as quantified by comproportionation constants $K_c > 10^3$. The problems associated with small K_c values of mixed-valence intermediates have been extensively discussed by Richardson and Taube,[3] and that equilibrium parameter can be considerably enhanced by changing the environment such as the temperature, the solvent (complex $3^{27,32}$) or the electrolyte.[33] This observation demonstrates the sensitivity of mixed-valence intermediates with respect to the medium.

The formation of mixed-valence intermediates is not limited to purely electron-transfer redox series, as pointed out before and as described below; however, the overall reversibility is typically checked by regenerating the starting spectrum to 100% after completing the backreaction. If the method is selective only for the intermediate state, as, *e.g.*, for EPR active odd-electron species in equilibrium with "EPR-silent" states, or if the two-step redox system can be analysed *via* stepwise monitoring,[34] the spectral features of an intermediate as distinct from those of the neighbouring redox states can be well identified. A case in point is discussed further below by example of intermediates $[(C_nR_n)M(\mu\text{-}L)M(C_nR_n)]^+$.

An important aspect in applying SEC methods to the study of mixed-valence intermediates concerns the timescale.[4b] The frequently asked question as to the valence situation, *i.e.* averaged (delocalised) or nonaveraged (localised, trapped),[1–5] must always be discussed with respect to the time frame Δt of the method used to identify the species in question. There is a huge difference between electronic absorption spectroscopy with $\Delta t < 10^{-15}$ s and NMR spectroscopy with a typical $\Delta t = 1$ s effective timescale.[4b] Accordingly, systems such as **5** have been found that apparently show a different "spectroscopic symmetry" *via* IR vibrational spectroscopy ($\Delta t \approx 10^{-13}$ s, trapped valence) and EPR spectroscopy ($\Delta t \approx 10^{-8}$ s, averaged valence).[35]

5

Temperature-dependent spectroelectrochemistry[36] not only serves to track and investigate less stable intermediates such as **3**[27] but also to study the dynamics of valence exchange.[16] Variable-temperature SEC cells have been described.[36] In order to effect a rapid exhaustive electrochemical conversion different cell designs have been used, employing thin-layer constructions with optically transparent grids or conducting materials (ITO, B-doped diamond) or waveguide arrangements.[31,37]

3.3 UV–Vis–NIR Absorption Spectroscopy

The most conspicuous feature of many mixed-valence compounds is their long-wavelength absorption, as illustrated so vividly by the intensely coloured Prussian Blue, formulated essentially as $\{Fe^{III}_4[Fe^{II}(CN)_6]_3\}$.[1,11] Simply speaking, this low-energy absorption results from an allowed transition ("intervalence charge transfer", IVCT) between electron-rich iron(II) to electron-deficient iron(III) *via* the bridging cyanide ligands. The asymmetry resulting from unsymmetrical CN^- and from the low-spin configuration of C-coordinated Fe^{II} *vs.* N-bound high-spin Fe^{III} causes the relatively high-energy absorption to occur in the visible range;[1,11] other mixed-valence coordination compounds such as the symmetrical Creutz–Taube ion $[(H_3N)_5Ru(\mu\text{-pz})Ru(NH_3)_5]^{5+}$ (**1**) with its assumed averaged valence of $+2.5$ for both ruthenium centres[2–5] have their absorptions in the near-infrared region, here at 1560 nm corresponding to $6400\,cm^{-1}$. The near-infrared (800–2500 nm, 12 500–4000 cm^{-1}) has become a very interesting spectral region due to its analytical potential and because of the relevance for information transfer *via* fibre optics.[29]

Typical responses of redox systems with a mixed-valence intermediate are shown in Figures 3.1 and 3.2, illustrating the rise (and decrease) of the IVCT absorption band as an often separated additional feature on (stepwise) electron transfer.[1–4] The spectral behaviour as in Figure 3.1 for the $Fe^{III}Fe^{II}$ (or, better,

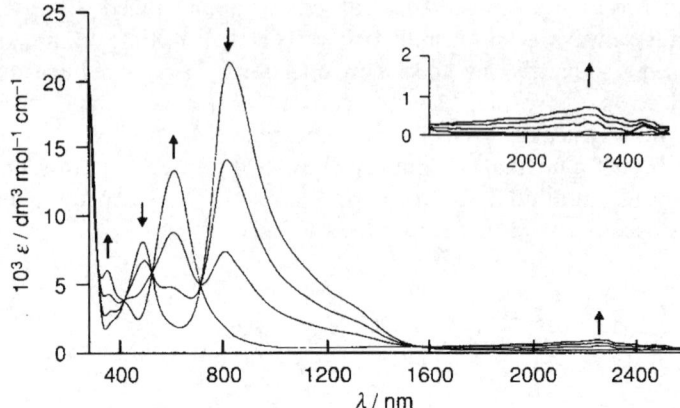

Figure 3.1 UV-VIS-NIR spectroelectrochemical response during the conversion $\{(\mu\text{-bmtz})[\text{Fe(CN)}_4]_2\}^{(4-) \to (3-)}$ (**6**) in $\text{CH}_3\text{CN}/0.1\,\text{M}$ Bu_4NPF_6 (insert: enhanced NIR region).

$\text{Fe}^{2.5}\text{Fe}^{2.5}$) species $\{(\text{NC})_4\text{Fe}(\mu\text{-bmtz})\text{Fe(CN)}_4\}^{3-}$ **6** ($K_c = 10^{14}$ in $\text{CH}_3\text{CN}/0.1\,\text{M}$ Bu_4NPF_6)[32] from an experiment with an optically transparent thin-layer electrolysis (OTTLE) cell[37] with Pt gauze working electrode is only one form of graphical representation, difference spectra or three-dimensional plots are also being used.[37]

3,6-bis(2-pyrimidyl)-1,2,4,5-
tetrazine (bmtz)

Although Figures 3.1 and 3.2 show a typical behaviour it is by no means compelling that the increasing long-wavelength absorption of an intermediate signifies a mixed-valence situation. Many organic ligands, serving, *e.g.*, as bridges in dinuclear or oligonuclear complexes, can undergo stepwise electron transfer that may give rise to low-energy absorbing radical intermediates.[38,39] An additional independent probe such as EPR (Section 3.4) is thus advised before finally attributing long-wavelength bands to IVCT transitions.[39]

On the other hand, the appearance of a near-infrared absorption for an electrochemically or otherwise generated intermediate can be viewed as indicating a *potentially* mixed-valence situation. For instance, the stepwise chloride-dissociative reduction of compounds $[\text{Cl}(\text{C}_n\text{R}_n)\text{M}(\mu\text{-L})\text{M}(\text{C}_n\text{R}_n)\text{Cl}]^{2+}$, $n = 5$ for $\text{M} = \text{Rh}$ or Ir, $n = 6$ for $\text{M} = \text{Ru}$, may lead, after addition of three electrons (and the loss of two Cl^- ions) to a species $[(\text{C}_n\text{R}_n)\text{M}(\mu\text{-L})\text{M}(\text{C}_n\text{R}_n)]^+$ with a distinct near-infrared absorption.[22-24] In conjunction with the EPR spectra showing large *g* factor anisotropy (see Section 3.4) it can be concluded that these are mixed-valence metal intermediates and not radical complexes, obtained

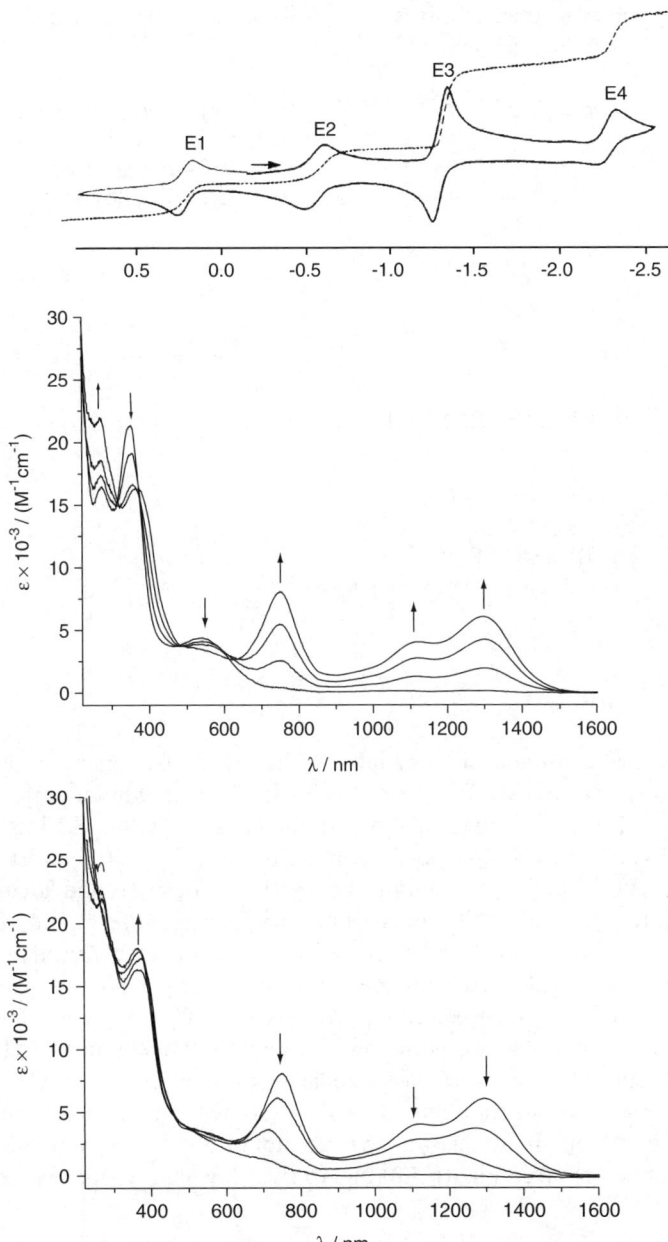

Figure 3.2 (Top) Cyclic voltammogram of $\{(\mu\text{-abpy})[MCl(C_5Me_5)]_2\}^{2+}$ in $CH_3CN/$ 0.1 M Bu_4NPF_6. (Center and bottom) UV–VIS–NIR spectroelectrochemical response during the conversions $\{(\mu\text{-abpy})[MCl(C_5Me_5)][M(C_5Me_5)]\}^+ \rightarrow \{(\mu\text{-abpy})[M(C_5Me_5)]_2\}^+$ and $\{(\mu\text{-abpy})[M(C_5Me_5)]_2\}^{(+)\rightarrow(0)}$ in $CH_3CN/0.1$ M Bu_4NPF_6.

through a process that involves atom displacement (*i.e.* chemical steps) in addition to electron transfer.[22–24]

$$\{(\mu\text{-BL})[M(C_nR_n)Cl]_2\}^{2+}$$

BL: bridging ligand

M = Rh, Ir; $n = 5$
M = Ru, Os; $n = 6$

$\downarrow$ +e⁻

$$\{(\mu\text{-BL})[M(C_nR_n)Cl]_2\}^+$$

$\downarrow$ +e⁻, −Cl⁻

$$\{(\mu\text{-BL})[M(C_nR_n)Cl][M(C_nR_n)]\}^+$$

$\downarrow$ +e⁻, −Cl⁻

$$\{(\mu\text{-BL})[M(C_nR_n)]_2\}^+$$

$\downarrow$ +e⁻

$$\{(\mu\text{-BL})[M(C_nR_n)]_2\}$$

The 3-electron-reduced organometallic mixed-valence intermediates $[(C_5Me_5)M(\mu\text{-L})M(C_5Me_5)]^+$, M = Rh or Ir, L = 2,5-diiminopyrazines, thus involve an unusual[40] oxidation state combination $M^I M^{II}$ with $d^7 d^8$ configuration and very small comproportionation constants $K_c < 50$. Nevertheless, the stepwise electrochemical reduction allowed us to observe and identify those intermediates *via* their IVCT bands around 1600 nm (Figure 3.2) and their EPR spectra[23] that signify mostly, albeit not exclusively, spin localisation at the metal centres. Similar results were obtained for the closely related organometallic diruthenium system $[(Cym)Ru(\mu\text{-abpy})Ru(Cym)]^+$, Cym = *p*-cymene (an arene ligand, 4-isopropyltoluene), with lower than the conventional[2–5] ($Ru^{III}Ru^{II}$) oxidation states.[24] The spectroelectrochemical method, here UV–Vis–NIR spectroscopy in conjunction with EPR, thus allows for the detection of intermediates which might otherwise even go unnoticed in conventional voltammetric experiments (see Figure 3.2) due to their low equilibrium concentration.

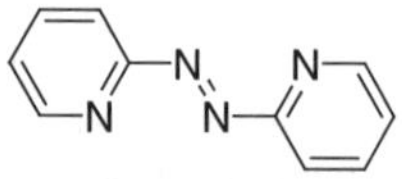

2,2′-azobispyridine
(abpy)

In general, however, the energy of the IVCT transition is not restricted to the near-infrared region. Depending on the electronic coupling[1–5] the corresponding band may be shifted either to higher energies, into the visible where it may be obscured by other bands, or to lower energies, into the midinfrared region where it may remain undetected among the generally much more narrow vibrational bands. Such an example has been observed for the Mo^IMo^0 combination[41] $\{(OC)_2(^nBu_3P)_2Mo(\mu\text{-bpym})Mo(P^nBu_3)_2(CO)_2\}^+$ (7) with 2,2′-bipyrimidine (bpym) as a bridge, which shows both CH and CO vibrational stretching absorptions in the presence of a broad underlying electronic band centred at about $2700\,cm^{-1}$.[41]

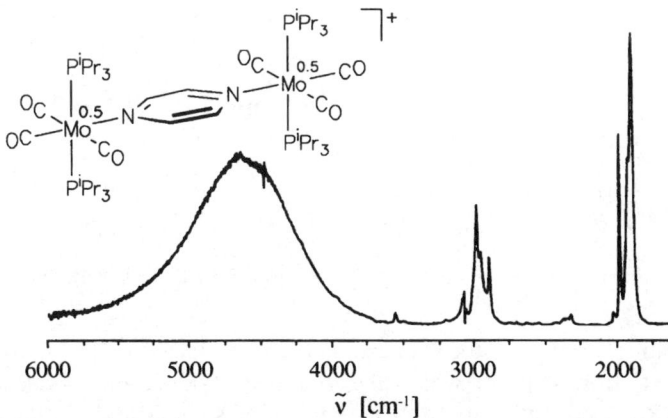

2,2′-bipyrimidine (bpym)

Figure 3.3, showing the SEC response in the formation of the related complex 4, illustrates the difference between the electronic transition (broad new band) and vibrational features (narrow CH and shifted CO bands) in the near- to mid-IR regions.[28]

Further, the intensities of the typically several IVCT transitions[5] may vary quite extensively, ranging from very strong to practically undetectable weak absorptions.[2–5,42–43] The typically lower sensitivity of spectrophotometers in the NIR region may further cause problems in the identification of weak IVCT bands, especially when a large-bandwidth $\Delta v_{1/2}$ makes detection even more difficult. Narrow bands with bandwidths $\Delta v_{1/2}$ of a few $100\,cm^{-1}$ at half-height as well as very broad absorptions with $\Delta v_{1/2} = 5000\,cm^{-1}$ such as for 2 have been observed.[25–28] The spectral width has been often used to estimate the extent of valence averaging according to a model derived for weakly coupled systems by Hush,[1,2–5] later approaches have extended and modified this

Figure 3.3   VIS–NIR–IR spectroelectrochemical result from the conversion $\{(\mu\text{-pz})[Mo(CO)_3(PR_3)_2]_2\}^{(0)\rightarrow(+)}$ (4) in $CH_2Cl_2/0.1\,M$ Bu_4NPF_6 (R = isopropyl).

Table 3.1 Characteristics of mixed-valence intermediates.[a]

	1	2	3	4
K_c	$10^{6.6}$	$10^{4.7}$	$10^{6.5}$	$10^{6.4}$
λ_{IVCT} (nm)	1560	1760	2475	2150
ε_{IVCT} ($M^{-1}cm^{-1}$)	5000	2600	3900	n.d.
$\Delta\nu_{1/2}$ (cm^{-1})	1250	4200	1500	700
g	2.799	n.o. (4 K)	2.45 ($g_\perp$)	2.096
	2.487		1.79 ($g_\parallel$)	2.066
	1.346			1.992
Ref.	2b,5	26	27	28

[a]For measurement conditions see references.

model.[13] In any case, broad absorption features signify substantial geometrical reorganisation after excitation to the IVCT excited state.

For comparison, the essential parameters of the near-infrared band for the pyrazine-bridged mixed-valence intermediates **1–4** are summarised in Table 3.1.

Table 3.1 demonstrates how charges play a large role, mostly *via* interactions with the environment. Thus, complex **3** has been studied in aqueous medium without appreciable metal–metal interaction[44] but was found to exhibit a much more pronounced effect in aprotic acetonitrile.[27]

Weakly coupled mixed-valence systems or those with certain coordination modes[6b] such as chelation by a bis- or tris-bidentate acceptor ligand (*e.g.* bptz, bmtz or dqp) can show very low intensity IVCT bands in the NIR (Figure 3.1) or even escape completely their detection by electronic absorption spectro-electrochemistry.[42,43]

3,6-bis(2-pyridyl)-1,2,4,5-   diquinoxalino[2,3-*a*:2′,3′-*c*]phenazine
tetrazine (bptz)   (dqp)

For instance, trinuclear species such as $(\mu_3\text{-L})[Ru(acac)_2]_3$ can be oxidised in three one-electron steps of which the first two produce mixed-valence inter-mediates ($Ru^{III}Ru^{II}Ru^{II}$ and $Ru^{III}Ru^{III}Ru^{II}$). For some ligands L, such as dqp, the EPR spectroscopically characterised first (*i.e.* odd-electron) intermediates with metal-centred spin do not exhibit a detectable band in the near- or midinfrared region.[43] The reason for this behaviour is not yet clear, especially since related

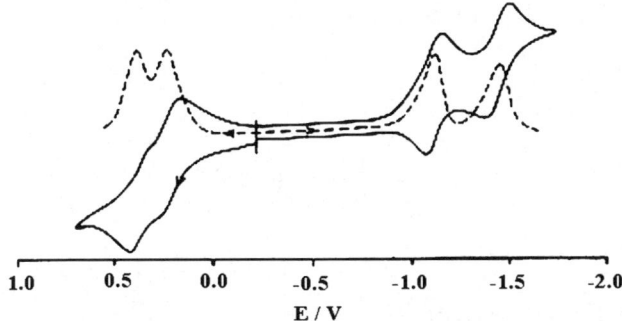

Figure 3.4   Cyclic voltammogram and differential pulse voltammogram of {(μ-tithb) [Ru(acac)$_2$]$_2$} (**8**) in CH$_3$CN/0.1 M Et$_4$NClO$_4$ (potentials *vs*. SCE).

compounds {(μ$_3$-L′)[Ru(bpy)$_2$]$_3$}$^{7+}$, L′ = 1,4,5,8,9,12-hexaazatriphenylene, show typical such IVCT bands in the near-infrared.[45]

While IVCT bands may thus not be detectable in the NIR region, NIR bands, on the other hand, may not be caused by IVCT transitions, as suggested by their appearance.

The complex {(μ-tithb)[Ru(acac)$_2$]$_2$} **8** with the noninnocent 3,3′,4,4′-tetra-imino-3,3′,4,4′-tetrahydrobiphenyl (tithb) ligand shows a clean double two-step oxidation and reduction behaviour (Figure 3.4).[46] One-electron oxidation to **8**$^+$ produces a weak NIR band at 1570 nm ($\varepsilon = 800$ M^{-1} cm^{-1}) while a more con-spicuous NIR absorption around 2160 nm ($\varepsilon = 4000$ M^{-1} cm^{-1}) occurs when **8** is reduced with one electron. According to EPR (see below) a mixed-valence oxidation state formulation is only appropriate for **8**$^+$ = {(μ-tithb)Ru^{III}Ru^{II}}$^+$ while the anion **8**$^-$ shows a radical type signal at $g = g$(electron) = 2.0023 that suggests the {(μ-tithb·$^-$)Ru^{II}Ru^{II}}$^-$ instead of the conceivable {(μ-tithb^{2-}) Ru^{III}Ru^{II}}$^-$ formulation.[46]

3,3′,4,4′-tetraimino-3,3′,4,4′-tetrahydrobiphenyl
(tithb)

In a related case, complexes {(μ-dih-R)[Ru(acac)$_2$]$_2$} with noninnocent 1,2-diiminohydrazido ligands dih-R (resulting from ring-opening of tetrazines)[47] show strong absorptions around 1500 nm when reduced with one electron. Although the mixed-valence oxidation state formulation (μ-dih-R^{2-})Ru^{III}Ru^{II} is tempting to interpret such a finding, the corresponding EPR results show ligand centred spin that clearly points to a description as (μ-dih-R·$^-$)Ru^{II}Ru^{II}.[47]

$$\begin{array}{c} R \\ \diagup \\ HN \diagup \diagdown N^{-} \diagdown N^{-} \diagdown \diagup NH \\ \diagdown R \end{array}$$

1,2-diiminohydrazido(2-)
(dih-R^{2-})

3.4 EPR Spectroscopy

Of the several less common spectroscopic methods to combine with electrochemical intermediate generation such as luminescence, Raman, NMR, or X-ray absorption spectroscopy,[48] the EPR method[49] is presented here because of its relative simplicity and pronounced selectivity. Only paramagnetic compounds with a certain, not too rapid relaxation rate from the spin-excited state give detectable signals for EPR spectroscopy, which helps to disregard many simultaneously present species. On the other hand, the rather slow time frame ($\Delta t \approx 10^{-8}$ s)[4b] and the sensitivity of the EPR method to electronic influences from the participating atoms *via* g-factor shift and hyperfine interaction can render EPR a very valuable method to determine the site of electron transfer (ligand or metal)[50] as well as the spin and thus valence distribution.

The EPR resonance condition $hv = g\beta H$ involves the g-factor as a substance-specific quantity that depends to a large extent on the admixture of higher excited states with nonzero orbital momentum.[49] Accordingly, spin–orbit coupling becomes crucial, and this effect is highly dependent on the spin–orbit coupling constant, a quantity increasing very strongly with the atomic number.[49] Therefore, it makes a huge difference whether heavy elements such as transition metals are dominating the spin and valence distribution or whether lighter (ligand) atoms such as C, N, O or H, *etc.* are prevalent.[38,50] Even in the absence of direct hyperfine information from individual nuclei the g-factor and the anisotropy of the corresponding tensor as measured from frozen solutions indicate the predominant site of residence of the spin and thus of the oxidation state situation.[6]

As an already-mentioned example, the diruthenium compound **8** can be converted (spectro)electrochemically to paramagnetic ions, cation and anion.[46] Absorption spectroscopy showed NIR bands for **8**$^{+}$ and for **8**$^{-}$. This ambiguous situation, metal-based IVCT absorptions for mixed-valence species or low-energy intraligand transitions for radical complexes, has been clarified by EPR spectroelectrochemistry: the cation **8**$^{+}$ showed a rhombic EPR signal of sizeable g component splitting in the frozen state (Figure 3.5), signifying considerable, if not exclusive spin density on the metals. In contrast, the anion **8**$^{-}$ was identified with a narrow signal at $g = 2.003$, suggesting an anion radical complex of two ruthenium(II) centres.[46]

In contrast to the above, there can be situations when the EPR method is not applicable (diamagnetism, *e.g.* in tetranuclear complexes, even-electron situations)[51] or when the signal is not observable even at low temperatures ("EPR

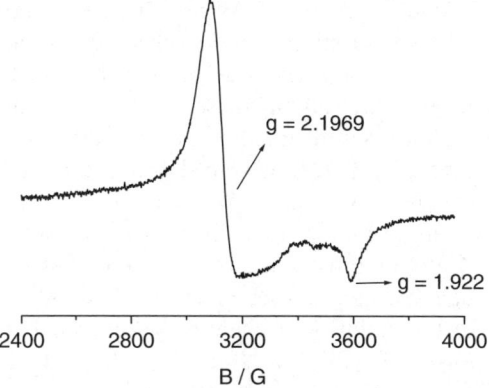

Figure 3.5 EPR spectrum resulting from the conversion $\{(\mu\text{-tithb})[Ru(acac)_2]_2\}^{(0)\rightarrow(+)}$ ($8 \rightarrow 8^+$) in $CH_3CN/0.1$ M Bu_4NPF_6 (measurement at 4 K).

silence").[26] Such cases are not uncommon for mixed-valence species because they often involve excited states lying very close to the paramagnetic ground state that, especially in the presence of heavy elements with their high spin-orbit coupling constants, allows for rapid EPR relaxation and thus line broadening up to the point of nonobservability. For instance, the decacyano analogue $\{(\mu\text{-pz})\text{-}[Ru(CN)_5]_2\}^{5-}$ (**2**) of the Creutz–Taube ion was reported as EPR-silent even at 4 K despite an IR and NIR spectroelectrochemical response.[26] The very broad IVCT band at 1760 nm, however, suggests considerable reorganisation energy and thus less pronounced valence delocalisation than in $\{(\mu\text{-pz})[Ru(NH_3)_5]_2\}^{5+}$.[2–5] Diosmium(III,II) species such as $\{(\mu\text{-adc-R})[Os(bpy)_2]_2\}^{3+}$ are also EPR-silent at 3.5 K despite an $S = 1/2$ state as determined by susceptibility measurements;[52] the ruthenium analogues $\{(\mu\text{-adc-R})[Ru(bpy)_2]_2\}^{3+}$ show $Ru^{III}Ru^{II}$-type EPR signals at 4 K which, however, begin to disappear at around 20 K.[53]

azodicarbonyl compound (adc-R)
R = alkyl, aryl, alkoxo or dialkylamino

1,2-diacylhydrazido(2-)
(adc-R^{2-})

Otherwise, the strong increase of the spin–orbit coupling within a group of the periodic table augments the *g*-component splitting as demonstrated for the previously mentioned $Rh^{III}Rh^{II}$ *vs.* $Ir^{III}Ir^{II}$ examples $[(C_5Me_5)M(\mu\text{-abpy})M(C_5Me_5)]^+$.[23]

The EPR method is particularly powerful with regard to the valence (de)-localisation question when hyperfine information can be obtained. This, unfortunately, requires not only the presence of suitable nuclei but also a sufficiently small linewidth, which is not always achieved for transition-metal species. Information on the *g*-factor anisotropy alone is not directly suitable to determine the symmetry of the mixed-valence intermediate.[25,27]

   Ideally, the hyperfine interaction between the free electron and the magnetically active nuclei in question should reflect quantitatively, or only by symmetry effects, what the spin and valence distribution looks like. A well-researched case concerns the dimanganese centres as found in enzymes that show the valence (de)localisation *via* the ^{55}Mn hyperfine coupling pattern.[9,54] With full valence averaging and thus spin delocalisation and a simple 11-line pattern with intensities following 1:2:3:4:5:6:5:4:3:2:1 should result,[54,55] whereas the experimental data show a more complex splitting, the so-called "16-line" signal.[9,54] Its analysis points to a situation with spin and valence asymmetry. Although spectroelectrochemical methods were not necessary for this example due to the stability of chemically generated species, it serves to demonstrate the power of EPR for establishing the extent of spin and thus valence distribution between electron-transfer sites.

   Using mixed-valence dimolybdenum compounds, *i.e.* involving an element with only about 25% of the naturally occuring isotope mixture bearing a nuclear spin (^{95,97}Mo: $I = 5/2$), one has to analyse the EPR spectra more carefully at the periphery. However, for bis(phenolate)-bridged $Mo^{VI}Mo^V$, $Mo^{II}Mo^I$,[56] and for the Mo^IMo^0 species **4**,[28] (the organometallic analogue $\{(\mu\text{–pz})[Mo(CO)_3(PR_3)_2]_2\}^+$ of the Creutz–Taube ion, Figure 3.6), the quantitative determination of satellite intensities as well as the approximately 50% diminished hyperfine splitting in comparison to related mononuclear compounds confirm the complete spin delocalisation and thus valence averaging on the EPR timescale of about 10^{-8} s.

   On the shorter timescale ($\approx 10^{-13}$ s) of vibrational spectroscopy the system **4** exhibits also full symmetry (see Figures 3.3 and 3.7), confirming the EPR results.

   Two additional aspects may be mentioned briefly with respect to EPR analysis: Hyperfine information on the symmetry of mixed-valence species can also be obtained through electron–nuclear double resonance (ENDOR). The main-group mixed-valence ($B^{III}B^{II}$) species **9** have thus been analysed despite the very large number of EPR lines with the consequence of unresolved spectra.[57]

   Secondly, the EPR method is also appropriate to study "inverse mixed-valency", involving metal-bridged ligands of different redox states.[58] In those cases, the "hopping" of unpaired electrons may be noted through linewidth effects as for system $[Ru(abpy)_2(bpy)]^+$ (**10**).[58a]

9   **10**

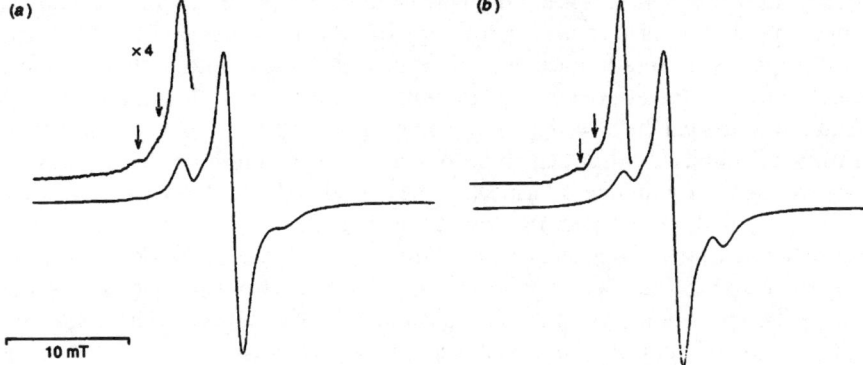

Figure 3.6 Room-temperature EPR spectrum resulting from the conversion $\{(\mu\text{-pz})[Mo(CO)_3(PR_3)_2]_2\}^{(0)\rightarrow(+)}$ (**4**) in $CH_2Cl_2/0.1$ M Bu_4NPF_6 ($R=$ isopropyl; $g = 2.0438$, $a(^{95,97}Mo) = 1.52$ mT). Experimental (top) and simulated spectrum (bottom, two equivalent Mo).

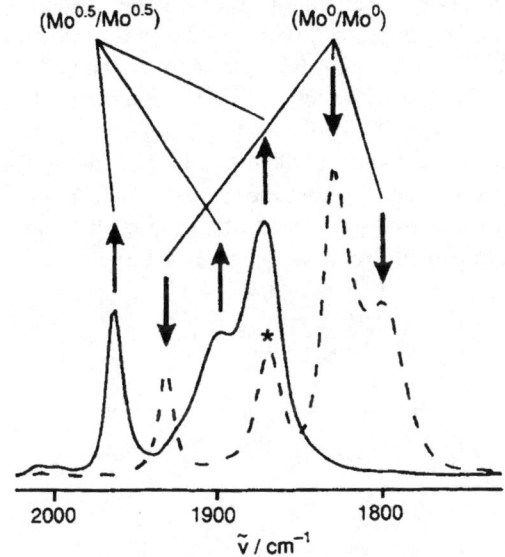

Figure 3.7 IR spectroelectrochemical response during the conversion $\{(\mu\text{-pz})[Mo(CO)_3(PR_3)_2]_2\}^{(0)\rightarrow(+)}$ (**4**) in $CH_2Cl_2/0.1$ M Bu_4NPF_6 ($R=$ isopropyl; *: band from decomposition product $Mo(CO)_4(PR_3)_2$).

3.5 IR Vibrational Spectroscopy

In contrast to EPR and to NIR absorption (electronic) spectroscopy, IR or Raman vibrational spectroscopies yield absorption bands for mixed-valence and homovalent states alike, mostly in similar spectral regions. Therefore, the shifts of bands associated with certain functional groups are often used in spectroelectrochemistry. Since the timescale (*ca.* 10^{-13} s) of this spectroscopy is longer than that of electronic spectroscopy but shorter than that of EPR,[4b] it is ideal for studying small-molecule mobility, so that the quintessential chemical question for localised or delocalised bonding and distinctly different or averaged oxidation states can be investigated by establishing a splitting or a coalescence of vibrational bands.[16] Although this approach seems convenient and straightforward there are a number of aspects to be considered in order to avoid pitfalls.

The use of infrared spectroscopy, either through "fingerprint" characterisation or by functional group identification, is well established. IR vibrational spectroscopy has thus been applied in spectroelectrochemistry for quite some time.[30] The possibility to establish the symmetry of a molecule has made IR-SEC a most valuable tool for mixed-valence chemistry,[16,31,59,60] allowing intramolecular electron-transfer rates in the picosecond region to be assessed and "electron-transfer isomers" to be established.[16a,i]

As in normal IR vibrational spectroscopy the typically strong bands of the CO group have featured prominently, either as metal carbonyl M–CO functions[16,28,60] or as organic (ligand) carbonyl bands in ligated species R_2CO.[43b,c] These intense, sensitive bands allow for differentiation between symmetrical situations with equivalent CO groups (valence averaging, see Figure 3.7) and those arrangements with nonequivalent CO groups, signifying at least partial valence trapping on the IR vibrational timescale despite formal molecular symmetry.[60] As an example for the former situation the diruthenium(II,III) complex [Cl(dpk)Ru(μ-tppz)-Ru(dpk)Cl]$^{3+}$ (**11**), tppz = 2,3,5,6-tetrakis(2-pyridyl)pyrazine, shows only one $v(CO)$ band for the two 2,2′-dipyridylketone (dpk) ancillary ligands[59] and, similarly, the three $v(CO)$ bands for the *mer*-configurated $Mo(CO)_3$ fragments in complex **4** shift uniformly and without splitting to higher energies after oxidation of the Mo^0Mo^0 homovalent precursor (Figures 3.3 and 3.7).[28]

2,3,5,6-tetrakis(2-pyridyl)pyrazine   2,2′-dipyridylketone
(tppz)   (dpk)

On the other hand, decoupling of redox-active entities with carbonyl functions may result in splitting of the $v(CO)$ bands.[16,31,35,59,60]

In more elaborate studies, Kubiak and coworkers have established experimentally the existence of dynamic equilibrium mixtures containing "charge-transfer isomers" of mixed-valence ruthenium complexes within [{Ru$_3$(μ-O)-(μ-CH$_3$COO)$_6$(^{12}C^{16}O)(L′)}](μ-BL)[{Ru$_3$(μ-O)(μ-CH$_3$COO)$_6$(^{13}C^{18}O)(L″)}] by monitoring the $v(CO)$ vibrations in the infrared (μ-BL = unsymmetrical pyrazine derivatives). The mixed-valence charge-transfer isomers differ in the location of the odd electron, and the asymmetric ligand environment around the metal centres facilitates their experimental identification.

12

The IR spectra of native $Ru_3^{III,III,II}$– $Ru_3^{III,III,II}$ and doubly reduced $Ru_3^{III,II,II}$– $Ru_3^{III,II,II}$ species exhibit two $\nu(CO)$ bands, separated by $90\,cm^{-1}$ because of the isotope substitution $\nu(^{12}C^{16}O)$ *vs.* $\nu(^{13}C^{18}O)$. In contrast, the monoanionic mixed-valence intermediate $Ru_3^{III,III,II}$– $Ru_3^{III,III,II}$ shows four $\nu(CO)$ bands at 1959/1861($^{13}C^{18}O$) and 1948/1859($^{13}C^{18}O$) cm^{-1} due to the presence of an equilibrium mixture of two charge-transfer isomers **12A** and **12B**, respectively.

Such investigations could also be carried out using molecular frameworks with symmetrical bridging ligands such as L = pyrazine in combination but different ancillary ligands L′ and L″ such as 4-dimethylaminopyridine/pyridine, pyridine/4-cyanopyridine or 4-dimethylaminopyridine/4-cyanopyridine.

The experimental evidence for the existence of bistable mixed-valence isomers **12A** and **12B** and their switching times of about one trillion per second as inferred from temperature-dependent coalescence of carbonyl stretching bands may be eventually employed to provide a solid basis for the fabrication of molecular electronic devices.

Not only well-separated and intense bands from carbonyl stretching but also other vibrational absorption features lying at lower frequencies may be used to investigate the molecular symmetry and thus the valence situation in complexes like the Creutz–Taube ion[61] or related osmium compounds.[5,62]

Strong unobscured bands in the $2000\,cm^{-1}$ region may also involve the stretching of the CN bond with its typically high bond order in metal cyanides MCN (Figure 3.8),[25,32] in metal-binding organopolynitriles such as TCNE or

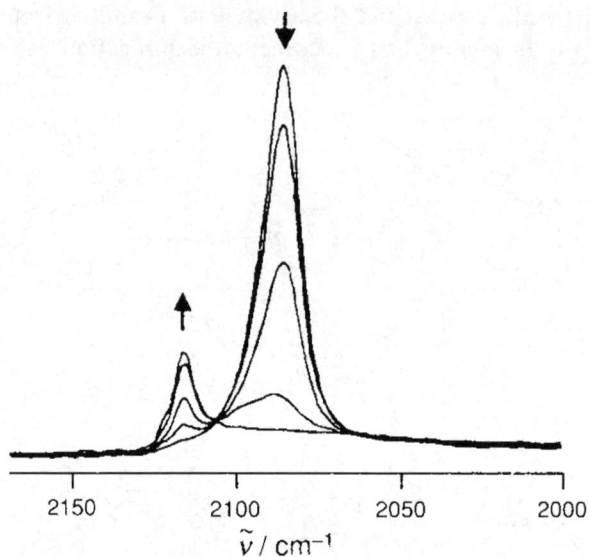

Figure 3.8 IR spectroelectrochemical response during the conversion $\{(\mu\text{-bmtz})$ $[\text{Fe(CN)}_4]_2\}^{(4-)\rightarrow(3-)}$ (**6**) in $\text{CH}_3\text{CN}/0.1$ M Bu_4NPF_6: shift, but no splitting of the cyanide stretching bands, indicating valence averaging.

TCNQ,[51] or in coordination compounds of π-conjugated organodicyanamides $^-$NCNRNCN$^-$.[63]

tetracyanoethene   7,7,8,8-tetracyano-p-quino-
(TCNE)   dimethane (TCNQ)

Metal compounds of the CN-containing ligands are often studied with respect to their potential for magnetic coupling.[12,64]

Further, the CC stretching around $2000\,\text{cm}^{-1}$ of oligoacetylene/cumulene redox systems[65,66] is well suitable for monitoring by IR-SEC. Such organometallic[65] or sometimes even Werner-type[66] systems have received attention as basic models of redox-active C_n materials.[67] An example with ligand system **13**$^{n-}$ showed the spectroelectrochemically characterised intermediate between a 1,3-diyne (diacetylene) and an 1,2,3-triene (cumulene) form with the respective C–C multiple bond-stretching features.[66]

13^{2-} **13**

3.6 Perspective

Summarising, spectroelectrochemical techniques are ideal for studying mixed-valence intermediates:

- UV–Vis–NIR spectroelectrochemistry normally allows for a convenient identification of the intervalence charge-transfer band that is crucial for the analysis of electronic interaction,[1–5]
- EPR serves to establish the mixed-valence character and may provide further insight into the electronic structure through the shift and anisotropy of the *g*-factor and through analysis of the electron–nuclear hyperfine interaction,[6]
- IR vibrational spectroelectrochemistry helps to determine the symmetry and thus the valence averaging (or trapping) on an otherwise experimentally challenging but chemically very valuable timescale.[16]

Ideally, the combination and integration of *several* spectroelectrochemical methods provides a most comprehensive picture of the valence situation and serves to avoid possible misinterpretation. Variable-temperature spectroelectrochemical setups[68] not only allow measurements of labile intermediates but also an assessment of electron-transfer dynamics in solution.

Acknowledgements

Our own research has been supported by the Land Baden-Württemberg, the Fonds der Chemischen Industrie, the Deutsche Forschungsgemeinschaft (DFG, Germany) and by the Department of Science and Technology (DST), New Delhi, India. Cooperation has been made possible by a personnel exchange grant from the Deutsche Akademische Austauschdienst (DAAD) and DST and by a Mercator Professorship (DFG). We thank Mrs. Angela Winkelmann for her contributions in preparing this article.

References

1. (a) M. B. Robin and P. Day, *Adv. Inorg. Chem. Radiochem.*, 1967, **10**, 247; (b) K. Prassides, (ed)., *Mixed Valency Systems – Applications in Chemistry, Physics and Biology*, Kluwer Academic Publishers, Dordrecht, 1991; (c) N. S. Hush, *Coord. Chem. Rev.*, 1985, **64**, 135; (d) F. A. Cotton, G. Wilkinson, C. A. Murillo and M. Bochmann, *Advanced Inorganic Chemistry*, Wiley, New York, 6th edn., 1999, p. 1017; (e) C. Housecroft and A. G. Sharpe, *Inorganic Chemistry*, Pearson, Harlow, 2nd edn., 2005, p. 679.
2. (a) H. Taube, *Angew. Chem.*, 1984, **96**, 315; *Angew. Chem. Int. ed. Engl.*, 1984, **23**, 329; (b) C. Creutz, *Prog. Inorg. Chem.*, 1983, **30**, 1.
3. D. E. Richardson and H. Taube, *Coord. Chem. Rev.*, 1984, **60**, 107.

4. (a) R. J. Crutchley, *Adv. Inorg. Chem.*, 1994, **41**, 273; (b) A. P. B. Lever, in *Comprehensive Coordination Chemistry II*, ed. J. A. McCleverty and T. J. Meyer, Elsevier Science, Amsterdam, 2003, p. 435; (c) J. T. Hupp, in *Comprehensive Coordination Chemistry II*, ed. J. A. McCleverty and T. J. Meyer Elsevier Science, Amsterdam, 2003, p. 709; (d) R. J. Crutchley, in *Comprehensive Coordination Chemistry II*, ed. J. A. McCleverty and T. J. Meyer, Elsevier Science, Amsterdam, 2003, p. 235.

5. K. D. Demadis, D. C. Hartshorn and T. J. Meyer, *Chem. Rev.*, 2001, **101**, 2655.

6. (a) W. Kaim, A. Klein and M. Glöckle, *Acc. Chem. Res.*, 2000, **33**, 755; (b) W. Kaim and B. Sarkar, *Coord. Chem. Rev.*, 2007, **251**, 584; (c) W. Kaim and G. K. Lahiri, *Angew. Chem. Int. ed.*, 2007, **46**, 1778.

7. E. C. Carson and S. J. Lippard, *J. Inorg. Biochem.*, 2006, **100**, 1109.

8. Y. Lu, *Angew. Chem. Int. ed.*, 2006, **45**, 5588.

9. C. Teutloff, K. O. Schaefer, S. Sinnecker, V. Barynin, R. Bittl, K. Wieghardt, F. Lendzian and W. Lubitz, *Magn. Reson. Chem.*, 2005, **43**, 51.

10. W. Kaim, W. Bruns, J. Poppe and V. Kasack, *J. Mol. Struct.*, 1993, **292**, 221.

11. (a) M. B. Robin, *Inorg. Chem.*, 1962, **12**, 337; (b) A. Ludi, *J. Chem. Educ.*, 1981, **58**, 1013.

12. (a) S. Ferlay, T. Mallah, R. Ouahes, P. Veillet and M. Verdaguer, *Nature*, 1995, **378**, 701; (b) J. S. Miller, *Inorg. Chem.*, 2000, **39**, 4392.

13. (a) J. Jortner and M. Bixon, *Adv. Chem. Phys.*, 1999, **106**, 35; (b) C. Creutz, M. D. Newton and N. Sutin, *J. Photochem. Photobiol. A.*, 1994, **882**, 47.

14. (a) M. D. Ward, *J. Chem. Educ.*, 2001, **78**, 321; (b) M. D. Ward, *Chem. Ind.*, 1996, 568; (c) M. D. Ward, *Chem. Ind.*, 1997, 640; (d) J.-P. Launay, *Chem. Soc. Rev.*, 2001, **30**, 386.

15. (a) S. B. Braun-Sand and O. Wiest, *J. Phys. Chem. B.*, 2003, **107**, 9624; (b) S. B. Braun-Sand and O. Wiest, *J. Phys. Chem. A*, 2003, **107**, 285; (c) C. S. Lent, B. Isaksen and M. Lieberman, *J. Am. Chem. Soc.*, 2003, **125**, 1056.

16. (a) J. C. Salsman and C. P. Kubiak, *J. Am. Chem. Soc.*, 2005, **127**, 2382; (b) C. H. Londergan and C. P. Kubiak, *Chem. Eur. J.*, 2003, **9**, 5962; (c) C. H. Londergan and C. P. Kubiak, *J. Phys. Chem. A.*, 2003, **107**, 9301; (d) C. H. Londergan, J. C. Salsman, S. Ronco, L. M. Dolkas and C. P. Kubiak, *J. Am. Chem. Soc.*, 2002, **124**, 6236; (e) T. Ito, T. Hamaguchi, H. Nagino, T. Yamaguchi, J. Washington and C. P. Kubiak, *Science*, 1997, **277**, 660; (f) C. H. Londergan, J. C. Salsman, S. Ronco and C. P. Kubiak, *Inorg. Chem.*, 2003, **42**, 926; (g) R. J. Crutchley, *Angew. Chem.*, 2005, **117**, 6610; *Angew. Chem. Int. ed.*, 2005, **44**, 6452; (h) R. C. Rocha, M. G. Brown, C. H. Londergan, J. C. Salsman, C. P. Kubiak and A. P. Shreve, *J. Phys. Chem. A.*, 2005, **109**, 9006; (i) J. C. Salsman and C. P. Kubiak, this volume, p. 124.

17. (a) A. F. Heyduk and D. G. Nocera, *Science*, 2001, **293**, 1639; (b) T. G. Gray and D. G. Nocera, *Chem. Commun.*, 2005, 1540.

18. J. Bonvoisin, J.-P. Launay, C. Rovira and J. Veciana, *Angew. Chem. Int. ed. Engl.*, 1994, **33**, 2106.
19. (a) S. F. Nelsen, *Chem. Eur. J.*, 2000, **6**, 581; (b) S. E. Bailey, J. I. Zink and S. F. Nelsen, *J. Am. Chem. Soc.*, 2003, **125**, 5939.
20. (a) C. Lambert, G. Nöll and J. Schelter, *Nature Mater.*, 2002, **1**, 69; (b) A. V. Szeghalmi, M. Erdmann, V. Engel, M. Schmitt, S. Amthor, V. Kriegisch, G. Noell, R. Stahl, C. Lambert, D. Leusser, D. Stalke, M. Zabel and J. Popp, *J. Am. Chem. Soc.*, 2004, **126**, 7834.
21. (a) A. Schulz and W. Kaim, *Chem. Ber.*, 1989, **122**, 1863; (b) W. Kaim and S. Zalis, *Main Group Chem.*, 2007, **6**, 267.
22. (a) W. Kaim, in *New Trends in Molecular Electrochemistry*, ed. A. J. L. Pombeiro, Fontis Media, Lausanne, 2004, p. 127; (b) W. Kaim, R. Reinhardt and J. Fiedler, *Angew. Chem.*, 1997, **109**, 2600; *Angew. Chem. Int. ed. Engl.*, 1997, **36**, 2493; (c) W. Kaim, S. Berger, S. Greulich, R. Reinhardt and J. Fiedler, *J. Organomet. Chem.*, 1999, **582**, 153.
23. (a) S. Berger, A. Klein, M. Wanner, J. Fiedler and W. Kaim, *Inorg. Chem.*, 2000, **39**, 2516; (b) S. Frantz, R. Reinhardt, S. Greulich, M. Wanner, J. Fiedler, C. Duboc-Toia and W. Kaim, *Dalton Trans.*, 2003, 3370.
24. B. Sarkar, W. Kaim, J. Fiedler and C. Duboc, *J. Am. Chem. Soc.*, 2004, **126**, 14706.
25. (a) M. Glöckle and W. Kaim, *Angew. Chem.*, 1999, **111**, 3262; *Angew. Chem. Int. ed.*, 1999, **38**, 3072; (b) M. Glöckle, W. Kaim, A. Klein, E. Roduner, G. Hübner, S. Zalis, J. van Slageren, F. Renz and P. Gütlich, *Inorg. Chem.*, 2001, **40**, 2256.
26. T. Scheiring, W. Kaim, J. A. Olabe, A. R. Parise and J. Fiedler, *Inorg. Chim. Acta.*, 2000, **300–302**, 125.
27. M. Ketterle, W. Kaim, J. A. Olabe, A. R. Parise and J. Fiedler, *Inorg. Chim. Acta.*, 1999, **291**, 66.
28. (a) W. Bruns, W. Kaim, E. Waldhör and M. Krejcik, *J. Chem. Soc., Chem. Commun.*, 1993, 1868; (b) W. Bruns, W. Kaim, E. Waldhör and M. Krejcik, *Inorg. Chem.*, 1995, **34**, 663.
29. (a) H. W. Siesler, Y. Ozaki, S. Kawata and H. M. Heise, ed., *Near-Infrared Spectroscopy*, Wiley-VCH, Weinheim, 2001; (b) Y. Ozaki, M. F. McClure and A. A. Christy, ed., *Near-Infrared Spectroscopy in Food Science and Technology*, Wiley-VCH, Weinheim, 2006; (c) Y. Qi, P. Desjardins and Z. Y. Wang, *J. Opt. A: Pure Appl. Opt.*, 2002, **4**, S273.
30. (a) K. Ashley and S. Pons, *Chem. Rev.*, 1988, **88**, 673; (b) Korzeniewski, in *Handbook of Vibrational Spectroscopy*, ed. J. M. Chalmers and P. R. Griffiths, Wiley, New York, 2002.
31. (a) S. P. Best, S. J. Borg and K. Vincent this volume, p. 1; (b) R. Holze, *J. Solid State Electrochem.*, 2004, **8**, 982.
32. M. Ketterle, J. Fiedler and W. Kaim, *J. Chem. Soc., Chem. Commun.*, 1998, 1701.
33. (a) R. LeSuer and W. E. Geiger, *Angew. Chem.*, 2000, **112**, 254; *Angew. Chem. Int. ed.*, 2000, **39**, 248; (b) R. LeSuer, C. Buttolph and W. E. Geiger,

Anal. Chem., 2004, **76**, 6395; (c) D. M. D'Alessandro and F. R. Keene, *Dalton Trans.*, 2004, 3950.

34. (a) S. Berger, A. Klein, W. Kaim and J. Fiedler, *Inorg. Chem.*, 1998, **37**, 5664; (b) W. Kaim, B. Schwederski, A. Dogan, J. Fiedler, C. J. Kuehl and P. J. Stang, *Inorg. Chem.*, 2002, **41**, 4025.

35. A. P. Meacham, K. L. Druce, Z. R. Bell, M. D. Ward, J. B. Keister and A. B. P. Lever, *Inorg. Chem.*, 2003, **42**, 7887.

36. (a) T. Mahabiersing, H. Luyten, R. C. Nieuwendam and F. Hartl, *Collect. Czech. Chem. Commun.*, 2003, **68**, 1687; (b) F. Hartl, H. Luyten, H. A. Nieuwenhuis and G. C. Schoemaker, *Appl. Spectrosc.*, 1994, **48**, 1522.

37. (a) A. M. Bond, *Broadening Electrochemical Horizons*, Oxford University Press, Oxford, 2002, p. 107; (b) J. W. Verhoeven (ed.), *Pure Appl. Chem.*, 1996, **68**, 2223; (c) W. Plieth, G. S. Wilson and C. Gutiérrez de la Fe, *Pure Appl. Chem.*, 1998, **70**, 1395; (d) M. H. Cheah, C. Tard, S. J. Borg, X. Liu, S. K. Ibrahim, C. J. Pickett and S. P. Best, *J. Am. Chem. Soc.*, 2007, **129**, 11085; (e) M. J. Shaw and W. E. Geiger, *Organometallics*, 1996, **15**, 13; (f) W. Kaim and J. Fiedler, *Chem. Soc. Rev.*, in preparation.

38. S. Ernst, P. Hänel, J. Jordanov, W. Kaim, V. Kasack and E. Roth, *J. Am. Chem. Soc.*, 1989, **111**, 1733.

39. S. Maji, B. Sarkar, S. Patra, J. Fiedler, S. M. Mobin, V. G. Puranik, W. Kaim and G. K. Lahiri, *Inorg. Chem.*, 2006, **45**, 1316.

40. (a) J. A. DeGray, P. H. Rieger, N. G. Connelly and G. Garcia Herbosa, *J. Magn. Reson.*, 1990, **88**, 376; (b) D. C. Boyd, N. G. Connelly, G. Garcia Herbosa, M. G. Hill, K. R. Namm, C. Mealli, A. G. Orpen, K. E. Richardson and P. H. Rieger, *Inorg. Chem.*, 1994, **33**, 960.

41. W. Kaim, S. Kohlmann, A. J. Lees, T. L. Snoeck, D. J. Stufkens and M. M. Zulu, *Inorg. Chim. Acta*, 1993, **210**, 159.

42. (a) J. Poppe, M. Moscherosch and W. Kaim, *Inorg. Chem.*, 1993, **32**, 2640; (b) S. Chellamma and M. Lieberman, *Inorg. Chem.*, 2001, **40**, 3177.

43. (a) S. Patra, B. Sarkar, S. Ghumaan, J. Fiedler, W. Kaim and G. K. Lahiri, *Dalton Trans.*, 2004, 754; (b) S. Ghumaan, B. Sarkar, N. Chanda, M. Sieger, J. Fiedler, W. Kaim and G. K. Lahiri, *Inorg. Chem.*, 2006, **45**, 7955; (c) M. Koley, B. Sarkar, S. Ghumaan, E. Bulak, J. Fiedler, W. Kaim and G. K. Lahiri, *Inorg. Chem.*, 2007, **46**, 3736.

44. (a) F. Felix, U. Hauser, H. Siegenthaler, F. Wenk and A. Ludi, *Inorg. Chim. Acta.*, 1975, **15**, L7; (b) F. Felix and A. Ludi, *Inorg. Chem.*, 1978, **17**, 1782.

45. (a) D. M. D'Alessandro and F. R. Keene, *Chem. Eur. J.*, 2005, **11**, 3679; (b) D. M. D'Alessandro and F. R. Keene, *Chem. Rev.*, 2006, **106**, 2270; (c) D. M. D'Alessandro and F. R. Keene, *Chem. Soc. Rev.*, 2006, **35**, 424.

46. S. Patra, B. Sarkar, S. Ghumaan, J. Fiedler, S. Zalis, W. Kaim and G. K. Lahiri, *Dalton. Trans.*, 2004, 750.

47. S. Berger, T. Scheiring, J. Fiedler and W. Kaim, *Inorg. Chem.*, 2004, **43**, 1530.

48. (a) R. P. Van Duyne, M. R. Suchanski, J. M. Lakovits, A. R. Siedle, K. D. Parks and T. M. Cotton, *J. Am. Chem. Soc.*, 1979, **101**, 2832; (b) L. R. Sharpe, *Chem. Rev.*, 1990, **90**, 705; (c) P. D. Prenzler, R. Bramley, S. R. Downing and G. A. Heath, *Electrochem. Commun.*, 2000, **2**, 516; (d) A. J. L. Pombeiro and C. Amatore, ed., *Trends in Molecular Electrochemistry*, Marcel Dekker, New York, 2004, p. 339.

49. (a) J. A. Weil and J. R. Bolton, *Electron Paramagnetic Resonance*, Wiley, New York, 2nd edn., 2007; (b) P. R. Murray and L. J. Yellowlees, this volume, p. 208.

50. W. Kaim, *Coord. Chem. Rev.*, 1987, **76**, 187.

51. (a) M. Moscherosch, E. Waldhör, H. Binder, W. Kaim and J. Fiedler, *Inorg. Chem.*, 1995, **34**, 4326; (b) F. Baumann, W. Kaim, J. A. Olabe, A. Parisse and J. Jordanov, *J. Chem. Soc., Dalton Trans.*, 1997, 4455.

52. (a) W. Kaim and V. Kasack, *Inorg. Chem.*, 1990, **29**, 4696; (b) M. Heilmann, S. Frantz, W. Kaim, J. Fiedler and C. Duboc, *Inorg. Chim. Acta*, 2006, **359**, 821.

53. (a) V. Kasack, W. Kaim, H. Binder, J. Jordanov and E. Roth, *Inorg. Chem.*, 1995, **34**, 1924; (b) A. Knödler, J. Fiedler and W. Kaim, *Polyhedron*, 2004, **23**, 701.

54. Cf. e.g. G. C. Dismukes in K. Prassides, ed., *Mixed Valency Systems: Applications in Chemistry, Physics and Biology*, NATO ACSI Series, Kluwer Academic Publishers, Dordrecht, 1991, p 137.

55. R. Gross and W. Kaim, *Inorg. Chem.*, 1986, **25**, 4865.

56. J. A. McCleverty and M. D. Ward, *Acc. Chem. Res.*, 1998, **31**, 842.

57. A. Lichtblau, W. Kaim, A. Schulz and T. Stahl, *J. Chem. Soc., Perkin Trans.*, 1992, 1497.

58. (a) M. Heilmann, F. Baumann, W. Kaim and J. Fiedler, *J. Chem. Soc., Faraday Trans.*, 1996, **92**, 4227; (b) P. H. Dinolfo, M. E. Williams, C. L. Stern and J. T. Hupp, *J. Am. Chem. Soc.*, 2004, **126**, 12989.

59. S. P. Best, S. J. Borg and K. Vincent this volume, p. 1.

60. C. G. Atwood and W. E. Geiger, *J. Am. Chem. Soc.*, 2000, **122**, 5477.

61. S. P. Best, R. J. H. Clark, R. C. S. McQueen and S. Joss, *J. Am. Chem. Soc.*, 1989, **111**, 548.

62. R. C. Rocha and A. P. Shreve, *Chem. Phys.*, 2006, **326**, 24.

63. (a) C. E. B. Evans, M. L. Naklicki, A. R. Rezvani, C. A. White, V. V. Kondratiev and R. J. Crutchley, *J. Am. Chem. Soc.*, 1998, **120**, 13096; (b) M. Al-Noaimi, G. P. A. Yap and R. J. Crutchley, *Inorg. Chem.*, 2004, **43**, 1770; (c) M. L. Naklicki and R. J. Crutchley, *Inorg. Chim. Acta*, 1994, **225**, 123.

64. (a) J. S. Miller, *Angew. Chem. Int. ed.*, 2006, **45**, 2508; (b) H. Miyasaka, T. Izawa, N. Takahashi, M. Yamashita and K. R. Dunbar, *J. Am. Chem. Soc.*, 2006, **128**, 11358.

65. (a) R. F. Winter and S. Zalis, *Coord. Chem. Rev.*, 2004, **248**, 1565; (b) J. Maurer, R. F. Winter, B. Sarkar and S. Zalis, *J. Solid State Electrochem.*, 2005, **9**, 738.

66. Y. Hoshino, S. Higuchi, J. Fiedler, C.-Y. Su, A. Knödler, B. Schwederski, B. Sarkar, H. Hartmann and W. Kaim, *Angew. Chem.*, 2003, **115**, 698; *Angew. Chem. Int. ed.*, 2003, **42**, 674.

67. S. I. Ghazala, F. Paul, L. Toupet, T. Roisnel, P. Hapiot and C. Lapinte, *J. Am. Chem. Soc.*, 2006, **128**, 2463.

68. J. Daschbach, D. Heisler and S. Pons, *Appl. Spectrosc.*, 1986, **40**, 489.

CHAPTER 4

Spectroelectrochemistry of Metalloporphyrins

AXEL KLEIN

Universität zu Köln, Department für Chemie, Bereich Anorganische Chemie, Greinstraße 6, D-50939 Köln, Germany

4.1 Introduction

Porphyrins and related tetrapyrrole macrocyclic ligands are one of the most prominent ligand types in coordination chemistry. Their occurrence in metallo-enzymes, their application as inorganic dyes and photosensitisers, their use as homogeneous catalysts, mainly for oxidation or oxygenation reactions, and their outstanding role in coordination chemistry is in general well documented.[1]

One of the most prominent properties of metalloporphyrins is the accessibility of several oxidation states, which is crucial in electron-transfer reactions, *e.g.* oxygenation catalysis or oxygen transport. In most cases a larger number of these oxidation states are stable on the timescale of cyclovoltammetric experiments or even longer, an example of such multiredox behaviour can be seen in Figure 4.1.

Since the porphyrin ligands themselves (*i.e.* their protonated forms H_2Por) are able to form several stable oxidation states and on the other hand many of the metals in metalloporphyrins (*e.g.* Fe or Ru) can exist in various oxidation states, a clear assignment of such oxidation states in metalloporphyrins is not easy to make, as exemplified in Scheme 4.1.

For a deeper understanding of the interplay of the porphyrin ligands, the metals and (if present) the axial coligands in these various oxidation states, thorough electrochemical and spectroscopic investigations, and of course spectroelectrochemical experiments are sought. This chapter will report on how

Spectroelectrochemistry
Edited by Wolfgang Kaim and Axel Klein
© Royal Society of Chemistry, 2008

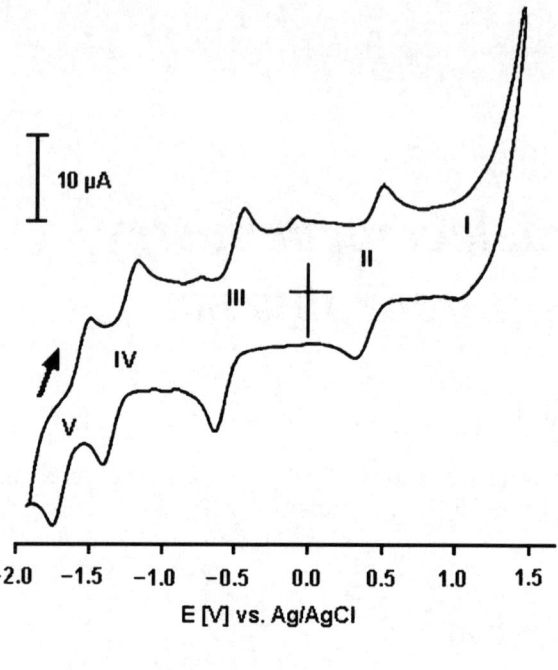

Figure 4.1 Cyclic voltammogram of [(CNOTPP)Co] in DMSO/0.1 M TBAP (scan
rate = 0.1 V/s). With permission from the RSC, ref. 2.

spectroelectrochemical methods can contribute significantly to the assignment
of metal–porphyrin oxidation states.

Since the number of reports on spectroelectrochemical investigations on
metalloporphyrins is very large and numerous reviews have been published,[1a,1b,3,4] this chapter will focus on a few selected examples.

We will start with the findings on the free porphyrin ligands, *i.e.* the protonated forms H$_2$Por to learn about the characteristic spectroscopy of pure
porphyrin-centred electron transfer (Section 4.2). In the same section

State		formal oxidation states				
$[(Por)M]^{2+}$	Ox2			$(Por)M(II)$	$(Por^{-})M(III)$	$(Por^{2-})M(IV)$
$[(Por)M]^{+}$	Ox1			$(Por^{-})M(II)$	$(Por^{2-})M(III)$	
$[(Por)M]$	Parent			$(Por^{2-})M(II)$		
$[(Por)M]^{-}$	Red1		$(Por^{2-})M(I)$	$(Por^{3-})M(II)$		
$[(Por)M]^{2-}$	Red2	$(Por^{2-})M(0)$	$(Por^{3-})M(I)$	$(Por^{4-})M(II)$		
		A	B	C	D	E

Scheme 4.1 Formal oxidation states of a divalent M and the porphyrin ligand Por in species generated from a neutral metalloporphyrin $[(Por)M(II)]$.

magnesium- and zinc-containing metalloporphyrins were set aside, since these two metals usually fit into the porphyrin cavity and are not redox-active. Therefore any redox chemistry should take place mainly at the porphyrin ligands and the spectroelectrochemical "performance" is not disturbed by protonation or deprotonation reactions, frequently observed for H_2Por systems.

Section 4.3 leads us into the redox chemistry of main group metals or metalloids especially phosphorus. These elements can exist in several oxidation states separated by two-electron steps. Electron transfer to this system might occur in one-electron steps (porphyrin-centred), two-electron steps (metalloid-centred, usually combined with ligand loss or addition) or mixed electron transfer/chemical reaction steps.

In contrast to the main-group elements, metals of the d-block usually show one-electron separated oxidation states. Thus, the discrimination between metal-centred and ligand-centred electron transfer is not easy and must often be based on several spectroscopic and electrochemical methods. As examples Section 4.4 reports on recent investigations on the nickel-containing cofactor F_{430} and Section 4.5 on the famous oxidation catalyst system $[(Por)Ru(CO)(L)]$. Sometimes the peculiar electrochemistry of porphyrins allows unusual oxidation states as shown with the divalent gold species $[(Por)Au(II)]$ in Section 4.6.

After having learned the lesson how to recognise typical spectroscopic patterns of the porphyrin ligand-centred electron transfer, and to discriminate them from metal-centred reaction, the assignment of oxidation states in multiredox arrays including metalloporphyrins is feasible as exemplified in Section 4.7, where covalently bound fullerene-$[(TPP)Co]$ dyads were explored by spectroelectrochemical means.

4.2 Electrochemistry and Spectroelectrochemistry of the Free Porphyrin Ligands H_2Por and Metalloporphyrins MPor (M = Mg, Zn, Cd)

Porphyrins H_2Por usually exhibit four reversible redox waves in cyclovoltammetric experiments (comparable to Figure 4.1) in the potential range from -3

to +3 V. Therefore, five distinguishable oxidation states can be characterised spectroscopically: $(H_2Por)^{2+} - (H_2Por)^+ - (H_2Por) - (H_2Por)^- - (H_2Por)^{2-}$.[5,6] The corresponding frontier orbitals can be described by the simple Gouterman four-orbital model (based on HMO calculations)[7] shown in Scheme 4.2.

The corresponding redox potentials are strongly dependent on the peripheral substituents and on the coligands on the metal, which are, in the case of complexes of the type [(Por)M(L)] the used solvent molecules[1b,6,8–10] but might also be influenced by ion pairing.[11]

Zinc- or magnesium-containing metalloporphyrins very often show similar behaviour as the free porphyrin ligands H_2Por. This is not unexpected, since both metals in their oxidation state +2 are electrochemically inert down to very low potentials and in this sense equivalent to the two protons in H_2Por. Furthermore, the M^{2+} ions ($M = Mg$ or Zn) fit well into the porphyrin cavity (ionic radii Mg^{2+}: 71 pm, Zn^{2+}: 74 pm),[12] therefore ligand distortion is minimal. Therefore zinc or magnesium porphyrins are well-suited references for systems with unperturbed ligand-centred electron transfer. Cadmium (ionic radius: 92 pm)[12] is too large for the porphyrin cavity and therefore leads to strain.[13]

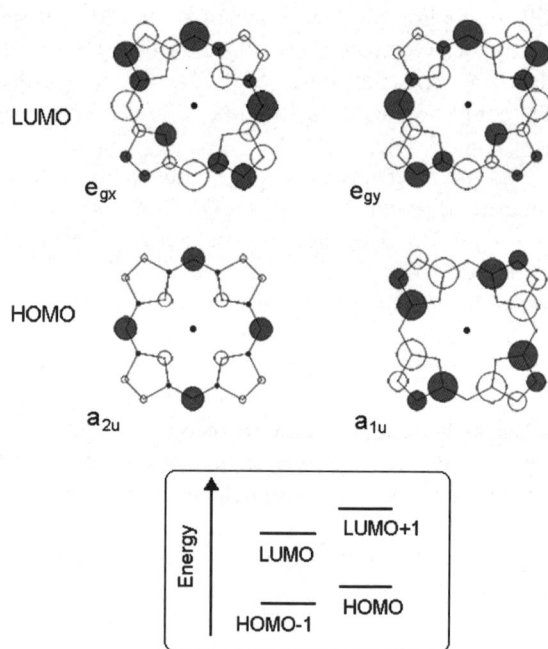

Scheme 4.2 Frontier orbitals of the porphyrins according to the Gouterman four-orbital model for D_4h point group (top). In the presence of a perturbation (substituents or ligand distortion) the degeneracy might be lifted (bottom). With permission from Wiley, ref. 7a.

4.2.1 Reduction

The reduction chemistry of free porphyrins and zinc or magnesium porphyrins was studied in detail using various spectroscopic techniques such as UV–Vis–NIR absorption,[5,6,9,10,14–16] EPR,[5,7a,16a,b,17–19] and resonance Raman.[20a,b,21a,b] The first reduction of porphyrins H_2Por is usually fully reversible. The absorption spectra of the thus-generated porphyrin π-anion radicals are characterised by a loss of intensity of the Soret band (sometimes together with a slight redshift) (see Figure 4.2). Also, the Q-bands are less intense and sometimes redshifted, compared to the parent molecules and an additional (sometimes structured) band system appears around 800 nm.[6,9,14] The same behaviour has been reported for zinc derivatives $[(Por)Zn]^-$.[5,6,10,16]

EPR spectra of the monoreduced species $[H_2Por]^-$ recorded at ambient temperature usually exhibit narrow, unstructured signals (no hyperfine splitting; HFS) with g values close to the "free-electron" value of 2.0023.[6,18] At low temperatures, H_2TPP^- exhibits an anisotropic signal of axial symmetry with $g_\perp = 2.0029$ and $g_\parallel = 2.0020$. For the metalloporphyrin species $[(Por)M]^-$ (M = Mg, Zn, Cd) HFS to N and H atoms could be resolved in many cases (occasionally by ENDOR or TRIPLE) and g values are usually slightly lower than found for the corresponding free ligand radical anions.[5,6,17,19,22] At low temperatures slightly higher anisotropy (Mg < Zn < Cd) and lower averaged g values were observed, as exemplified in Figure 4.3.

These phenomena can be reliably traced back to Jahn–Teller and strain distortion combined to unquenched orbital angular momentum due to spin-orbit-coupling in the metalloporphyrins. The Jahn–Teller distortion has been also studied by resonance Raman spectroscopy.[20,21]

In contrast to these extensive studies on the first reduced state, reports on the species obtained after the second one-electron reduction step were not that frequent, due to chemical reactions following the second reduction.[9] For metalloporphyrins $[(Por)M]$ with zinc or magnesium the dianions resulting from the second reduction were usually more stable. The absorption spectra of the dianions $[(Por)M]^{2-}$ show Soret bands that are blueshifted compared to those of the anion radicals and the 800-nm band system is replaced by a comparable one at higher energy (around 600 nm).[4,15,16]

An outstanding work by Baumgarten *et al.* even went beyond the second reduction step for the zinc porphyrins $[(Por)Zn]$ (Por = TPP, OEP, THPP and TDPP).[5] Absorption data for Soret and Q bands for the tri-, tetra-, penta-, and hexa-anions and EPR, ENDOR and TRIPLE data for tri- and penta-anions were recorded (Figure 4.4). The reduced species were generated using elemental potassium, therefore an influence of the K^+ ion might be assumed. However, comparison of the obtained spectra with those generated by reduction using sodium[16] or of electrochemically generated species[15] (*e.g.* Figure 4.2) reveals that the cation influence is marginal at least for the first two reduction steps.

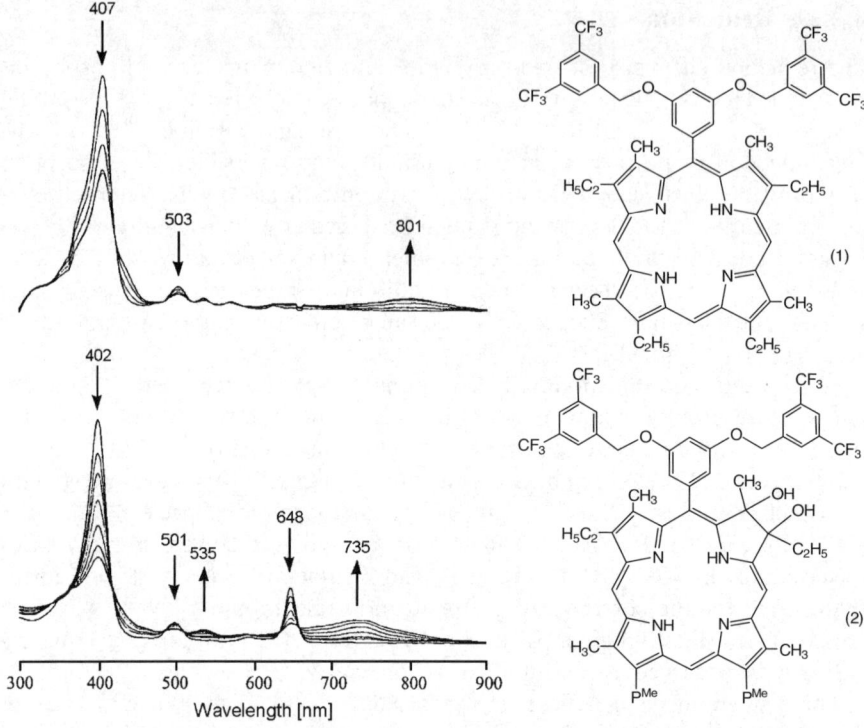

Figure 4.2 Thin-layer UV–Vis spectral changes of porphyrin (1) (top) and of the *vic-*
dihydroxychlorin (2) ($P^{Me} = -CH_2-CH_2-COO_2CH_3$) (bottom) upon the
first reduction at $-1.50\,V$ (*vs.* SCE) in benzonitrile/0.2 M TBAP. With
permission from Pergamon Press, ref. 9.

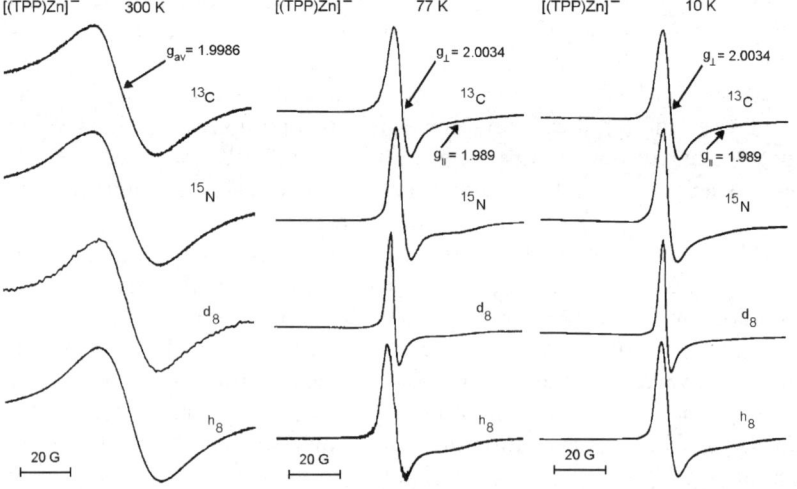

Figure 4.3 EPR spectra of electrochemically generated [(TPP)Zn]$^{•-}$ in THF/, butyro-
nitrile/, and DMF/0.1 M TBAP at 300, 77, and 20 K. With permission
from the American Chemical Society, ref. 18.

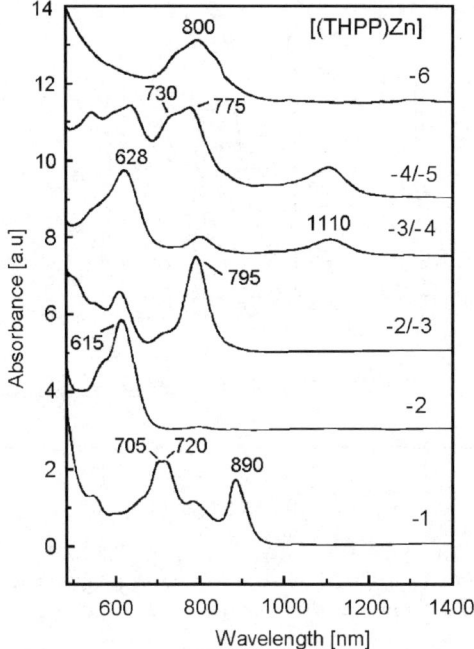

Figure 4.4 Vis–NIR optical absorption spectra of the six anionic redox stages of [(THPP)Zn] in THF taken at room temperature, obtained by reduction with potassium. The spectra are shown for the visible to NIR region from 480 to 1400 nm since the Soret region is far too intense. Acquisition time was 230 s per spectrum. With permission from Elsevier, ref. 5.

4.2.2 Oxidation

The first oxidation of the porphyrins H_2Por usually leads to corresponding π-radical cations that exhibit a hypsochromic shift for the Soret band at half the intensity of the original band. The Q bands are replaced by a structured band system with a maximum at around 650 nm[6,11,23] (see Figure 4.5). The same behaviour was reported from transient absorption spectra of chemically generated H_2TPP^+ cations,[24] and from spectroelectrochemical investigation of zinc porphyrins.[15]

Usually, it is already difficult to observe the first oxidised state in a pure form since these cations undergo rapid proton-transfer reaction, leading to species like H_4Por^{2+}.[11,15] In the spectroelectrochemical experiment this is revealed by the incomplete backconversion to the starting species. For zinc or magnesium containing derivatives the first oxidation is fully reversible, which has been recently used to create multizinc porphyrin arrays [(μ-(4-diphenyl-ethinyl)$_m$-{(Por)Zn}$_n$] (with n up to 20) capable of releasing up to n electrons or creating up to n holes, respectively.[25] EPR and ENDOR spectra of [(TPP)M]$^{•+}$ (M = Mg, Zn) have been studied and reveal hyperfine splitting to N and H atoms of the porphyrin ligand[19,22,26] (see Figure 4.6).

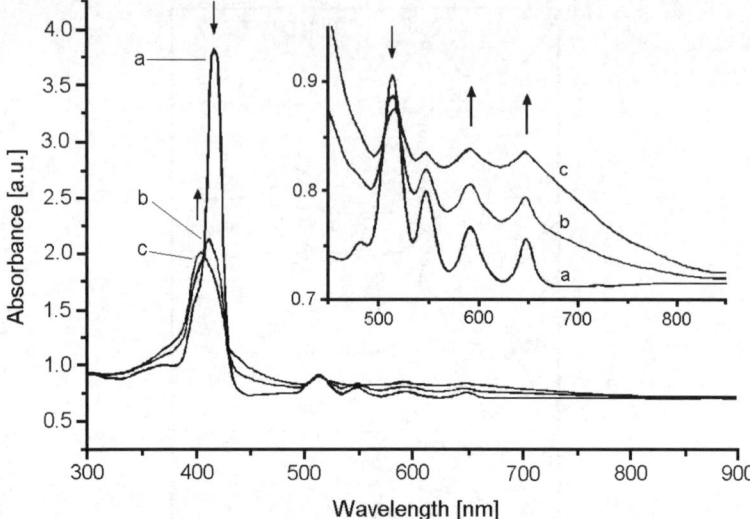

Figure 4.5 Spectroelectrochemistry of H_2TPP in 1,2-difluorobenzene/0.1 M TBAPF$_6$. Open circuit (a) to 0.812 V (b and c) (*vs.* Ag/Ag$^+$ reference electrode). With permission from the RSC, ref. 11.

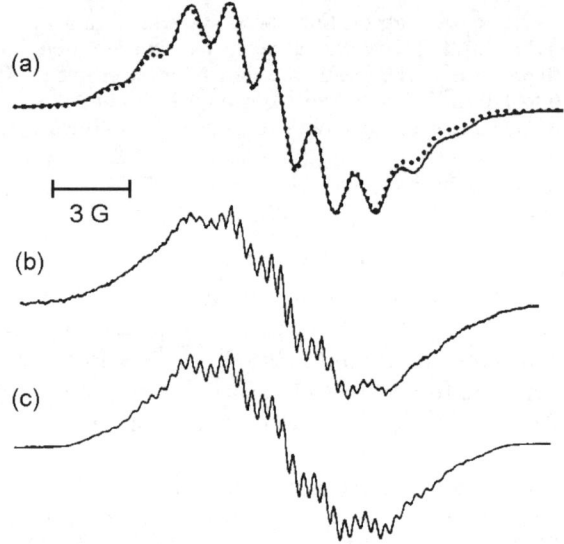

Figure 4.6 EPR spectra of electrochemically generated [(TPP)Zn]$^{\bullet+}$ in CH_2Cl_2/0.1 M TBAP at room temperature (a) and at 200 K (b). Computer simulation of spectrum (a) (dotted curve) involves four equivalent nuclei of $I=1$ with HFS $a=1.46$ G and a Gaussian single component linewidth of $\Delta H_{pp}=1.42$ G that contains the unresolved proton hyperfine structure (1 G = 0.1 mT = 2.80 MHz). Computer simulation of spectrum (b) [curve (c)] involves four nuclei of $I=1$, $a=1.50$ G; eight nuclei of $I=1/2$, $a=0.07$ G; eight nuclei of $I=1/2$, $a=0.28$ G; eight nuclei of $I=1/2$, $a=0.33$ G; four nuclei of $I=1/2$, $a=0.41$ G; $\Delta H_{pp}=0.20$ G. With permission from the RSC, ref. 22.

The second oxidation product of the porphyrins H_2Por^{2+} is not accessible by the usual techniques. DFT calculations for H_2TPP in the presence of various anions have shown that the second oxidation leads to a saddled dication with N–H localised positive charge and strong ion pairing.[11] Unfortunately, also for the zinc porphyrins the second oxidation product is unstable on the timescale of spectroelectrochemical experiments.

4.3 Metalloporphyrins of the Group V Elements P, As, Sb, Bi

Almost any main group metal or metalloid can be coordinated by porphyrin ligands.[27,28] For the group V metals in the oxidation state $+V$ (with the exception of Bi) a number of hexacoordinated complexes $[(Por)M(L^1)(L^2)]X$ (L = anionic axial ligands like halogen, OH^-, OR^-, alkyl or aryl; X = various anions) have been synthesised and investigated electrochemically and spectroelectrochemically.[1a,27,29–33] For bismuth the oxidation state $+V$ is not accessible. In full agreement with the "inert pair effect" only bismuth(III) porphyrin complexes $[(Por)Bi(X)]$ were obtained.[31] In turn, porphyrin complexes of the trivalent elements decrease in stability within the series Sb > As > P.[27,32,33] A series of bismuth(III) porphyrin complexes $[(Por)Bi(X)]$ (Por = TMP, TPP, TpTP, OEP; X = $SO_3CF_3^-$, NO_3^-) has been studied by spectroelectrochemistry in detail.[31] The results suggest that the oxidation reactions occur exclusively at the porphyrin ligand with no evidence for the formation of a Bi(V) species. For the other elements the electron-transfer reactions should be comparable, although the small size of P(V) leads to a ruffling of the porphyrin ligand.[1a,28,30,34,35] Within the accessible electrochemical range such complexes show one reversible one-electron oxidation and two more or less reversible one-electron reductions (see Figure 4.7).

The spectroelectrochemical investigation by UV–Vis–NIR absorption spectroscopy and in part EPR reveals that all three processes occur at the porphyrin ligand as can be seen from the typical shifts of Soret and Q bands or narrow EPR lines, centred at around $g = 2$ (see Section 4.2). From EPR spectra some metal contribution can be deduced from hyperfine coupling to the nuclei (^{31}P, $I = 1/2$ (nat. abundance = 100%); ^{75}As, $I = 3/2$ (100%); ^{121}Sb, $I = 5/2$ (57.21%); ^{123}Sb, $I = 7/2$ (42.79). For phosphorus the coupling only leads to line broadening of isotropic signals (even at T down to 115 K),[30,33] whereas for $[(TpTP)SbCl_2]^{\cdot}$ hyperfine splitting was resolved with $a^{121}Sb = 63.5$ G and $a^{123}Sb = 35.1$ G[33] (Figure 4.8).

Interestingly, upon cooling the latter sample to 77 K the HFS is lost and the signal exhibits a slightly axial shape. For related samples with unresolved HFS a decrease of the total linewidth was observed. It seems that the metal contribution increases with temperature suggesting equilibria like in Eqn (4.1).

$$[(TpTP^-)Sb(V)Cl_2]^{\cdot} \rightleftharpoons [(TpTP)Sb(IV)Cl_2]^{\cdot} \qquad (4.1)$$

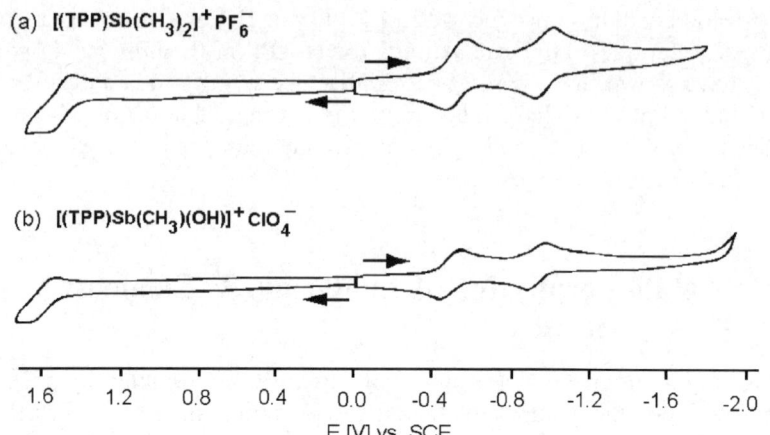

(a) [(TPP)Sb(CH$_3$)$_2$]$^+$PF$_6^-$

(b) [(TPP)Sb(CH$_3$)(OH)]$^+$ClO$_4^-$

E [V] vs. SCE

Figure 4.7 Cyclic voltammograms of (a) [(TPP)Sb(CH$_3$)$_2$]$^+$PF$_6^-$ and (b) [(TPP)Sb(CH$_3$)(OH)]$^+$ClO$_4^-$ in benzonitrile/0.1 M TBAP. With permission from the American Chemical Society, ref. 32.

The total metal contribution to the unpaired spins in the reduced states of the metalloporphyrins is definitely very small, not exceeding a few per cent.

As already mentioned above, some of the reduction processes are not fully reversible due to follow-up chemical reactions (EC and ECE mechanisms). Although not all chemical reactions could be clarified to date for all three elements the partial formation of trivalent species [(Por)M(L)] was observed, as exemplified in Figure 4.9 for the reduction of [(TpTP)SbCl$_2$]$^+$. Such species were in most cases unambiguously detected by their typical two p-hyper-type absorption bands, which occur at 350–380 and 450–470 nm, respectively.[27,30,35,36]

4.4 Exploring the Oxidation States of Methyl-Coenzyme M Reductase (MCR) and the Cofactor F$_{430}$

Methyl-coenzyme M reductase (MCR) catalyses the reaction of methyl-coenzyme M (H$_3$C–SCoM) and coenzyme B (HS–CoB) to methane and the corresponding disulfide CoM–S–S–CoB. This unique reaction proceeds under strictly anaerobic conditions in the presence of the cofactor F$_{430}$, also-called factor F$_{430}$, see Figure 4.10.[37–44]

Despite numerous efforts to gain an insight into this unusual enzymatic reaction, the catalytic mechanism remains largely unknown. In the last few years different forms of the enzyme and interconversions between various MCR states are known and have been investigated.[46–51] In the crystallographically characterised, inactive forms of MCR, the cofactor is EPR-silent and a nickel(II) oxidation state is proposed.[42,47,52,53] Three oxidised EPR-active states of

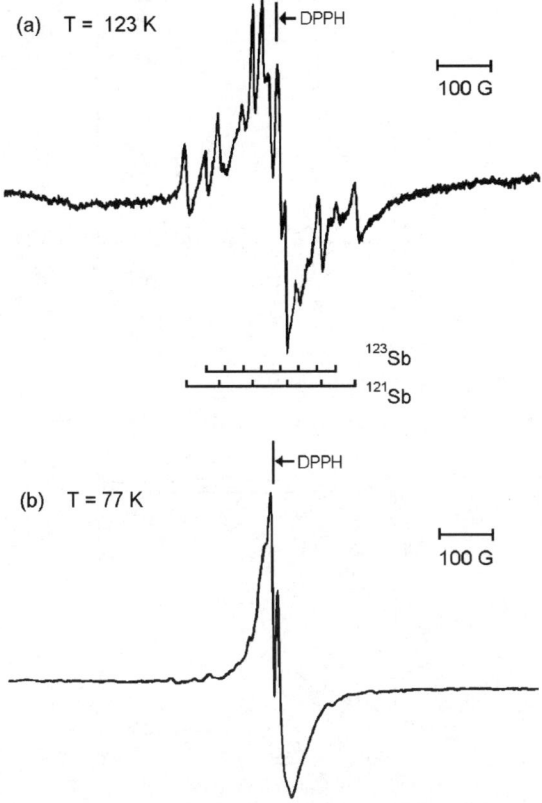

Figure 4.8 EPR spectra obtained at (a) 123 K and (b) 77 K after the bulk controlled-potential reduction of $[(TpTP)SbCl_2]^+$ at -0.40 V (*vs.* SCE) for 3 min in $CH_2Cl_2/0.2$ M TBAP. With permission from the American Chemical Society, ref. 33.

MCR have been detected so far (MCR$_{OX1-3}$).[50] Additionally, there is an EPR-silent oxidised form MCR$_{OX1-SILENT}$.[47,52] The EPR-active states can be either assigned to paramagnetic Ni(I) or Ni(III) species or to Ni(II) bound to an oxidised hydrocorphin ligand. A recent EPR study has revealed that for MCR$_{OX1}$ two resonance forms should be considered: **A**: a Ni(III) centre bound to a thiolate ligand (Ni(III)-[S$^{(-)}$—CoM]) and **B**: high-spin Ni(II) with a thiyl radical (Ni(II)-[S·—CoM]) with an estimated contribution of $7 \pm 3\%$ spin density on the thiolate sulfur atom and the main density of the unpaired electron residing in the Ni $d_{x^2-y^2}$ orbital and the four hydropyrrolic nitrogen.[47] Results from UV–Vis absorption, MCD and X-ray spectroscopy (XAS) reveal a high resemblance of MCR$_{OX1}$ with the EPR-silent Ni(II) states thus proposing a high contribution of the corphin ligand to the SOMO.[51,52] The metal, corphin ligand and thiolate contributions in MCR$_{OX2}$ and MCR$_{OX3}$ probably do not differ much from MCR$_{OX1}$ since the EPR spectra are quite similar.[50]

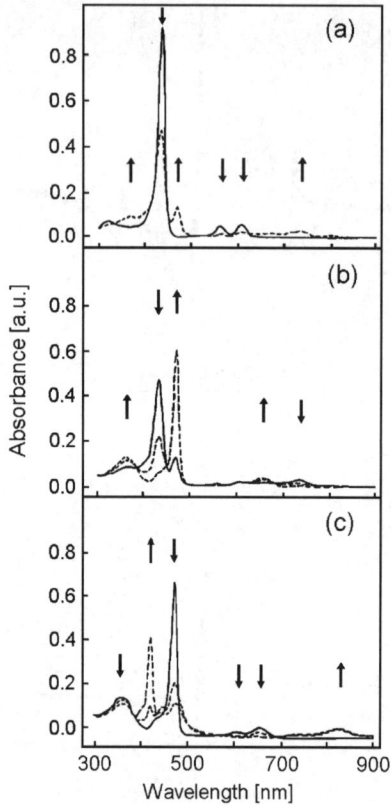

Figure 4.9 Time-dependent UV–Vis spectral changes obtained during the controlled-potential reduction of $[(TpTP)SbCl_2]^+$ in $CH_2Cl_2/0.2$ M TBAP (a) during the first 10 s at -0.40 V ($vs.$ SCE), (b) at longer time scales at -0.40 V and (c) at -0.80 V ($vs.$ SCE). With permission from the American Chemical Society, ref. 33.

The active form MCR$_{RED1}$ and the second reduced form MCR$_{RED2}$ both accumulate in the cells when the gas mixture normally used for cell growth (80% H_2/20% CO_2) is changed to 100% H_2 gas. They are both EPR-active and usually assigned to formal Ni(I) centres.[46,47,49–51,53,54]

The assignment of the MCR$_{RED1}$ oxidation states was recently under debate when, based on XAS and resonance Raman (RR) investigations on MCR and the isolated cofactor F$_{430}$, it was proposed that on reduction of the EPR-silent forms of MCR to MCR$_{RED1}$ the electron density at the nickel atom does not vary significantly (XAS) and secondly the 2-electron conversion of MCR$_{OX1}$ to MCR$_{RED1}$ can be described as a 2-electron reduction of the corphin ligand.[51] Hereupon, Piskorski and Jaun[48] started a spectroelectrochemical study using MCR, the free coenzyme (NiF$_{430}$) and its pentamethyl ester (NiF$_{430}$M). Figure 4.11 shows the UV–Vis spectra during the reduction of NiF$_{430}$M, revealing a reversible, clean one-electron reduction.

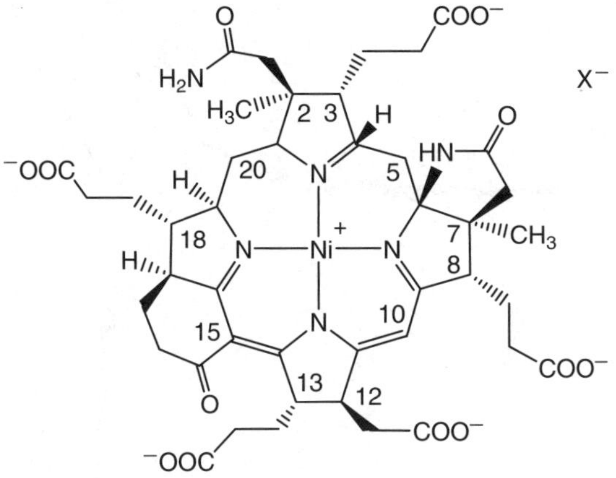

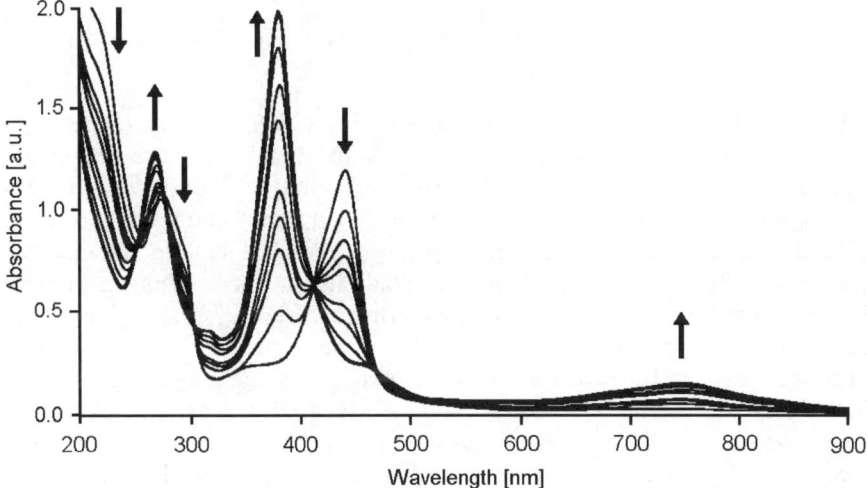

Figure 4.10 Structure of F_{430}, the prosthetic group of methyl-coenzyme M reductase (MCR), which is involved in the formation of methane in methanogenic bacteria. The highly reduced macrocyclic structure, related to porphyrins, is termed a corphin.[45]

Figure 4.11 Changes in the UV–Vis spectrum of $Ni(II)F_{430}M$ in acetonitrile/0.1 M $TBAPF_6$ during reduction in an OTTLE cell. With permission from the American Chemical Society, ref. 48.

The obtained spectrum of the reduced form is almost identical to spectra of MCR_{RED1}, whereas the long-wavelength band around 750 nm is blue-shifted for the NiF_{430} system. This is in keeping with a charge-transfer character of this band[52] and the highly hydrophobic environment of F_{430} in the enzyme.[42] EPR spectroscopy on reduced $NiF_{430}M$ (Figure 4.12) revealed an axial spectrum ($g_{||} = 2.2201$, $g_{\perp} = 2.0722$) with a high g anisotropy ($\Delta g = 0.1479$) consistent with a high nickel contribution to the SOMO.

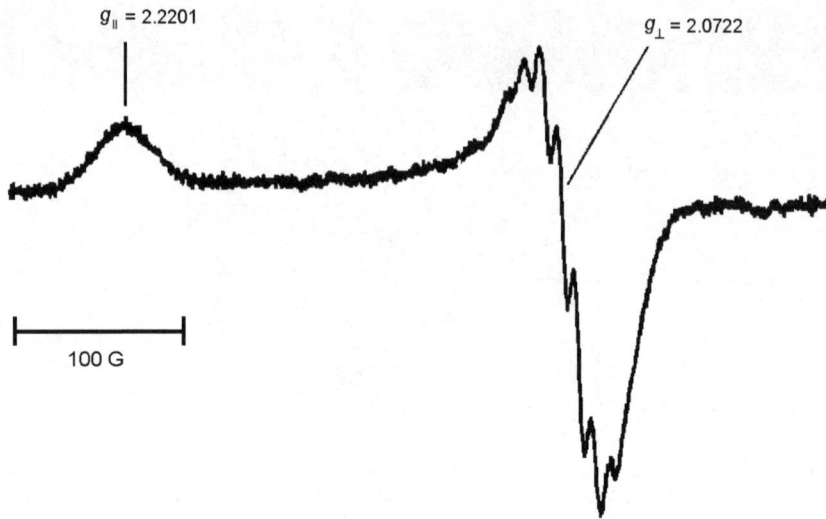

Figure 4.12   X-Band EPR spectrum of reduced $NiF_{430}M$ generated by electrolysis in acetonitrile/0.1 M $TBAPF_6$ (frozen solution at $T = 130\,K$, microwave frequency = 9.512 GHz). With permission from the American Chemical Society, ref. 48.

For MCR_{RED1}[49,53] the reduced free coenzyme NiF_{430}[53] and related isobacteriochlorin[55] similar values have been obtained. In view of the markedly different EPR behaviour (narrow EPR signals around $g = 2$) and strongly deviating UV–Vis absorption spectra (usually redshifted Soret band upon reduction) of related reduced nickel-porphyrin, -hydroporphyrin, or -chlorin systems that are best described as tetrapyrrolic π anion radicals,[55,56] the assignment to nickel(I) species in MCR_{RED1} and the reduced form of NiF_{430} or $NiF_{430}M$ is justified. However an appreciable contribution of the corphin ligand cannot be ruled out since the EPR spectra reveal HFS to the N atoms of the magnitude of 10 G. Investigation of the higher reduced states of MCR, NiF_{430} or $NiF_{430}M$ might reveal species with distinct π radical anionic or π dianionic character. The EPR spectra of MCR_{RED2}, which is produced from MCR_{RED1} in the presence of coenzyme M (HS—CoM) reveal even higher g anisotropy, indicative of an even higher nickel contribution to the SOMO[46,49,50,54] but its character is not yet fully understood and very probably MCR_{RED2} does not represent a higher reduced state but an isomeric form of MCR_{RED1}. For $NiF_{430}M$ these highly reduced states lie beyond $-2.4\,V$ (*vs.* the ferrocene/ferrocenium couple), the corresponding redox waves are irreversible, which hampers their investigation.[48]

4.5   The Oxidation Catalyst [(Por)Ru(CO)(L)]

A very prominent class of metalloporphyrins are the tetra(aryl)porphyrin ruthenium carbonyl complexes [(Aryl)$_4$Por)Ru(CO)(L)] (L = solvent) that, in recent years have gained great importance in oxygenation catalysis.[57–62] For a deeper understanding of the role the porphyrin ligands, the metals and the axial coligands play in the oxidation catalysis it is necessary to explore their electronic states. Tetra(aryl)porphyrin ruthenium carbonyl complexes have therefore been the subject of thorough electrochemical and spectroscopic investigation, including spectroelectrochemical methods.[4,63–74]

The electrochemistry of the complexes [(Por)Ru(CO)(L)] (Por = TPP and substituted derivatives or OEP; L = diverse ligands including solvent molecules) reveals that within the usual range (+2 V to −3 V) four further redox states are accessible, corresponding to two individual one-electron reductions and two individual one-electron oxidations (see Figure 4.13).

For this class of compounds, IR spectroelectrochemistry should be well suited to monitor the change in electron density in the ruthenium centre upon oxidation or reduction of the metalloporphyrins since the C–O stretching frequency has proved in many investigations of carbonyl ruthenium compounds to be a powerful probe for such changes.[74–76,77a,b,78,79] In Figure 4.14 a typical IR spectroelectrochemical monitoring is shown, Table 4.1 lists some collected data.

In all cases an increase in the energies of the stretching vibration is observed upon oxidation of the ruthenium porphyrins, whereas upon reduction the energies decrease. The free CO ligand absorbs at $2140 \, cm^{-1}$, so in a ruthenium porphyrin complex the metal-to-ligand π backbonding into antibonding orbitals has already lowered the C–O bond strength markedly. Upon oxidation of the complex this backbonding is reduced, leading to an enhancement of the C–O bond strength. Reduction leads consequently to the reverse effect. Therefore, the generally observed trend is fully in line with our expectations.

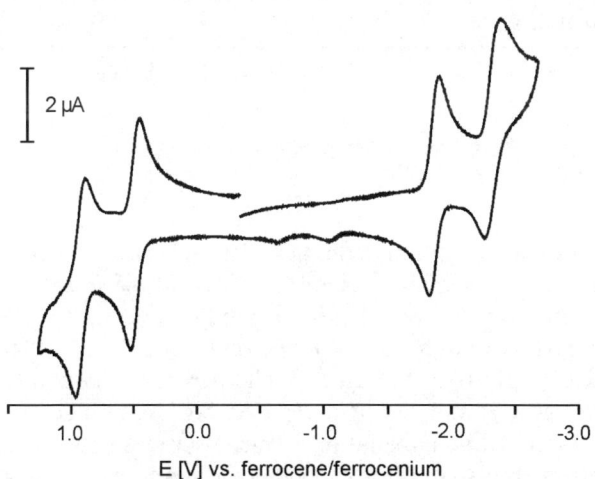

Figure 4.13 Cyclic voltammogram of [(TMP)Ru(CO)] in acetonitrile/0.1 M TBAPF$_6$ solution (scan rate = 0.1 V/s). Potentials are given *vs.* ferrocene/ferrocenium from ref. 74.

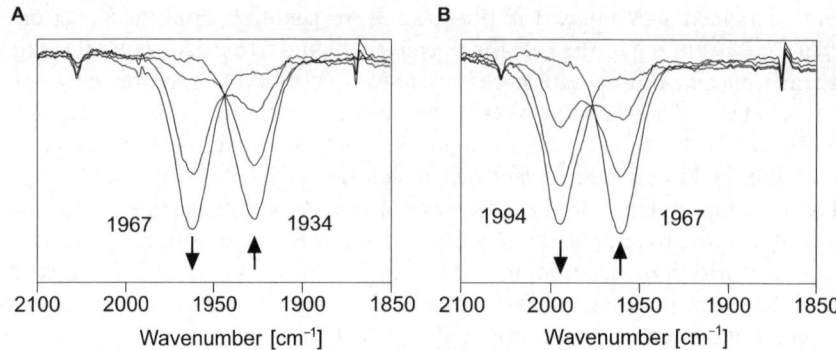

Figure 4.14 IR spectroelectrochemistry of [(TMP)Ru(CO)] in CH$_2$Cl$_2$/0.1 M TBAPF$_6$ solution. **A**: first oxidation, **B**: second oxidation, from ref. 74.

Table 4.1 Spectroelectrochemical data: IR absorption maxima ν of C≡O stretching vibration for ruthenium complexes [(Por)Ru(CO)(L)].^{a}

			ν_{max}	*[cm^{-1}]*		*Ref.*
Complex	[Ru]$^{2-}$	[Ru]$^-$	[Ru]	[Ru]$^+$	[Ru]$^{2+}$	
[(TMP)Ru(CO)(CH$_2$Cl$_2$)]	1853	1893	1934	1967	1994	74
[(TDMAP)Ru(CO)(CH$_2$Cl$_2$)]	1851	1895	1934	1961	1991	74
[(TPP)Ru(CO)(THF)]^b	1853	1898	1941	1971	–	67
[(TPP)Ru(CO)(DMF)]^c	–	1896	1923	1962	–	67
[(TPP)Ru(CO)(Py)]	1853	1905	1942	1985	–	67
[(TPP)Ru(CO)(CH$_2$Cl$_2$)]	–	–	1930	1945	–	73
[(TPP)Ru(CO)(Py)]	–	–	1943	1970	–	73
[(OEP)Ru(CO)(Py)]	–	1896	1931	–	–	65
[(OEP)Ru(CO)(THF)]^b	–	1878^d	1931	1946	–^{e}	65
[(OEP)Ru(CO)(CH$_2$Cl$_2$)]	–	–	1917	1950	–	73

^{a} Wave numbers ν in cm^{-1} from measurements in CH$_2$Cl$_2$ unless stated otherwise.
^{b} in THF.
^{c} in DMF.
^{d} Further peak at 1896 cm^{-1} after prolonged electrolysis.
^{e} Not observed due to decomposition.

In the trinuclear ruthenium complex [Ru$_3$(μ^3-O)(μ-CH$_3$COO)$_6$(CO)(pyridine)$_2$] which belongs to the well-established class of the so-called triruthenium clusters, described in more detail in Chapter 5, a spectroelectrochemical investigation gave νCO frequencies of 1900 cm^{-1} for [Ru$_3$]$^-$, 1940 cm^{-1} [Ru$_3$], 2055 cm^{-1} [Ru$_3$]$^+$, 2110 cm^{-1} [Ru$_3$]$^{2+}$.[79] The parent compound [Ru$_3$] formally consists of one Ru(II) and two Ru(III). The biggest shift for the CO stretch (+115 cm^{-1}) occurs upon oxidation, which is presumably a highly metal-centred electron transfer. The second oxidation leads to a delocalised Ru(III)Ru(III)Ru(IV) state,[80] therefore the shift is far smaller (+55 cm^{-1}). The same argument holds for the reduction (−40 cm^{-1}) also leading to a delocalised Ru(II)Ru(II)Ru(III) state.[80] A further look at other IR spectroelectrochemical studies of ruthenium carbonyl complexes reveals that shifts of 20–50 cm^{-1} are typical for coligand-centred electron transfer,[76,77,81,82a,b] whereas metal-centred electron transfer should lead to much larger shifts.[75,83,84] From this we can already conclude that for the complexes [(Por)Ru(CO)L] the site of electron transfer is mainly on the porphyrin ligand. We will now combine this with further spectroscopic evidence for porphyrin-centred electron transfer.

Thorough spectroscopic investigation of the first oxidised states using UV–Vis absorption and EPR spectroscopy lead to the assignment of mainly porphyrin-centred oxidations.[65,67–70,72–74,85–87a,b,88a,b,c,89] The typical UV–Vis–NIR spectrum of the corresponding π cation radicals exhibits a split Soret band, each component with half the initial intensity, structured long-wavelength band systems around 650 nm and further weak absorptions around 900 nm. EPR signals so far reported for such π cation radicals are narrow, unstructured with g values around the "free-electron" value of 2.0023. Interestingly, upon exchange of the CO ligand by other ligands, *e.g.* OEt$^-$ in the recently reported [(TPFPP)Ru(OEt)(EtOH)], the corresponding cation radical [(TPFPP)Ru(OEt)(EtOH)]$^{\cdot+}$ generated by chemical reduction of the Ru(VI) precursor complex using Zn(Hg) exhibits a rhombic EPR signal at low temperatures with a g anisotropy ($\Delta g = g_1 - g_3$) of 0.64,[57] which points to a marked metal contribution,[90a,b,c,d] *e.g.* Δg for the [(bpy)$_3$Ru(III)]$^{\cdot 3+}$ radical lies at 1.5.[91] Such remarkable "gain of metal character" of corresponding cation radicals upon CO dissociation has been reported earlier by others.[73,88]

The species generated during the second oxidation were not studied so far, which is due to decomposition reactions of the formed dications,[65,67,68] or even the monocations,[88a,b,c,89] either through loss of CO or disproportionation reactions. In 1981 Rillema *et al.*[71] proposed a mainly metal-based second oxidation on the basis of the potential separation (between first and second oxidation) of 0.42 V for [(TPP)Ru(CO)(CH$_2$Cl$_2$)] whereas the usual separation for the porphyrin should be around 0.29 V. However, this discrepancy might be due to other mechanisms and to date no spectroscopic evidence has been provided for such a metal-centred oxidation reaction that should lead to a biradical [(Por$^{\cdot+}$)Ru(III)(CO)(L)]$^{2+}$ or a ruthenium(IV) species [(Por)Ru(IV)(CO)(L)]$^{2+}$.

For the first reduction the IR shifts point to a porphyrin-centred electron transfer. This is supported by further spectroscopy on the anion radical complexes [(Por)Ru(CO)(L)]$^{\cdot-}$. The observed EPR lines are narrow, unstructured, with g values around 2.[67] The UV–Vis–NIR spectra of the radical anions are characterised by redshifted Soret bands of reduced intensity, a weak structured band system around 600 nm and weak broad absorptions around 800 or 900 nm (see Figure 4.15).[65,67] Further support comes from resonance Raman investigations on [(OEP)Ru(CO)(THF)]$^{\cdot-}$ for which the observed Raman bands fit perfectly to those of the [(OEP)VO]$^{\cdot-}$ radical anion.[63] There is some evidence that if the spectroelectrochemistry is not carried out in very aprotic and unpolar solvents or traces of water are present, the radical anionic complexes are readily transformed. This has been investigated for the [(OEP)Ru(CO)(L)] system, where the use of solvents like MeOH or nitriles for the electrochemical reduction leads to altered species with unreduced porphyrin ligands (see Figure 4.15).[4,65]

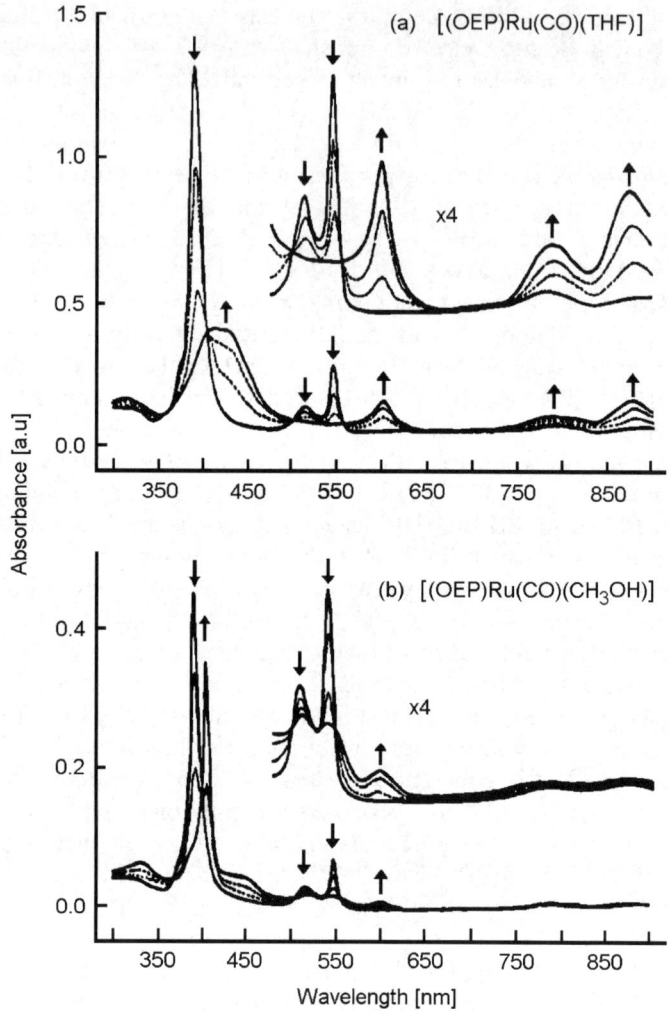

Figure 4.15   Time-resolved thin-layer UV-Vis spectra obtained during controlled-potential reduction of (a) [(OEP)Ru(CO)(THF)] and (b) [(OEP)-Ru(CO)(CH$_3$OH)] in THF/0.2 M TBAP at −1.9 V (*vs.* SCE). With permission from the American Chemical Society, ref. 65.

However, the interpretation of Ru(I) species being the products of the transformation is not very plausible, since the CO-stretching frequency of the resulting species is closer to the parent complex than the primary radical anions. Another study revealed that the radical anionic [(TPP)Ru(CO)(THF)]$^{\cdot-}$ readily undergoes formation of the corresponding chlorine complex in the presence of MeI.[66] Such electrophilic attack might also be the reason for the above-described transformations in protic or highly polar media.

For the species after the addition of a second electron, *i.e.* $[(Por)Ru(CO)(L)]^{2-}$, the reported shift of the CO-stretching vibration suggests again a porphyrin-centred reduction. However, probably due to the above-described follow-up chemical reactions for the monoreduced species there is only a report on $[(TPP)-Ru(CO)(THF)]^{2-}$. The Soret band undergoes a further redshift, compared to the monoanion, the band system around 600 nm has also moved to lower energy.[67]

To summarise, the first oxidation step for $[(Por)Ru(CO)(L)]$ can surely be regarded as almost exclusively porphyrin ligand-centred, a conclusion which is important for the understanding of the various catalytic transformations of such systems.[61,62,74] Also well characterised is the first reduced state (ligand-centred). The assignment for the two further states (dianionic or dicationic) is less clear. IR data (Table 4.1) also suggests a ligand-centred electron transfer but the lack of stability of the generated species leaves some uncertainty.

4.6 Searching for Divalent Gold

Gold porphyrins have received enormous interest in the last decade due to their versatile applications *e.g.* in photoactive dyads or triads,[92a,b,c,93] as building blocks in supramolecular aggregates,[93–95] in organic catalysis,[96] and as potential anticancer drugs.[97a,b,c] Of more academic interest is the fact that porphyrin ligands have played an important role in efforts to extend the rather short series of mononuclear divalent gold compounds: gold(II) porphyrins $[(Por)Au]$;[98–101] complexes with thio-macrocyclic ligands such as $[Au([9]aneS_3)_2]^{2+}$ or $[Au([12]aneS_4)_2]^{2+}$;[102a,b,c,103] $[AuXe_4]^{2+}$;[104] $[Au(SO_3F)_{3-x}]$ species;[105] complexes of gold(II) with $N_2S_2X_1$ or $N_2S_2X_2$ ligand sets (ligands like *o*-methylthio)aniline and 1,2-bis((*o*-aminophenyl)thio)ethane);[106a,b,c] complexes with an S_4 ligand set (ligands like mnt = 1,2-dicyanoethene-1,2-dithiolate or Et_2dtc = diethyldithiocarbamate);[107–109] or the ill-defined $[Au(phthalocyanin)]$.[110]

In 1978 Jamin and Iwamoto[111] reported that $[(TPP)Au(III)](AuCl_4)$ undergoes three reversible one-electron reduction steps but from the potential separations and the UV–Vis absorption data of the species after the first two reductions they concluded that the gold atom remains in the oxidation state + III. Only recently Kadish *et al.*[99,101] investigated the UV–Vis and EPR spectroelectrochemistry of a number of gold porphyrin systems $[(Por)Au(III)](PF_6)$ (see Figure 4.16, to the right) and found strong evidence for divalent gold species.

The first three reductions could be monitored by UV–Vis absorption spectroscopy (Figure 4.17) and indicate a first metal-centred reduction (small bathochromic shift of the Soret band with only marginal loss of intensity) followed by two porphyrin-based reductions (*e.g.* strong decrease in intensity for the Soret band and broad bands in the visible region).

Strong evidence for a Au(II) formulation came from EPR spectroscopy (Figure 4.18) showing broad (linewidth $\Delta H \sim 300$ G), unstructured signals with *g* values around 2.06. During the same experiments (reduction with naphthalene radical anion) narrow signals centred around 2.006 were also detected and

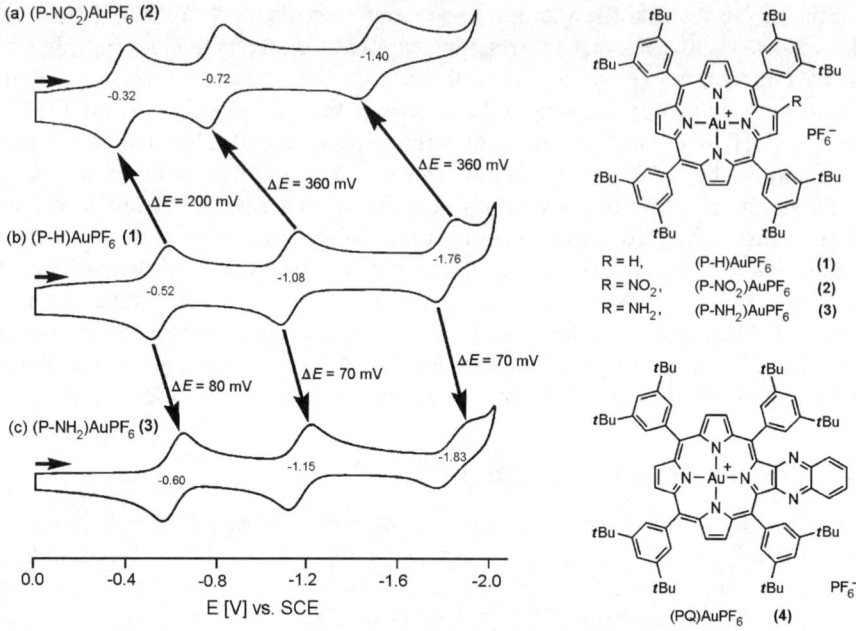

Figure 4.16   Cyclic voltammograms of (a) [(P-NO$_2$)Au]PF$_6$ (**2**), (b) [(P-H)Au]PF$_6$ (**1**), and (c) [(P-NH$_2$)Au]PF$_6$ (**3**) in pyridine/0.1 M TBAP. With permission from the American Chemical Society, ref. 99.

assigned to corresponding species, where the single electron resides mainly in the porphyrin ligand (π radical anions). For the Por = P-NO$_2$ complex the reduction gave exclusively this type of signal (ligand-centred) that can be explained by the electron-withdrawing nature of the NO$_2$-substituent. This is fully in line with a study of Lobstein and coworkers[98] in which for electrogenerated [(1-NpyridiniumTPP)Au]$^+$ a narrow signal with $g = 2.006$ was observed down to 15 K (ligand centred) and at 4 K an additional broad signal ($g = 2.06$, $\Delta H \sim 300$ G, Au(II) species) appeared, and agrees also with a study by Gencheva et al.[100] on a hematoporphyrinIX gold(II) complex, where at 130 K from a polycrystalline sample of the isolated complex both an unresolved narrow and isotropic signal ($g = 2.003$) and an axial signal ($g_\perp = 2.035$, $g_\parallel = 1.97$) were observed, whereas in glassy frozen solution only the narrow signal could be detected.

   A number of Au(II) complexes with sulfur-containing macrocyclic ligands have been investigated by EPR showing g values around 2.02,[102,103] whereas (*o*-methylthio)aniline and 1,2-bis((*o*-aminophenyl)thio)ethane complexes of gold(II), which represent a N$_2$S$_2$X$_1$ or N$_2$S$_2$X$_2$ coordination sphere, exhibit EPR signals centred around $g = 2.003$.[106] For [Au(mnt)$_2$]$^{2-}$ a g_{iso} value of 2.009 in glassy frozen solution[109] but $g_{iso} = 1.9947$ was measured for a solid sample of [Au(mnt)$_2$]$^{2-}$ doped in [Ni(mnt)$_2$]$^{2-}$,[108] indicating a strong dependency on the environment.[105] A very similar g value compared to the porphyrin systems was

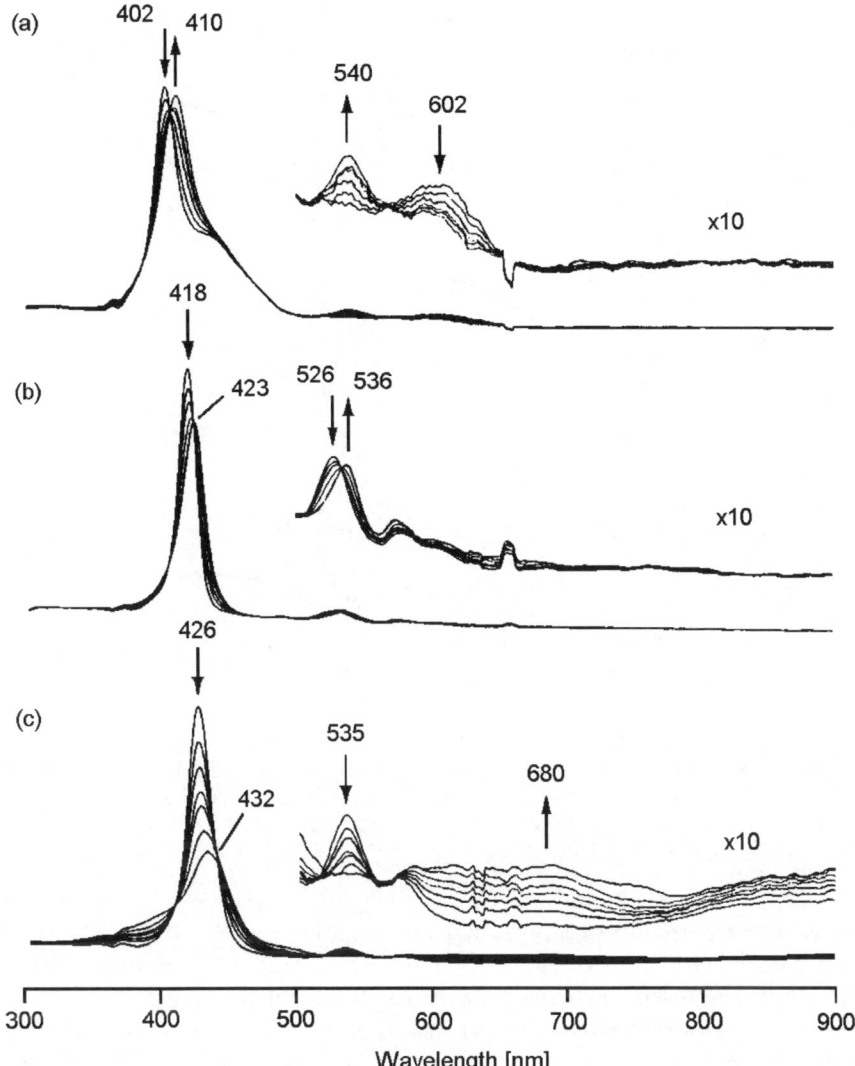

Figure 4.17 UV–Vis spectral changes upon the first electroreduction of (a) [(P-NH₂)Au]PF₆ (**3**) at an applied potential of −0.80 V, (b) [(P-H)Au]PF₆ (**1**) at −0.65 V, and (c) [(P-NO₂)Au]PF₆ (**2**) at −0.60 V in pyridine/0.2 M TBAP, all values given *vs.* SCE. With permission from the American Chemical Society, ref. 99.

reported very early by MacCragh and Koski for chemically generated solid gold phthalocyanine ($g_{iso} = 2.0065$).[110]

Apart from the g value, the coupling a (hyperfine splitting = HFS) of the unpaired electron to the gold nucleus ([197]Au, $I = 3/2$, 100% natural abundance) should reveal the metal contribution to the unpaired spin. The highest HFS of about 30 to 60 G have been observed for the S-macrocycle coordinated species,

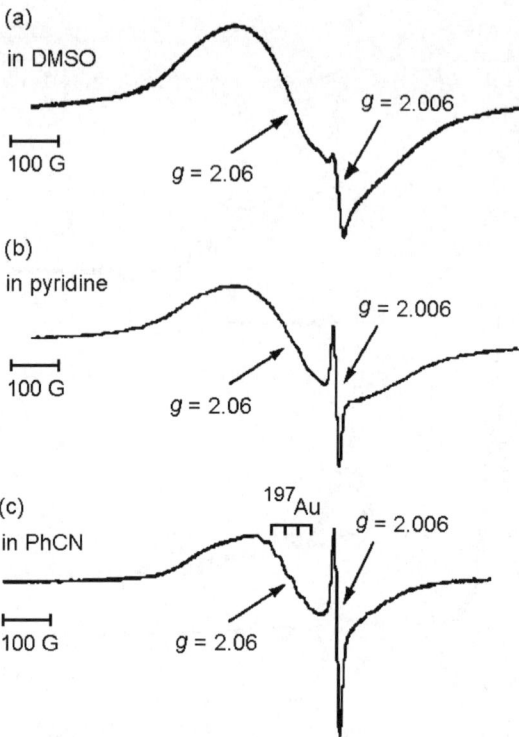

Figure 4.18 EPR spectra of singly reduced [(P-H)Au]PF$_6$ (1.0 mM) generated by the reduction with one equivalent of naphthalene radical anion (1.0 mM) in deaerated (a) DMSO, (b) pyridine and (c) benzonitrile at $-160\,^{\circ}$C. With permission from the RSC, ref. 101.

the values obtained for the S$_2$N$_2$X$_2$ systems are far lower (4–10 G). For [(P-H)Au]$^-$ measured in benzonitrile (Figure 4.18(c)) a^{197}Au HFS of 27 G was found, placing the porphyrin systems between these two groups. For the hematoporphyrinIX gold(II) complex HFS from a^{14}N ($a_\perp$ and $a_\parallel$ both 17 G) and a^{197}Au (both 15 G) were obtained from spectra simulation.[100] Larger values have been measured for solid gold phthalocyanine with a^{197}Au$_\perp = 65$ G, a^{14}N$_\perp = 16.9$ G,[110] for [Au(mnt)$_2$]$^{2-}$ ($a_{iso} = 83$ G (in glassy frozen solution),[109] or 121 G for [Au(mnt)$_2$]$^{2-}$ doped in [Ni(mnt)$_2$]$^{2-}$),[108] and for HSO$_3$F solvated Au^{2+} ions ($a_{iso} = 133.7$ G).[105]

An estimation of the virtual spin density on the metal gave $\sim25\%$ for [Au([9]Sane)]$^{2+}$ the rest mainly localised on the S atoms.[102a] Therefore for the porphyrin systems a slightly smaller contribution of the gold atom might be deduced (if such correlation is permissive). Although these complexes are far from being "pure" Au(II) species, porphyrin ligands provide the base to generate radical species with relatively high gold contribution to the unpaired electron. With the use of electron-donating substituents the metal contribution might even be enhanced.[98]

4.7 Multiredox-Arrays: Cobalt(II)Porphyrin-Fullerene Dyads

Within the large number of multiredox arrays containing metalloporphyrins,[112a,b] covalently bound (conjugate) fullerene-metalloporphyrin dyads have gained enormous interest in the last ten years, mainly due to their potential application as artificial antennae[113,114a,b,115,116] Due to the multiredox behaviour of the fullerenes (up to six reversible one-electron reductions and at least one reversible one-electron oxidation),[117a,b,118] the porphyrin ligands and the incorporated metals, the assignment of electron-transfer steps in such systems is difficult. Recently, spectroelectrochemical characterisation has been carried out on a number of fullerene-[(TPP)Co] dyads shown in Scheme 4.3,[119] which exhibit rather complex redox behaviour (Figure 4.19).

The first oxidation is quasireversible (peak-to-peak separation is scan-rate dependent) in such systems and is commonly assigned to a cobalt-centred Co(II)/Co(III) redox couple.[120] The quasireversible behaviour is attributed to changes in axial coordination. Steps two and three correspond to reversible one-electron oxidations of the porphyrin ligand, the fourth and fifth one-electron oxidations involve the C_{60} moiety.[117a,119,121] The first one-electron reduction is assigned to a C_{60}-centred electron transfer, the huge second reduction wave is dependent on the type of conjugate (**M1** to **M3**). Therefore it can be assumed that it is formed by two overlapping one-electron processes that are assigned to one-electron reductions of the fullerene and the porphyrin ligand, respectively. To support the assignment of the redox waves the first two oxidation steps were investigated by UV–Vis spectroelectrochemistry as shown in Figure 4.20.

Comparison with analogous investigations on [(TPP)Co] reveals that for the cobalt porphyrin complex [(T*t*BuP)Co] the first electrochemical oxidation occurs metal-centred, leading to a Co(III) species [(T*t*BuP)Co(III)]$^{+}$.[120,122] The second

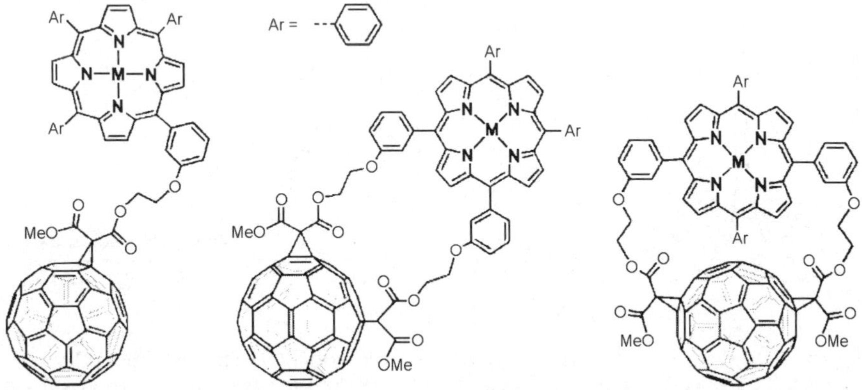

M1   **M2**   **M3**

Scheme 4.3 Representation of three fullerene-[(TPP)Co] dyads. With permission from the American Chemical Society, ref. 119.

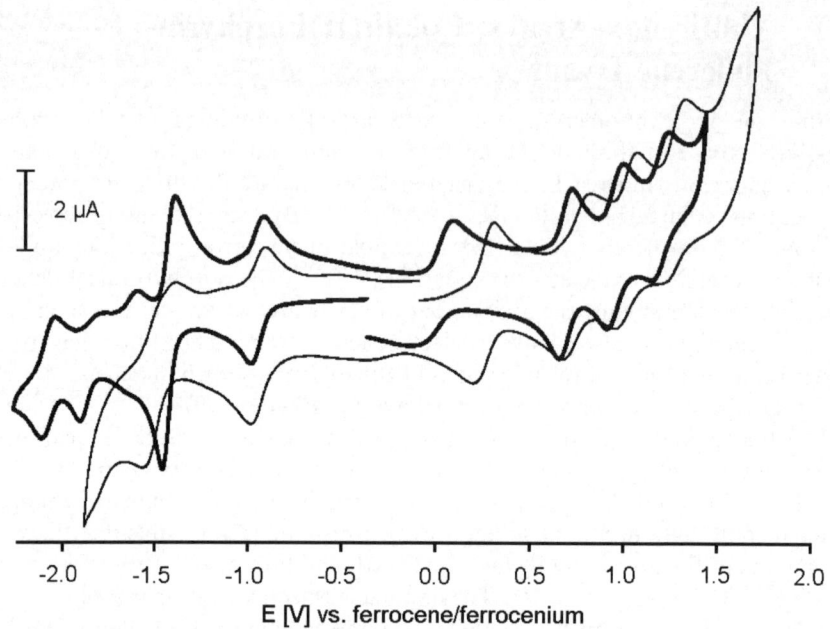

Figure 4.19 Cyclic voltammetry (scan rate = 0.1 V/s) of **M1** (bold line) and **M3** (thin line) in benzonitrile/0.1 M TBAPF$_6$. With permission from the American Chemical Society, ref. 119.

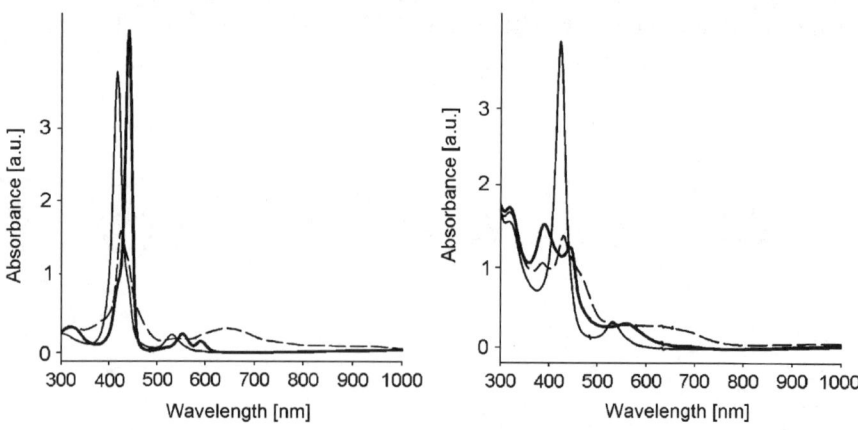

Figure 4.20 UV–Vis absorption spectra of the cobalt(II) species (line), the one-electron oxidised species (bold line), and the two one-electron oxidised species (dashed line) obtained during spectroelectrochemical oxidation of the cobalt porphyrin [(TtBuP)Co] (left) and the dyad **M3** (right) in benzonitrile/0.1 M TBAPF$_6$. With permission from the American Chemical Society, ref. 119.

oxidation, in contrast, leads to a dication $[(TtBuP^{\cdot+})Co(III)]^{2+}$ with a π cationic porphyrin ligand. The same behaviour was also reported for $[(TPP)Co]$.[120,122]

For the dyad **M3** the conclusion from the UV–Vis spectra contrasts with the free Co porphyrin complexes. The marked decrease of the Soret band (while split into two) strongly suggests the formation of a porphyrin π cation, *i.e.* $[full-(TtBuP^{\cdot+})Co(II)]^{\cdot+}$. The second oxidation then probably leads to the same type of species $[full-(TtBuP^{\cdot+})Co(III)]^{2+}$ as observed for the free Co-porphyrin complexes.

4.8 Conclusions

The electrochemistry of metalloporphyrins $[(Por)M]$ is very often assigned on the basis of the potential separations from electrochemical measurements (usually CV or related methods). If the measured separations agree more or less quantitatively with those found for the free porphyrin ligands H_2Por the electron-transfer steps were assigned to occur within the porphyrin π-system, leading to π radical cationic $[(Por^{\cdot+})M]$, dicationic $[(Por^{2+})M]$, π radical anionic $[(Por^{\cdot-})M]$ or dianionic species $[(Por^{2-})M]$. If the potential separations found for the metalloporphyrins do not fit to those of the free porphyrin ligands, a metal-centred electron-transfer step can be assumed. However, the potential separation is not a reliable tool in many cases, as we have seen especially in Section 4.6. Therefore, spectroelectrochemical characterisation of the redox states is necessary to discriminate between purely ligand-based electron transfer and metal-centred redox chemistry. In the preceding sections we have seen that, usually based on several spectroscopic methods, such discrimination is possible even for rather complicated systems (*e.g.* Section 4.7). Also, we have learned that in some systems there is a smooth transition between porphyrin- and metal-centred electron transfer (Sections 4.3 and 4.6) and the spectroelectrochemical characterisation allows the individual contribution of metal and porphyrin ligand to be estimated.

Abbreviations

TPP	*meso*-Tetraphenylporphyrin
TMP	*meso*-Tetramesitylporphyrin
T*p*TP	*meso*-Tetra(*p*-tolyl)porphyrin
THPP	*meso*-Tetra[*p*-(*n*hexyl)phenyl]porphyrin
TDPP	*meso*-Tetra[(3,5-*t*Bu)$_2$C$_6$H$_3$]porphyrin
TDMAP	*meso*-Tetra(9-[anti-(1,2,3,4,5,6,7,8-octahydro-1,4:5,8-dimethanoanthracenyl)]porphyrin
TPTBP	*meso*-Tetraphenyltetrabenzoporphyrin
TPFPP	*meso*-Tetra(pentafluorophenyl)porphyrin
T*t*BuP	*meso*-Tetra(*p*-*t*Buphenyl)porphyrin
OEP	2,3,7,8,12,13,17,18-Octaethylporphyrin
F$_{430}$	Cofactor F$_{430}$
MCR	Methyl-coenzyme M reductase

TBAP   Tetra-*n*-butylammonium perchlorate
TBAPF$_6$ Tetra-*n*-butylammonium hexafluorophosphate
OTTLE cell   Optically transparent thin-layer electrochemical cell

Numbering scheme for porphyrin ligands. 5, 10, 15 and 20 are the *meso* positions, sometimes referred to as α, β, γ and δ.

References

1. Due to their importance in many fields in chemistry, numerous monographs have been devoted to porphyrin chemistry. Comprehensive information is available in (a) *The Porphyrin Handbook*, Volumes 1–10, K. M. Kadish, K. M. Smith and R. Guilard (ed.), Academic Press, San Diego, 2000. (b) *The Porphyrin Handbook*, Volumes 11–20, K. M. Kadish, K. M. Smith and R. Guilard (ed.), Academic Press, Amsterdam, 2003, whereas detailed scientific work can be retrieved in (c) the *Journal of Porphyrins and Phthalocyanines* (JPP), Society of Porphyrins & Phthalocyanines from 2002 to date, (d) *ibid.*, John Wiley & Sons, Chichester, 1996–2001. (e) K. M. Smith, *Porphyrins*, in *Comprehensive Coordination Chemistry II*, J. A. Mc Cleverty, T. J. Meyer (ed.), Elsevier, Oxford, UK, 2004, Vol. 1, 493–506. (f) S. J. A. Van Gisbergen, E. J. Baerends, *Scientific Computing and Modelling*, in *Comprehensive Coordination Chemistry II*, J. A. Mc Cleverty and T. J. Meyer (ed.), Elsevier, Oxford, UK, 2004, **2**, 511–517. (g) C. M. Drain, I. Goldberg, I. Sylvain, A. Falber, *Top. Curr. Chem.* 2005, **245**, 55–88.
2. Z. Zhao, K. I. Ozoemena, D. M. Maree and T. Nyokong, *Dalton Trans.*, 2005, 1241–1248.
3. (a) J. A. Crayston, *Spectroelectrochemistry*, in *Comprehensive Coordination Chemistry II*, J. A. Mc Cleverty and T. J. Meyer, (ed.), Elsevier, Oxford UK, 2004, **1**, 775–789; (b) J. A. Crayston, *Spectroelectrochemistry In situ UV-Vis Spectroscopy*, in *Encyclopedia of Electrochemistry*, A. J. Bard, M. Stratmann and P. R. Unwin, (ed.), Wiley-VCH, Weinheim, 2003, Vol. 3, 491–529.

4. K. Kadish and X. Mu, *Pure Appl. Chem.*, 1990, **62**, 1051–1054.

5. J. Pawlik, L. Gherghel, S. Karabunarliev and M. Baumgarten, *Chem. Phys.*, 1997, **221**, 121–133.

6. (a) R. H. Felton, in *The Porphyrins-Physical Chemistry*, D. Dolphin (ed.), Part C, Vol. 5, Academic Press, New York, 1978, 53–125.; (b) R. H. Felton and H. Linschitz, *J. Am. Chem. Soc.*, 1966, **88**, 1113–1116.

7. (a) M. Huber, *Eur. J. Org. Chem.*, 2001, 4379–4389; (b) M. Gouterman, *J. Chem. Phys.*, 1959, **30**, 1139–1161.

8. A. Brisach-Wittmeyer, S. Lobstein, M. Gross and A. Giraudeau, *J. Electroanal. Chem.*, 2005, **576**, 129–137.

9. S. K. Pandey, A. L. Gryshuk, A. Graham, K. Ohkubo, S. Fukuzumi, M. P. Dobhal, G. Zheng, Z. Ou, R. Zhan, K. M. Kadish, A. Oseroff, S. Ramaprasad and R. K. Pandey, *Tetrahedron*, 2003, **59**, 10059–10073.

10. K. M. Kadish, L. R. Shiue, R. K. Rhodes and L. A. Bottomley, *Inorg. Chem.*, 1981, **20**, 1274–1277.

11. H. Sun, J. C. Biffinger and S. G. DiMagno, *Dalton Trans.*, 2005, 3148–3154.

12. So-called effective ionic radii. (a) R. D. Shannon and C. T. Prewitt, *Acta Crystallogr.*, 1969, **B25**, 925–946; (b) R. D. Shannon, *Acta Crystallogr.*, 1976, **A32**, 751–767.

13. (a) A. Hazell, *Acta Crystallogr.*, 1986, **C42**, 296–299; (b) H. J. Jakobsen, P. D. Ellis, R. R. Inners and C. F. Jensen, *J. Am. Chem. Soc.*, 1982, **104**, 7442–7452.

14. I. Mayer, A. L. B. Formiga, F. M. Engelmann, H. Winnischofer, P. V. Oliveira, D. M. Tomazela, M. N. Eberlin, H. E. Toma and K. Araki, *Inorg. Chim. Acta*, 2005, **358**, 2629–2642.

15. F. X. Redl, M. Lutz and J. Daub, *Chem. Eur. J.*, 2001, 5350–5358.

16. G. L. Closs and L. E. Closs, *J. Am. Chem. Soc.*, 1963, **85**, 818–819.

17. (a) M. T. Barton, N. M. Rowley, P. R. Ashton, C. J. Jones, N. Spencer, M. S. Tolley and L. J. Yellowlees, *New J. Chem.*, 2000, **24**, 555–560; (b) M. T. Barton, N. M. Rowley, P. R. Ashton, C. J. Jones, N. Spencer, M. S. Tolley and L. J. Yellowlees, *J. Chem. Soc., Dalton Trans.*, 2000, 3170–3175.

18. J. Seth and D. F. Bocian, *J. Am. Chem. Soc.*, 1994, **116**, 143–153.

19. N. S. Hush and J. R. Rowlands, *J. Am. Chem. Soc.*, 1967, **89**, 2976–2979.

20. (a) J.-H. Peng and D. F. Bocian, *J. Phys. Chem.*, 1992, **96**, 4804–4811; (b) M. Atamian, R. J. Donohoe, J. S. Lindsey and D. F. Bocian, *J. Phys. Chem.*, 1989, **93**, 2236–2243.

21. (a) K. Prendergast and T. G. Spiro, *J. Phys. Chem.*, 1991, **95**, 9728–9736; (b) R. A. Reed, R. Purrello, K. Prendergast and T. G. Spiro, *J. Phys. Chem.*, 1991, **95**, 9720–9727.

22. M. Huber, H. Kurreck, B. von Maltzan, M. Plato and K. Möbius, *J. Chem. Soc., Faraday Trans.*, 1990, **86**, 1087–1094.

23. J. Fajer, D. C. Borg, A. Forman, D. Dolphin and R. H. Felton, *J. Am. Chem. Soc.*, 1970, **92**, 3451–3459.

24. K. Okamoto, K. Ohkubo, K. M. Kadish and S. Fukuzumi, *J. Phys. Chem. A*, 2004, **108**, 10405–10413.

25. M. del Rosario Benites, T. E. Johnson, S. Weghorn, L. Yu, P. D. Rao, J. R. Diers, S. I. Yang, C. Kirmaier, D. F. Bocian, D. Holten and J. S. Lindsey, *J. Mater. Chem.*, 2002, **12**, 65–80.

26. H. van Willigen and M. H. Ebersole, *J. Am. Chem. Soc.*, 1987, **109**, 2299–2302.

27. P. Sayer, M. Gouterman and C. R. Connell, *Acc. Chem. Res.*, 1982, **15**, 73–79.

28. P. J. Brothers, *J. Porphyrins Phthalocyanines*, 2002, **6**, 259–267.

29. K. M. Kadish, Z. Ou, X. Tan, W. Satoh, Y. Yamamoto and K.-Y. Akiba, *J. Porphyrins Phthalocyanines*, 2002, **6**, 325–335.

30. K.-Y. Akiba, R. Nadano, W. Satoh, Y. Yamamoto, S. Nagase, Z. Ou, X. Tan and K. M. Kadish, *Inorg. Chem.*, 2001, **40**, 5553–5567.

31. L. Michaudet, D. Fasseur, R. Guilard, Z. Ou, K. M. Kadish, S. Dahaoui and C. Lecomte, *J. Porphyrins Phthalocyanines*, 2000, **4**, 261–270.

32. K. M. Kadish, M. Autret, Z. Ou, K.-Y. Akiba, S. Masumoto, R. Wada and Y. Yamamoto, *Inorg. Chem.*, 1996, **35**, 5564–5569.

33. Y. H. Liu, M.-F. Bénassy, S. Chojnacki, F. D'Souza, T. Barbour, W. J. Belcher, P. J. Brothers and K. M. Kadish, *Inorg. Chem.*, 1994, **33**, 4480–4484.

34. T. Vangberg and A. Ghosh, *J. Am. Chem. Soc.*, 1999, **121**, 12154–12160.

35. P. Sayer, M. Gouterman and C. R. Connell, *J. Am. Chem. Soc.*, 1977, **99**, 1082–1087.

36. K. S. Suslick and R. A. Watson, *New J. Chem.*, 1992, **16**, 633–642.

37. R. K. Thauer and S. Shima, *Nature*, 2006, **440**, 878–879.

38. U. Ermler, *Dalton Trans.*, 2005, 3451–3458.

39. W. Kaim and B. Schwederski, *Pure Appl. Chem.*, 2004, **76**, 351–364.

40. S. W. Ragsdale, Biochemistry of methyl-CoM reductase and coenzyme F_{430}, in: *The Porphyrin Handbook*, K. M. Kadish, K. M. Smith, R. Guilard, (ed.), Academic Press, Amsterdam, 2003, **11**, 205–228.

41. W. Grabarse, S. Shima, F. Mahlert, E. C. Duin, R. K. Thauer and U. Ermler, *Handbook of Metalloproteins*, John Wiley & Sons, Chichester, 2001, Vol. 2, 897–914.

42. U. Ermler, W. Grabarse, S. Shima, M. Goubeaud and R. K. Thauer, *Science*, 1997, **278**, 1457–1462.

43. W. Kaim and B. Schwederski, *Bioinorganic Chemistry: Inorganic Elements in the Chemistry of Life*, John Wiley & Sons, Chichester, 1994, 180–186.

44. A. Pfaltz, B. Jaun, A. Fässler, A. Eschenmoser, R. Jaenchen, H. H. Gilles, G. Diekert and R. K. Thauer, *Helv. Chim. Acta*, 1982, **65**, 828–865.

45. J. L. Sessler and S. J. Weghorn, in: *Expanded, Contracted & Isomeric Porphyrins*, J. E. Baldwin, (ed.), Tetrahedron Organic Chemistry Series, Vol. 18, Pergamon, New York, 1997, 11–120.

46. M. Goenrich, E. C. Duin, F. Mahlert and R. K. Thauer, *J. Biol. Inorg. Chem.*, 2005, **10**, 333–342.

47. J. Harmer, C. Finazzo, R. Piskorski, C. Bauer, B. Jaun, E. C. Duin, M. Goenrich, R. K. Thauer, S. van Doorslaer and A. Schweiger, *J. Am. Chem. Soc.*, 2005, **127**, 17744–17755.
48. R. Piskorski and B. Jaun, *J. Am. Chem. Soc.*, 2003, **125**, 13120–13125.
49. F. Mahlert, W. Grabarse, J. Kahnt, R. K. Thauer and E. C. Duin, *J. Biol. Inorg. Chem.*, 2002, **7**, 101–112.
50. F. Mahlert, C. Bauer, B. Jaun, R. K. Thauer and E. C. Duin, *J. Biol. Inorg. Chem.*, 2002, **7**, 500–513.
51. Q. Tang, P. E. Carrington, Y.-C. Horng, M. J. Maroney, S. W. Ragsdale and D. F. Bocian, *J. Am. Chem. Soc.*, 2002, **124**, 13242–13256.
52. J. L. Craft, Y.-C. Horng, S. W. Ragsdale and T. C. Brunold, *J. Am. Chem. Soc.*, 2004, **126**, 4068–4069.
53. C. Holliger, A. J. Pierik, E. J. Reijerse and W. R. Hagen, *J. Am. Chem. Soc.*, 1993, **115**, 5651–5656.
54. S. Rospert, M. Voges, A. Berkessel, S. P. Albracht and R. K. Thauer, *Eur. J. Biochem.*, 1992, **210**, 101–107.
55. M. W. Renner, L. R. Furenlid, K. M. Barkigia, A. Forman, H.-K. Shim, D. J. Simpson, K. M. Smith and J. Fajer, *J. Am. Chem. Soc.*, 1991, **113**, 6891–6898.
56. K. M. Kadish, M. M. Franzen, B. C. Han, C. Araullo-McAdams and D. Sazou, *J. Am. Chem. Soc.*, 1991, **113**, 512–517.
57. C. Wang, K. V. Shalyaev, M. Bonchio, T. Carofiglio and J. T. Groves, *Inorg. Chem.*, 2006, **45**, 4769–4782.
58. R. Zhang, W.-Y. Yu and C.-M. Che, *Tetrahedron: Asymmetry*, 2005, **16**, 3520–3526.
59. E. Rose, B. Andrioletti, S. Zrig and M. Quelquejeu-Ethève, *Chem. Soc. Rev.*, 2005, **34**, 573–583.
60. Q.-H. Xia, H.-Q. Ge, C.-P. Ye, Z.-M. Liu and K.-X. Su, *Chem. Rev.*, 2005, **105**, 1603–1662.
61. A. Berkessel, P. Kaiser and J. Lex, *Chem. Eur. J.*, 2003, **9**, 4746–4756.
62. S. Funyu, T. Isobe, S. Takagi, D. A. Tryk and H. Inoue, *J. Am. Chem. Soc.*, 2003, **125**, 5734–5740.
63. S. E. Vitols, R. Kumble, M. E. Blackwood Jr., J. S. Roman and T. G. Spiro, *J. Phys. Chem.*, 1996, **100**, 4180–4187.
64. K. M. Kadish, Y. Hu, P. Tagliatesta and T. Boschi, *J. Chem. Soc., Dalton Trans.*, 1993, 1167–1172.
65. K. M. Kadish, P. Tagliatesta, Y. Hu, Y. J. Deng, X. H. Mu and L. Y. Bao, *Inorg. Chem.*, 1991, **30**, 3737–3743.
66. Y. J. Deng, X. H. Mu, P. Tagliatesta and K. M. Kadish, *Inorg. Chem.*, 1991, **30**, 1957–1960.
67. X. H. Mu and K. M. Kadish, *Langmuir*, 1990, **6**, 51–66.
68. T. Malinski, D. Chang, L. A. Bottomley and K. M. Kadish, *Inorg. Chem.*, 1982, **21**, 4248–4253.
69. K. M. Kadish, D. J. Leggett and D. Chang, *Inorg. Chem.*, 1982, **21**, 3618–3622.
70. K. M. Kadish and D. Chang, *Inorg. Chem.*, 1982, **21**, 3614–3618.

71. D. P. Rillema, J. K. Nagle, L. F. Barringer Jr. and T. J. Meyer, *J. Am. Chem. Soc.*, 1981, **103**, 56–62.

72. G. M. Brown, F. R. Hopf, T. J. Meyer and D. G. Whitten, *J. Am. Chem. Soc.*, 1975, **97**, 5385–5390.

73. G. M. Brown, F. R. Hopf, J. A. Ferguson, T. J. Meyer and D. G. Whitten, *J. Am. Chem. Soc.*, 1973, **95**, 5939–5942.

74. A. Berkessel, P. Kaiser, E. Ertürk, A. Klein, R. M. Kowalczyk and B. Sarkar, *Dalton Trans.*, 2007, 3427–3434.

75. B. Han, J. Shao, Z. Ou, T. D. Phan, J. Shen, J. L. Bear and K. M. Kadish, *Inorg. Chem.*, 2004, **43**, 7741–7751.

76. A. P. Meacham, K. L. Druce, Z. R. Bell, M. D. Ward, J. B. Keister and A. B. P. Lever, *Inorg. Chem.*, 2003, **42**, 7887–7896.

77. (a) J. van Slageren, F. Hartl and D. J. Stufkens, *Eur. J. Inorg. Chem.*, 2000, 847–855; (b) M. P. Aarnts, F. Hartl, K. Peelen, D. J. Stufkens, C. Amatore and J.-N. Verpeaux, *Organometallics*, 1997, **16**, 4686–4695.

78. M. J. Shaw and W. E. Geiger, *Organometallics*, 1996, **15**, 13–15.

79. S. Ye, H. Akutagawa, K. Uosaki and Y. Sasaki, *Inorg. Chem.*, 1995, **34**, 4527–4528.

80. J. A. Baumann, D. J. Salmon, S. T. Wilson, T. J. Meyer and W. E. Hatfield, *Inorg. Chem.*, 1978, **17**, 3342–3350.

81. T. Tomon, T.-A. Koizumi and K. Tanaka, *Eur. J. Inorg. Chem.*, 2005, 285–293.

82. (a) C. Carriedo and N. G. Connelly, *J. Organomet. Chem.*, 1991, **403**, 359–363; (b) N. G. Connelly, I. Manners, J. R. C. Protheroe and M. W. Whiteley, *J. Chem. Soc. Dalton Trans.*, 1984, 2713–2717.

83. G. Cripps, A. Pellissier, S. Chardon-Noblat, A. Deronzier and R. J. Haines, *J. Organomet. Chem.*, 2004, **689**, 484–488.

84. J. Halpern, B. R. James and A. L. W. Kemp, *J. Am. Chem. Soc.*, 1966, **88**, 5142–5147.

85. R.-J. Cheng, S.-H. Lin and H.-M. Mo, *Organometallics*, 1997, **16**, 2121–2126.

86. J. T. Groves, M. Bonchio, T. Carofiglio and K. Shalyaev, *J. Am. Chem. Soc.*, 1996, **118**, 8961–8962.

87. (a) I. Morishima, Y. Shiro and K. Nakajima, *Biochemistry*, 1986, **25**, 3576–3584; (b) I. Morishima, Y. Shiro and Y. Takamuki, *J. Am. Chem. Soc.*, 1983, **105**, 6168–6170.

88. (a) B. R. James, S. R. Mikkelsen, T. W. Leung, G. M. Williams and R. Wong, *Inorg. Chim. Acta*, 1984, **85**, 209–213; (b) B. R. James, D. Dolphin, T. W. Leung, F. W. B. Einstein and A. C. Willis, *Can. J. Chem.*, 1984, **62**, 1238–1245; (c) M. Barley, D. Dolphin, B. R. James, C. Kirmaier and D. Holten, *J. Am. Chem. Soc.*, 1984, **106**, 3937–3943.

89. P. D. Smith, D. Dolphin and B. R. James, *J. Organomet. Chem.*, 1981, **208**, 239–248.

90. (a) R. Raveendran and S. Pal, *Inorg. Chim. Acta*, 2006, **359**, 3212–3220; (b) M. Heilmann, S. Frantz, W. Kaim, J. Fiedler and C. Duboc, *Inorg.*

Chim. Acta, 2006, **359**, 821–829; (c) S. Ghumaan, S. Kar, S. M. Mobin, B. Harish, V. G. Puranik and G. K. Lahiri, *Inorg. Chem.*, 2006, **45**, 2413–2423; (d) M. J. Ingleson, M. Pink, J. C. Huffman, H. Fan and K. G. Caulton, *Organometallics*, 2006, **25**, 1112–1119.

91. (a) W. H. Quayle and J. H. Lunsford, *Inorg. Chem.*, 1982, **21**, 97–103; (b) R. E. DeSimone and R. S. Drago, *J. Am. Chem. Soc.*, 1970, **92**, 2343–2352.

92. (a) K. Pettersson, J. Wiberg, T. Ljungdahl, J. Martensson and B. Albinsson, *J. Phys. Chem. A*, 2006, **110**, 319–326; (b) M. P. Eng, T. Ljungdahl, J. Andréasson, J. Martensson and B. Albinsson, *J. Phys. Chem. A*, 2005, **109**, 1776–1784; (c) K. Kilsa, J. Kajanus, A. N. Macpherson, J. Martensson and B. Albinsson, *J. Am. Chem. Soc.*, 2001, **113**, 3069–3080.

93. S. Saha, E. Johansson, A. H. Flood, H.-R. Tseng, J. I. Zink and J. F. Stoddart, *Chem. Eur. J.*, 2005, **11**, 6846–6858.

94. K. M. Roth, D. T. Gryko, C. Clausen, J. Li, J. S. Lindsey, W. G. Kuhr and D. F. Bocian, *J. Phys. Chem. B*, 2002, **106**, 8639–8648.

95. N. Solladié, J.-C. Chambron and J.-P. Sauvage, *J. Am. Chem. Soc.*, 1999, **121**, 3684–3692.

96. C.-Y. Zhou, P. W. H. Chan and C.-M. Che, *Org. Lett.*, 2006, **8**, 325–328.

97. (a) Y. Wang, Q.-Y. He, C.-M. Che and J.-F. Chiu, *Proteomics*, 2006, **6**, 131–142; (b) Y. Wang, Q.-Y. He, R. W.-Y. Sun, C.-M. Che and J.-F. Chiu, *Cancer Res.*, 2005, **65**, 11553–11564; (c) C.-M. Che, patent 2004, WO2004024146.

98. A. Brisach-Wittmeyer, S. Lobstein, M. Gross and A. Giraudeau, *J. Electroanal. Chem.*, 2005, **583**, 109–115.

99. Z. Ou, K. M. Kadish, W. E. J. Shao, P. J. Sintic, K. Ohkubo, S. Fukuzumi and M. J. Crossley, *Inorg. Chem.*, 2004, **43**, 2078–2086.

100. G. Gencheva, D. Tsekova, G. Gochev, D. Mehandjiev and P. R. Bontchev, *Inorg. Chem. Commun.*, 2003, **6**, 325–328.

101. K. M. Kadish, W. E. Z. Ou, J. Shao, P. J. Sintic, K. Ohkubo, S. Fukuzumi and M. J. Crossley, *Chem. Commun.*, 2002, 356–357.

102. (a) M. Kampf, J. Griebel and R. Kirmse, *Z. Anorg. Allg. Chem.*, 2004, **630**, 2669–2676; (b) M. Kampf, R.-M. Olk and R. Kirmse, *Z. Anorg. Allg. Chem.*, 2002, **628**, 34–36; (c) L. Ihlo, M. Kampf, R. Böttcher and R. Kirmse, *Z. Naturforsch. B*, 2002, **57 b**, 171–176.

103. A. J. Blake, J. A. Greig, A. J. Holder, T. I. Hyde, A. Taylor and M. Schröder, *Angew. Chem., Int. Ed. Engl.*, 1990, **29**, 197–198.

104. S. Seidel and K. Seppelt, *Science*, 2000, **290**, 117–118.

105. F. G. Herring, G. Hwang, K. C. Lee, F. Mistry, P. S. Phillips, H. Willner and F. Aubke, *J. Am. Chem. Soc.*, 1992, **114**, 1271–1277.

106. (a) A. P. Koley, S. Purohit, L. S. Prasad, S. Ghosh and P. T. Manoharan, *Inorg. Chem.*, 1992, **31**, 305–311; (b) A. P. Koley, L. S. Prasad, P. T. Manoharan and S. Ghosh, *Inorg. Chim. Acta*, 1992, **194**, 219–225; (c) A. P. Koley, S. Purohit, S. Ghosh, L. S. Prasad and P. T. Manoharan, *J. Chem. Soc., Dalton Trans.*, 1988, 2607–2613.

107. L. Ihlo, R. Böttcher, R.-M. Olk and R. Kirmse, *Inorg. Chim. Acta*, 1998, **281**, 160–164.

108. R. L. Schlupp and A. H. Maki, *Inorg. Chem.*, 1974, **13**, 44–51.

109. J. H. Waters and H. B. Gray, *J. Am. Chem. Soc.*, 1965, **87**, 3534–3535.

110. A. MacCragh and W. S. Koski, *J. Am. Chem. Soc.*, 1965, **87**, 2496–2497.

111. M. E. Jamin and R. T. Iwamoto, *Inorg. Chim. Acta*, 1978, **27**, 135–143.

112. (a) T. S. Balaban, *Acc. Chem. Res.*, 2005, **38**, 612–623; (b) D. Kim and A. Osuka, *Acc. Chem. Res.*, 2004, **37**, 735–744.

113. P. D. W. Boyd and C. A. Reed, *Acc. Chem. Res.*, 2005, **38**, 235–242.

114. (a) D. M. Guldi, G. M. A. Rahman, F. Zerbetto and M. Prato, *Acc. Chem. Res.*, 2005, **38**, 871–878; (b) D. M. Guldi and H. Imahori, *J. Porphyrins Phthalocyanins*, 2004, **8**, 976–983.

115. M. Marcacchio, F. Paolucci and S. Roffia, *Supramolecular Electrochmistry of Coordination Compounds and Molecular Devices*, in: *Trends in Molecular Electrochemistry*, A. J. Pombeiro, C. Amatore, (ed.), Fontis Media, Lausanne, Marcel Dekker, New York, 2004, 223–281.

116. T. Da Ros, M. Prato, D. M. Guldi, M. Ruzzi and L. Pasimeni, *Chem. Eur. J.*, 2001, **7**, 816–827.

117. (a) Q. Xie, F. Arias and L. Echegoyen, *J. Am. Chem. Soc.*, 1993, **115**, 9818–9819; (b) Q. Xie, E. Pérez-Cordero and L. Echegoyen, *J. Am. Chem. Soc.*, 1992, **114**, 3978–3980.

118. D. Dubois, G. Moninot, W. Kutner, M. T. Jones and K. M. Kadish, *J. Phys. Chem.*, 1992, **96**, 7137–7145.

119. L. R. Sutton, M. Scheloske, K. S. Pirner, A. Hirsch, D. M. Guldi and J.-P. Gisselbrecht, *J. Am. Chem. Soc.*, 2004, **126**, 10370–10381.

120. K. M. Kadish, X. Q. Lin and B. C. Han, *Inorg. Chem.*, 1987, **26**, 4161–4167.

121. T. F. Guarr, M. S. Meier, V. K. Vance and M. Clayton, *J. Am. Chem. Soc.*, 1993, **115**, 9862–9863.

122. A. Wolberg and J. Manassen, *J. Am. Chem. Soc.*, 1970, **92**, 2982–2991.

CHAPTER 5

Infrared Spectroelectrochemical Investigations of Ultrafast Electron Transfer in Mixed-Valence Complexes

J. CATHERINE SALSMAN AND CLIFFORD P. KUBIAK

Department of Chemistry and Biochemistry, University of California San Diego, 9500 Gilman Drive, La Jolla, California 92093-0358, USA

5.1 Introduction

The introduction of spectroelectrochemical (SEC) cells in the 1960s made possible for the first time the spectroscopic characterisation of unstable oxidation states, either in thin layers or at the electrode surface.[1,2] Early versions of SEC cells were typically optically transparent thin-layer electrodes (OTTLEs) which involved the use of platinum or gold optically transparent minigrid electrodes, or germanium crystals.[3–13] The limitations of these SEC cells included leaking and the difficulty of maintaining an oxygen-free environment. Furthermore, many complexes become unstable in various oxidation states at room temperature or display interesting temperature-dependent properties; therefore, control of temperature is often desirable. Here, we review the design of the most recent generation of a variable-temperature reflectance spectroelectrochemical cell developed in our laboratories. We also review its use in infrared (IR) spectroelectrochemical studies of ultrafast ($\sim 10^{12}\,\text{s}^{-1}$) electron transfer in inorganic and organic mixed-valence complexes.

Spectroelectrochemistry
Edited by Wolfgang Kaim and Axel Klein
© Royal Society of Chemistry, 2008

5.2   Cell Design

With temperature control and anaerobic considerations in mind, we designed a variable-temperature thin-layer specular (external) reflectance spectroelectro-chemical cell (Figure 5.1). The cell design is such that the incident light from the spectrometer reflects off the working electrode, ensuring that the species detected

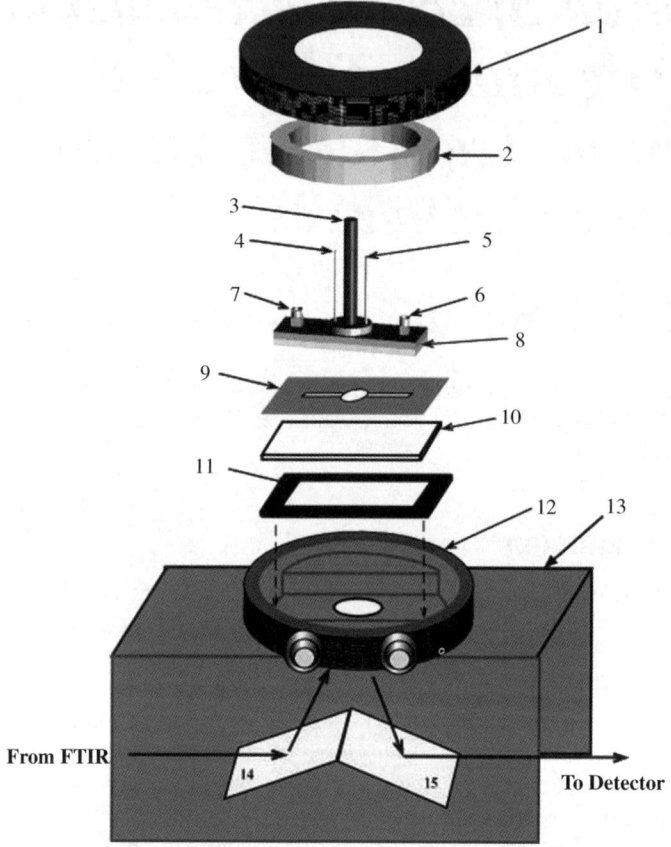

Figure 5.1   Disassembled view of the spectroelectrochemical cell. (1) Tightening brass cap (threaded inside). (2) Brass ring required to tighten the cell. (3) Working electrode (brass rod with platinum soldered to the base). (4) Auxiliary electrode: platinum wire with the tip made flush to the teflon base of the cell. (5) Pseudoreference electrode: silver wire, also made flush to the teflon. (6,7) Luer-lock-type injection ports. (8) Cell body, top part aluminium, lower part teflon. (All three electrodes and both filling ports are press fitted into the cell body, so that they can be replaced if needed.) (9) Teflon spacer, determines the pathlength of the cell and masks the reference and counter electrodes from the incident beam. (10) Calcium fluoride window (Wilmad, standard $38.5 \times 19.5 \times 4$ mm). (11) Rubber gasket. (12) Hollow brass cell body with threaded inlet and outlet ports (Swagelock) for connection to circulating bath. (13) Two-mirror reflect-ance accessory (Thermo-SpectraTech FT-30). (14,15) Mirrors.

is in the immediate vicinity of the working electrode. A teflon mask (spacer) is layered underneath the cell window to block the portion of the solution nearest the counter electrode.

A concentric three-ring electrode geometry optimises the amount of product visible to the spectroscopic beam and results in a small and symmetric voltage drop across the working electrode (Figure 5.2). The temperature is controlled by a hollow brass heat-exchanger ring that surrounds the cell and is connected to a high-capacity low-temperature circulating bath. The spectrometer body was fitted with a custom faceplate to allow the insulated hoses connected to the circulator to pass through. At low temperatures (*ca.* −30 °C) a sufficient flow of purge gas in the spectrometer is required to keep the windows free of frost.

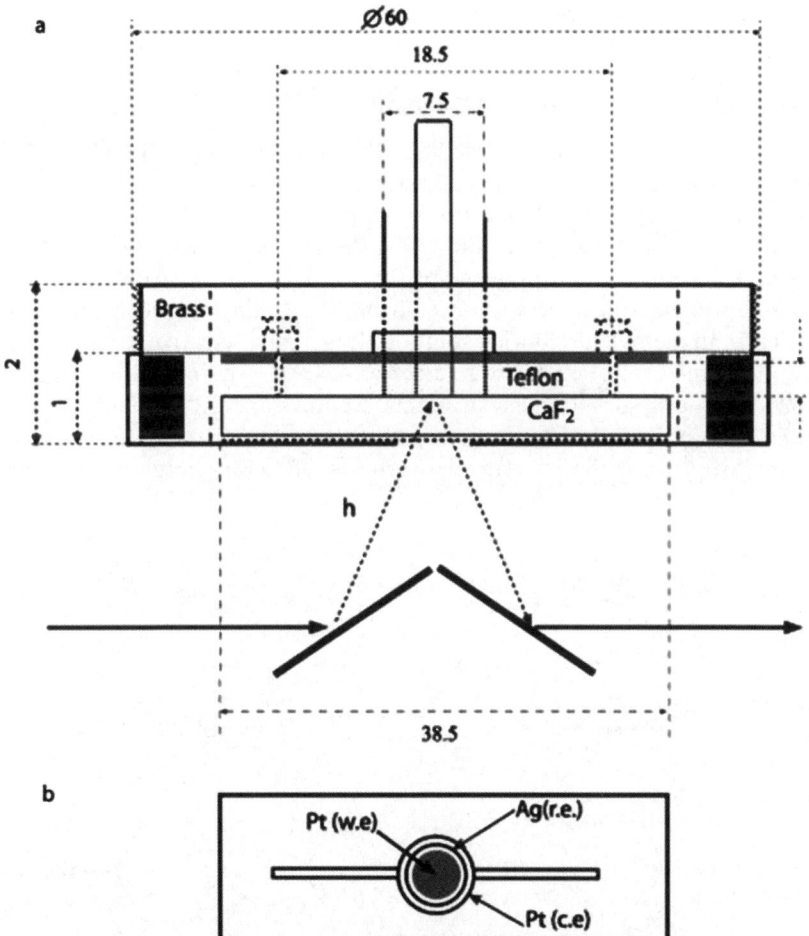

Figure 5.2 (a) Cross-sectional view of cell. (b) View of electrode geometry. "In" and "out" represent the inlet and outlet solvent ports, respectively.

5.3   Applications

The spectroelectrochemical cell described here has been used for infrared spectroscopy, UV-Vis electronic spectroscopy, and resonance Raman experiments. As designed, it can be used in a standard specular reflectance cell accessory holder in an IR spectrometer at room temperature. For temperature-dependence studies, a custom faceplate was fabricated for the spectrometer body to allow cooling hoses from a circulating bath to pass through and connect to the hollow brass housing surrounding the cell. For electronic spectroscopy, the same standard IR accessory holder was installed into the chamber of a UV–Vis–NIR spectrometer. For low-temperature electronic spectra, a custom plexiglass purge box equipped with CaF_2 windows was fitted over the SEC cell to prevent fogging of the mirrors. The SEC cell head can also be removed from the specular reflectance accessory to be used in laser experiments, such as resonance Raman spectroscopy.

5.3.1   Trinuclear Ruthenium Cluster Dimers

Trinuclear ruthenium cluster dimers of the type $[Ru_3O(CH_3COO)_6(CO)(L)]_2$ (μ-BL), where $L = 4$-dimethylaminopyridine (dmap) **1**, pyridine (py) **2**, or 4-cyanopyridine (cpy) **3**; $BL =$ pyrazine (Figure 5.3), first synthesised by Ito et al.,[14-16] are ideal candidates for spectroelectrochemical studies due to their rich electrochemistry and presence of the CO ligand as a strong IR chromophore. In the neutral dimers, the CO–ligand stretching frequency appears *ca.* 1935 cm^{-1} in methylene chloride solution. It is highly sensitive to the oxidation state of the metal clusters due to its strong π-acceptor nature. Although the CN groups of isocyanides are also strong π-acceptors and IR chromophores, isocyanides are often not well suited to spectroelectrochemistry since combination bands arising from the overlap of overtones of lower-energy normal modes

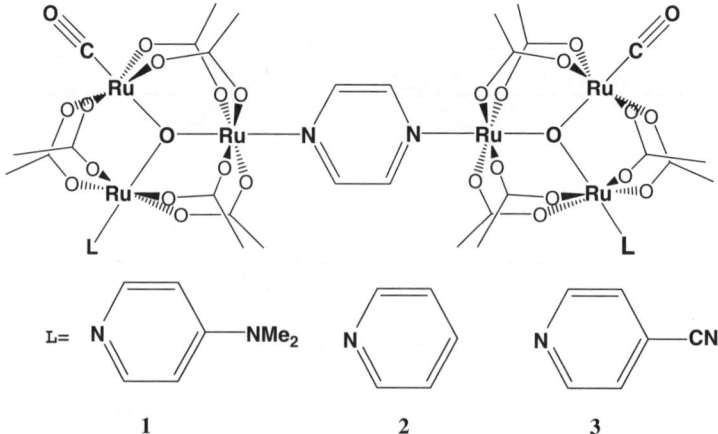

Figure 5.3   Structures of complexes **1–3**.

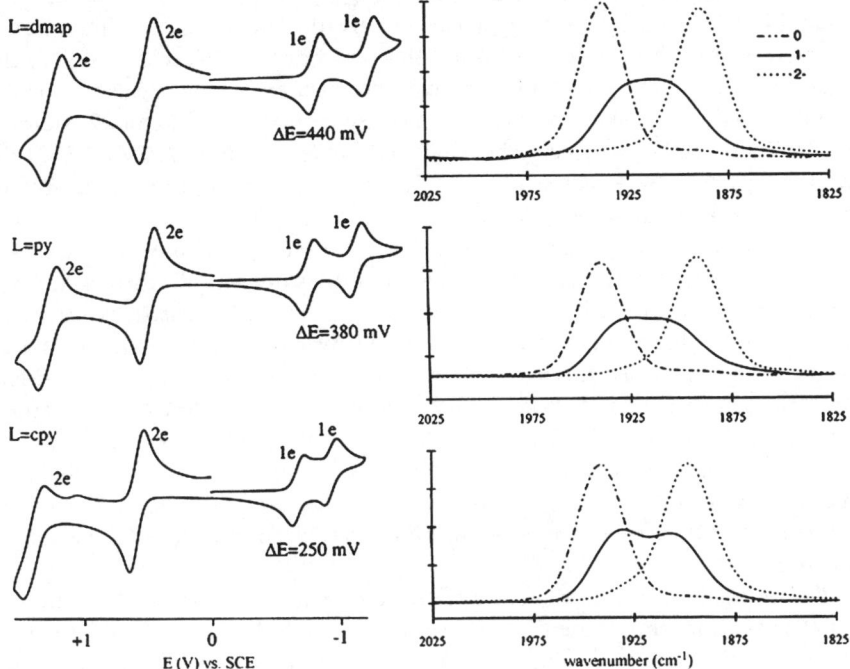

Figure 5.4   Cyclic voltammetry (left) and IR-SEC (right) of **1–3**.

with the fundamental C–N stretching frequency (Fermi resonances) can complicate the spectra.[17]

The neutral Ru_3 dimers display four reversible oxidation waves corresponding to $(+4)$, $(+2)$, (0), (-1), and (-2) redox states (Figure 5.4). It is important to note here that although the oxidation waves are each two-electron processes, the reduction waves involve a single electron. The splitting between these one-electron reduction waves, $\Delta E_{1/2}$, is a composite of five factors that determine the magnitude of the splitting:[18–20]

$$\Delta E_{1/2} = \Delta E_{deloc} + \Delta E_{coul} + \Delta E_{stat} + \Delta E_{induct} + \Delta E$$

where ΔE_{deloc} is the resonance stabilisation energy imparted to the complex in the mixed-valence state, ΔE_{coul} is the electrostatic repulsion term, ΔE_{stat} is the statistical thermodynamic contribution to the stability $(-RT\ln2)$, ΔE_{induct} is the inductive and backbonding effects, and ΔE, the intrinsic energy difference between donor and acceptor reduction potentials $(\Delta E = 0$ in the case of the symmetric dimers **1–3** discussed here). Since the dimers are similar in structural framework, but have different degrees of ligand–metal cluster overlap, the only terms that should vary significantly from dimer to dimer are ΔE_{deloc} and ΔE_{induct}. Therefore, the splitting in the electrochemistry can be used as a guideline to predict the degree of electronic coupling ($\Delta E_{\text{deloc}} = H^2_{\text{AB}}/\lambda$, where λ is the total reorganisational energy, usually measured by the energy of the intervalence charge-transfer absorption)[21] between the two bridged cluster moeities in the mixed-valence state. A comproportionation constant, K_c, is calculated for the stability of the (-1) state relative to a mixture of the neutral and (-2) states.[22] Dimer **1** has the largest electrochemical splitting ($\Delta E = 400\,\text{mV}$; $K_c = 2.3 \times 10^7$), while **3** cpy had the smallest ($\Delta E = 255\,\text{mV}$; $K_c = 1.7 \times 10^4$), and dimer **2** falls between the two ($\Delta E = 380\,\text{mV}$; $K_c = 2.7 \times 10^6$).

Cyclic voltammetry indicates that the dmap dimer is the most stable in the (-1) state compared to the other dimers, but does not provide insight into the detailed nature of the redox states observed. Is the charge localised, either at one cluster site or the ligands, or is it delocalised over the molecule? If the charge is at least quasilocalised, what is the rate of intramolecular electron transfer? Spectroelectrochemistry provides an additional level of electronic structural information required to investigate the electronic (and geometric) structure of these complexes. We will show that in favourable cases the lineshape of the IR spectroelectrochemical response can be used to obtain dynamical information regarding the rates of intramolecular electron transfer in the mixed-valence states.

The ruthenium cluster dimers were dissolved in 0.1 M tetra-n-butylammonium hexafluorophosphate (TBAH) solutions. TBAH was recrystallised from hot absolute ethanol, vacuum dried at 150 °C for 18 h, and stored under a nitrogen atmosphere. Solvents were dried over alumina in a custom solvent-purification system (Glass Contour). Spectroelectrochemistry was carried out at temperatures below 10 °C to prevent decomposition of the dimer in the (-1) and (-2) states, as the reduced dimers tend to break apart into their constituent monomeric species over prolonged periods above 10 °C.

The $v(CO)$ band can be observed in each of the five available redox states, and it shifts to progressively lower frequencies upon reduction from the ($+4$) state to the (-2) state due to the increasing π-backbonding ability of the metal centre. All of the redox states can be considered "symmetric" in that both cluster monomers are in the same oxidation state, with the exception of the singly reduced (-1) mixed-valence state. As such, the $v(CO)$ band in each of the symmetric states is sharp (FWHM = $16\,\text{cm}^{-1}$), while the $v(CO)$ band in the (-1) state is broad (FWHM ~ 50–$60\,\text{cm}^{-1}$) and is midway (with half-intensity) between the peaks for the (0) and (-2) states (Figure 5.4).

This behaviour is reminiscent of dynamical averaging seen in NMR spectroscopy, which is treated using a Bloch-equation analysis. Turner and co-workers[23] were the first to use this type of analysis of dynamic infrared spectra to calculate rate constants and self-isomerisation barrier heights in the case of a "turnstile"-type exchange of CO ligands observed in trigonal bipyramidal [(η^4-diene)Fe(CO)$_3$] complexes. In a similar fashion, rate constants for electron transfer in the ruthenium cluster dimers were calculated from the lineshapes of the (-1) states. As expected from the electrochemical data, the rate constant was fastest for the most electronically coupled dimer (**1**) and slowest for the

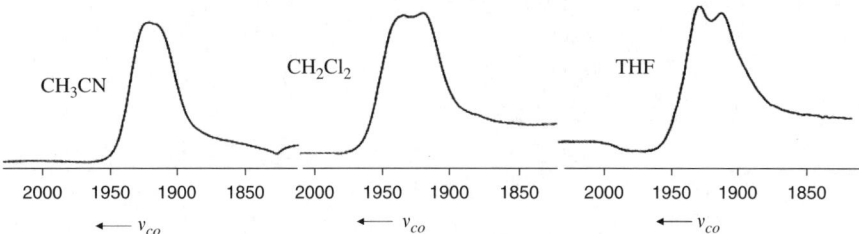

Figure 5.5 Spectra of **2** in the mixed valent (-1) state in three of the solvents studied: acetonitrile, methylene chloride, and tetrahydrofuran.

least coupled (**3**) (rate constants of all compounds discussed here are summarised in Table 5.2, page 142).

The SEC cell is versatile for use with a wide range of solvents, which allows for the determination of the role of the solvent in the electron-transfer process.[24] Many factors are thought to contribute to the electron-transfer rate, such as choice of ancillary and bridging ligands, electrolyte, temperature, and solvent. IR-SEC spectra of **2** in three representative solvents are shown in Figure 5.5.

The rate of electron transfer scales not with dielectric constant, solvent polarity, or calculated "outer-sphere reorganisation energies",[21] as might be expected, but with the "inertial" solvent relaxation time (t_{1e}) as measured and discussed by Maroncelli and coworkers[25] (Figure 5.6). It can be concluded from this study that for a given set of ancillary and bridging ligands, the role of the solvent is the single most important factor influencing the rate of electron transfer. Deviations from the linear correlation of $(k_{et})^{-1}$ with t_{1e} (particularly for complex **3** in the fastest solvents) indicate where solvent is no longer the limiting factor for the electron transfer. This is significant in that the solvent-relaxation times measured by Maroncelli were ultrafast time-resolved emission measurements of Coumarin 153. These data represent the first connection between our static IR-SEC rate estimates and a direct ultrafast time-resolved measurement of solvent dynamics.

In spite of the strong solvent dependence observed, no appreciable temperature dependence of the CO spectrum is observed in the temperature range available in the spectroelectrochemical cell ($25\,°C$ to *ca.* $-60\,°C$). In cryogenic studies reported elsewhere, we have observed that electron transfer rates *increase* in frozen solvents, indicating that the frozen solvent no longer plays a dynamical role in limiting the electron transfer rate.[26,27]

Small modifications to the bridging or ancillary ligands of these complexes result in large changes in electronic communication between the dimers, as evidenced by the measurement of comproportionation constants over a range of five orders of magnitude and the dramatic changes in appearance of the spectra of the mixed-valence state (Figure 5.7). By choosing the appropriate combination of bridging (BL = pyrazine, 2-methylpyrazine, 2,5-dimethylpyrazine, or 2-chloropyrazine) and ancillary (L = 4-(dimethylamino)pyridine, pyridine, or 4-cyanopyridine) ligands, the electronic coupling can essentially be

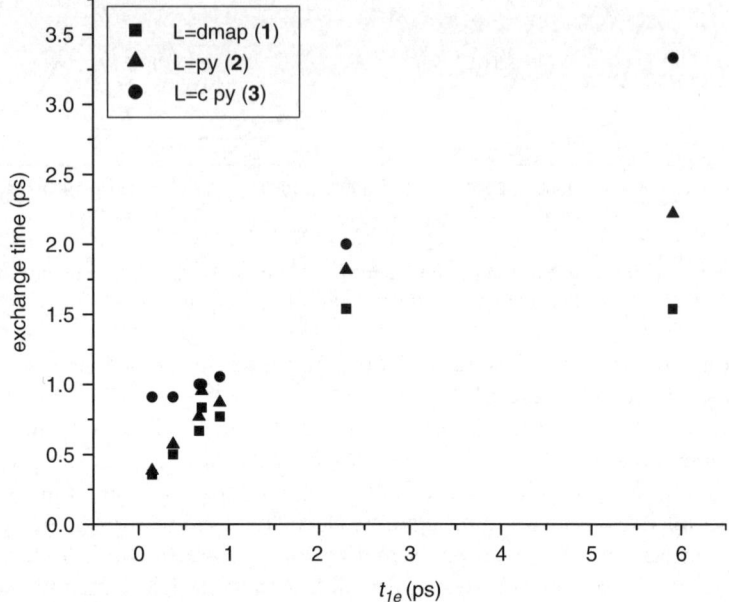

Figure 5.6 Graph of the relation between exchange time estimated by lineshape analysis of IR-SEC data for **1–3** and solvent relaxation time, t_{1e}.

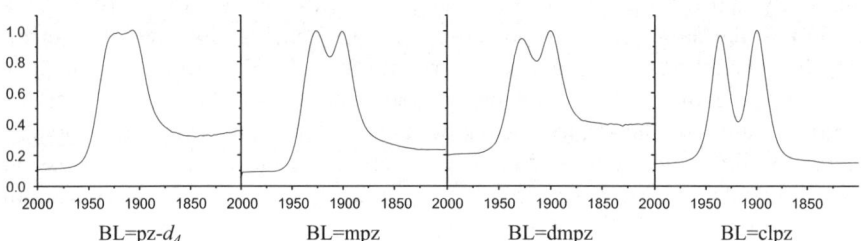

BL=pz-d_4 BL=mpz   BL=dmpz   BL=clpz

Figure 5.7 Spectra of [Ru$_3$(μ_3-O)(μ-CH$_3$CO$_2$)$_6$(CO)(py)]$_2$(μ-BL) in the mixed valent (−1) state; where BL = pyrazine-d_4 (pz-d_4), 2-methylpyrazine (mpz), 2,5-dimethylpyrazine, or 2-chloropyrazine (clpz).

"fine tuned" (Table 5.1). The rates of all of the complexes studied were estimated from lineshape analysis and plotted against the electrochemical splitting. The relationship was found to be linear and, within experimental error, the intercept is 0 (Figure 5.8).

The ν(CO) mode is not the only mode that provides insight into the electronic structure of these complexes. The *symmetric* ν_{8A} mode of pyrazine (Figure 5.9) appears only in the spectrum of the (−1) state.[28]

This phenomenon was originally thought to be indicative of a localised mixed-valence state, *i.e.* that the appearance of this mode is proof of electronic asymmetry on the infrared timescale.[29] However, control experiments with the cluster monomers that have pronounced structural asymmetry, were carried

Table 5.1   Electrochemical data for $[Ru_3(\mu_3\text{-}O)(\mu\text{-}CH_3CO_2)_6(CO)(L)]_2(\mu\text{-}BL)$.[a]

	BL	L	$E_{1/2}(0/-1)$, V	$E_{1/2}(-1/-2)$, V	ΔE (mV)	K_c
1	Pz	dmap	−1.216	−1.649	435	2.27×10^7
	d_4-pz	dmap	−1.216	−1.649	435	2.27×10^7
	mpz	dmap	−1.249	−1.605	355	1.01×10^6
	dmpz	dmap	−1.291	−1.613	320	2.58×10^5
2	pz	py	−1.133	−1.507	380	2.67×10^6
	d_4-pz	py	−1.133	−1.507	380	2.67×10^6
	mpz	py	−1.169	−1.503	334	4.63×10^5
	dmpz	py	−1.207	−1.492	285	5.44×10^4
	clpz	py	−1.198	−1.438	240	1.39×10^4
3	pz	4-cpy	−1.058	−1.313	255	1.69×10^4
	d_4-pz	4-cpy	−1.058	−1.313	255	1.69×10^4
	mpz	4-cpy	−1.110	−1.310	200	2.41×10^3
	dmpz	4-cpy	−1.150	−1.300	150	3.44×10^2

[a]In dichloromethane at 25 °C; 0.1 M tetra-*n*-butylammonium hexafluorophosphate supporting electrolyte; potentials referenced to ferrocene.

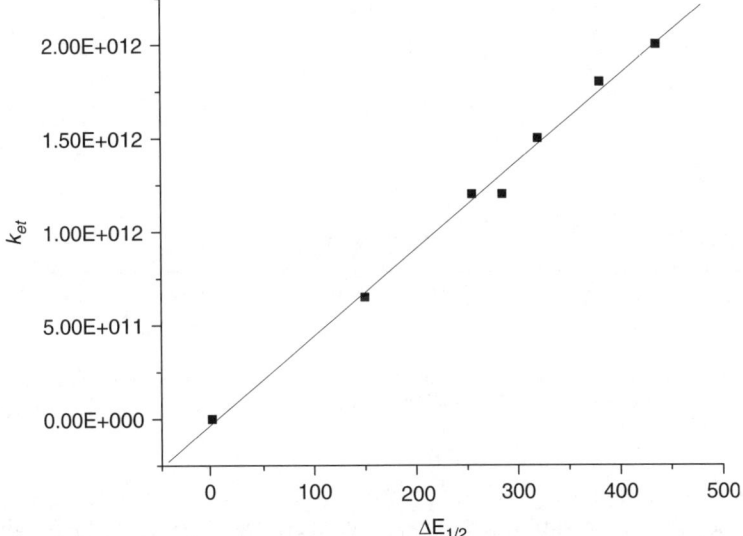

Figure 5.8   Graph of the relationship between the electrochemical splitting, $\Delta E_{1/2}$ between the (-1) and (-2) states and the rate constants estimated from $\nu(CO)$ IR lineshapes of the symmetric dimers: $k_{et} = 4.7 \times 10^9(\Delta E_{1/2}) - 2.6 \times 10^{10}$.

out and the symmetric ν_{8A} mode was not observed (Figure 5.10). This led to the interpretation of the appearance of the symmetric mode not as an indicator of electronic asymmetry, but as an indicator of strong vibronic coupling in the mixed-valence state. In inorganic chemistry, vibronic coupling is most frequently discussed with respect to the fact that Co(III) complexes are not

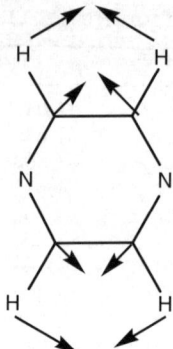

Figure 5.9   The symmetric ν_{8A} mode of pyrazine.

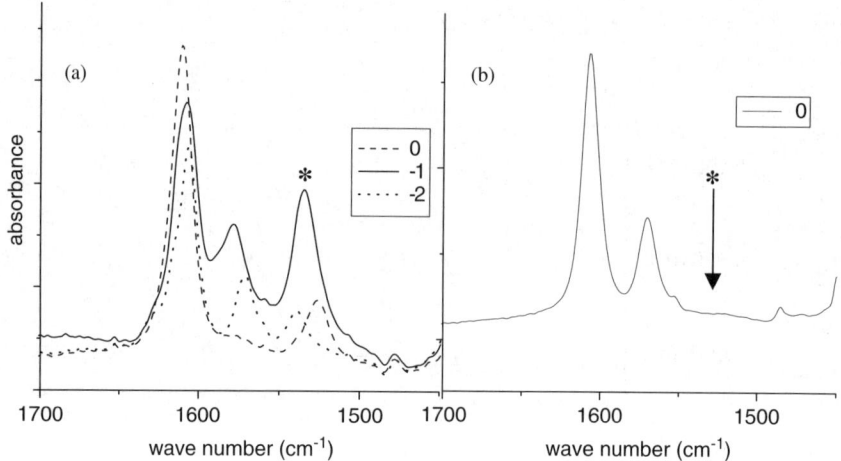

Figure 5.10   Comparison spectra for 1 (a) and its monomeric analogue: $Ru_3O(CH_3COO)_6(CO)(dmap)(pz)$ (b).

colorless. Although $[Co(NH_3)_6]^{3+}$ with a $^1A_{1g}$ ground electronic state should have electronic transitions that are symmetry (g→g) forbidden, vibronic coupling of normal modes of vibration that are asymmetric give some over-all symmetric character to the integrand of the transition moment integral, $\int \Psi_{el}^* \Psi_{vib}^* \bar{\mu}_{x,y,z} \Psi_{el} \Psi_{vib} d\tau$, producing nonzero intensities of vibronic d–d transitions in the electronic spectra. In the case of inorganic mixed valency, vibronic involvement of normally IR forbidden symmetric vibrations with intervalence charge-transfer (IVCT) electronic transitions that have strong g→u character, produces nonzero intensities of the vibrational transitions in the IR spectra.[30] This follows the same line of reasoning as vibronic enhancements of forbidden d–d transitions, but applies to vibronic enhancement of forbidden IR bands instead.

5.3.2 Resonance Raman Experiments

Resonance Raman spectroscopy is a valuable tool for probing the electronic structure of mixed-valence complexes. The Creutz–Taube ion, $[(NH_3)_5Ru]_2(pz)^{5+}$, was the subject of much controversy surrounding whether the charge was essentially localised on one ruthenium cluster (Robin–Day Class II) or delocalised over the molecule (Robin–Day Class III). After much investigation into the electronic structure, the compelling results of Stark (electroabsorption)[31] and resonance Raman[32] spectroscopies led researchers to understand the Creutz–Taube ion as an average-valence complex with vibronically enhanced bridging ligand modes.

The trinuclear ruthenium cluster dimers (**1–3**) were also studied by spectro-electrochemical resonance Raman spectroscopy.[33,34] Raman spectra were obtained in a backscattering geometry from solution samples in the SEC cell. Raman excitation was carried out using $\lambda_{exc} = 801$ nm (in resonance with the bands at 830 nm, 850 nm, and 930 nm of **1**, **2**, and **3**, respectively) from an Ar+ -pumped Ti:sapphire laser (Spectra Physics model 3900) operating at approximately 7–8 mW power. Scattered light was collected and passed through a triple spectrometer (SPEX 1877) onto a liquid-nitrogen-cooled charge-coupled device (CCD) detection system (Princeton Instruments) interfaced to a PC. Typical integration times to acquire spectra were 5–10 min (average of at least 30 scans per 10 s). Data were collected and processed using Princeton Instruments WinSpec32 software. Spectra were calibrated using known spectral lines from low-intensity Ne calibration lamps. *In-situ* controlled-potential electrolyses were performed in the SEC cell at –20 °C.

In each case an enhancement of totally symmetric pyrazine modes (v_1, v_{9a}, v_{8a}) was observed in the spectra of the mixed-valence state (Figure 5.11). Enhancement of nontotally symmetric modes was not observed. Interestingly, the degree of resonance Raman enhancement scales with the degree of electronic coupling in the cluster dimers (as estimated by $\Delta E_{1/2}$), *i.e.* **1** showed the greatest enhancement, while **3** showed the least, although it should be mentioned here that **1** has the greatest overlap with the excitation wavelength. The enhancement of these modes is significant in that it provides further compelling evidence that the totally symmetric normal modes of the bridging pyrazine ligand in the mixed-valence states of **1**, **2**, and **3** are vibronically involved with the IVCT bands. This adds strong support to the vibronic interpretation of the anomalously high intensities of some of these same modes, v_{8A} in particular, in IR-SEC measurements.

5.3.3 Individual Cluster "Monomers"

The isolated trinuclear clusters are also suitable for spectroelectrochemical studies of intracluster mixed-valence states. A new trinuclear cluster was synthesised by treatment of $Ru_3O(CH_3COO)_6(CO)(CH_3OH)_2$ with excess 4-acetylpyridine (acpy) in a 1 : 1 mixture of methanol and methylene chloride. After stirring overnight, the desired product, $Ru_3O(CH_3COO)_6(CO)(acpy)_2$

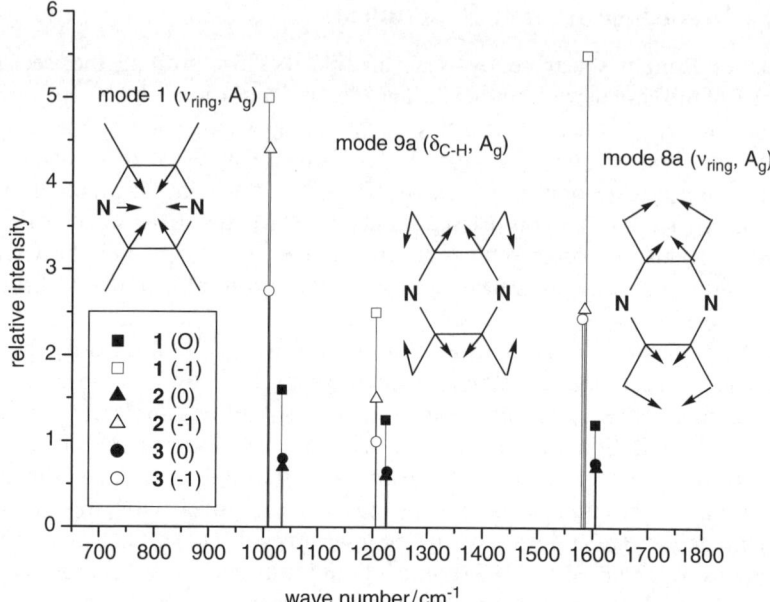

Figure 5.11   Stick graph displaying Raman frequencies of observed totally symmetric pyrazine modes (v_1, v_{8a}, and v_{9a}) and peak intensities (relative to CH$_2$Cl$_2$ solvent peaks at 1157 cm^{-1}, for v_1 and v_{9a}, and 1424 cm^{-1} for v_{8a}).

was isolated by column chromatography (eluent: 2% MeOH in CH$_2$Cl$_2$) and precipitated with hexanes from a methylene chloride solution.

The synthesis of this cluster provided the addition of the two acetyl IR chromophores to aid in characterisation of the reduced state. In the neutral state, this monomer contains three ruthenium atoms, one formally in the ($+2$) (bonded to CO) and two in the ($+3$) (bonded to acpy ligands) redox states. However, this formal description of the charge may not accurately represent the actual charge distribution over the cluster. Infrared spectroelectrochemistry was carried out on this monomer to determine whether the charge was in fact localised in this manner.

IR spectra were obtained for each of the four readily available oxidation states: ($+2$), ($+1$), (0), and (-1). At $-10\,°C$, the spectrum of the (-1) state indicates rapid, irreversible decomposition of the cluster. The experiment was repeated at $-30\,°C$ with no apparent decomposition of the (-1) state (Figure 5.12).

The v(CO) band redshifts 220 cm^{-1} on going stepwise from the ($+2$) redox state down to the (-1) state (Figure 5.13). The acetyl carbonyl band, however, does not shift quite so dramatically (*ca.* 30 cm^{-1}). It is clear that the most dramatic shift in the CO stretching frequency occurs between the ($+1$) and (0) oxidation states, while the most dramatic shift in the acetyl stretching frequency occurs between the (0) and (-1) oxidation states. Since the formal electronic distribution of the neutral state is thought to be Ru$_{CO}$(II), Ru$_{acpy}$(III), Ru$_{acpy}$(III), (Figure 5.14) reduction of the cluster would produce Ru$_{CO}$(II), Ru$_{acpy}$(II),

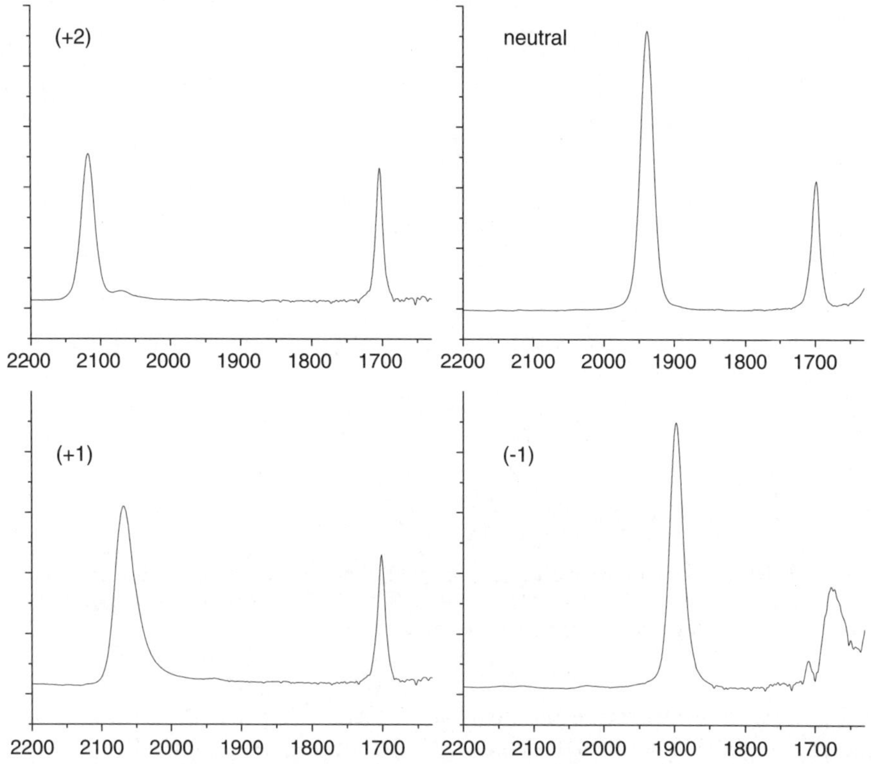

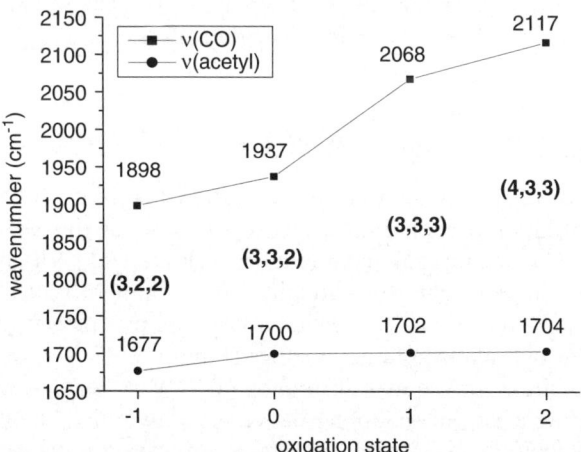

Figure 5.12 Spectra of $Ru_3O(CO)(acpy)_2$ in methylene chloride at $-30\,°C$.

Figure 5.13 Summary of IR-SEC data for $Ru(CO)(acpy)_2$. Spectral peaks are denoted in wave numbers and the formal charges on the ruthenium atoms for each oxidation state are indicated in bold.

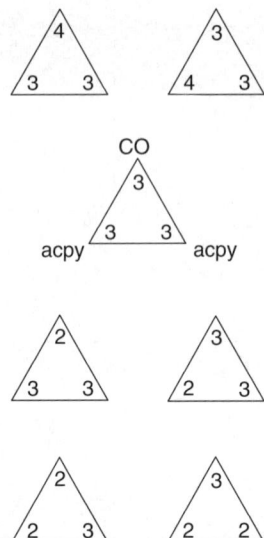

Figure 5.14 Schematic of the possible formal charges on the ruthenium clusters in
each of the available oxidation states. The points of the triangles represent the positions of the three ruthenium atoms.

$Ru_{acpy}(III)$; however, since the peak at $1677\,cm^{-1}$ appears broad, the actual
picture may be a Class III delocalised description: $Ru_{CO}(II)$, $Ru_{acpy}(II.5)$,
$Ru_{acpy}(II.5)$. Clearly, the oxidation of the neutral cluster to the singly oxidised
state produces the symmetric $Ru_{CO}(III)$, $Ru_{acpy}(III)$, $Ru_{acpy}(III)$ state, but it
appears that the Ru(IV) character is mainly on the ruthenium atom bearing the
CO ligand in the ($+2$) oxidation state, *i.e.* $Ru_{CO}(IV)$, $Ru_{acpy}(III)$, $Ru_{acpy}(III)$,
since $\nu(CO)$ shifts approximately $40\,cm^{-1}$, while the acetyl band shifts only
$2\,cm^{-1}$ between the ($+1$) and ($+2$) oxidation states.

5.3.4 Organic Intervalence Radicals

Although much of the research of our laboratory involves electron transfer in
inorganic coordination compounds, the applications of the SEC cell are not
limited to studies of inorganic compounds. Nelsen *et al.*[35] discovered an organic radical complex that was thought to be comparable to a traditional
intervalence M-B-M^{+} complex in which a bridging-oxo played the role of the
bridge between two dialkylaniline moeities (Figure 5.15). Two reversible one-
electron waves are observed at $+0.70$ and $+0.95$ V *vs.* SCE by cyclic voltammetry, indicating electronic coupling between the two dialkylaniline moeities,
and a broad ($7000\,cm^{-1}$) intervalence charge-transfer band is observed in the
electronic spectrum at $10\,600\,cm^{-1}$. These properties suggest the presence of a
strongly coupled charge-transfer state; and the presence of a ketone moeity
allows for further characterisation by infrared spectroelectrochemistry.

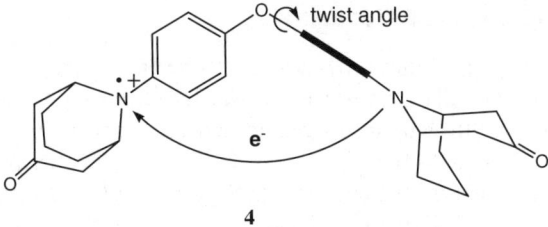

4

Figure 5.15   Structure of the oxo-bridged dialkylaniline dimer.

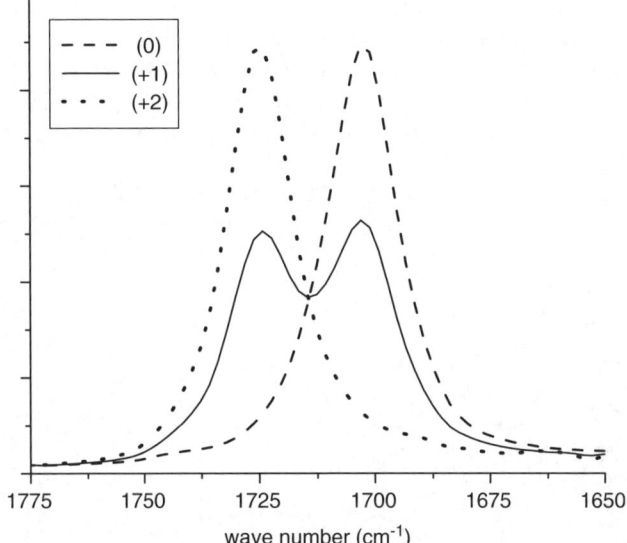

Figure 5.16   Spectra of **4** in a 0.1 M tetra-*n*-butyl ammonium hexafluorophosphate solution in methylene chloride at −25 °C.

The organic complex was dissolved in a 0.1 M tetra-*n*-butyl ammonium hexafluorophosphate (TBAH) solution in methylene chloride and introduced to the SEC cell at −25 °C to ensure stability of the oxidised species. The ketone stretch was examined in the (0), (+1), and (+2) states. Single sharp v(CO) bands were observed in the (0) and (+2) oxidation states, while two peaks of half-intensity were observed in the (+1) state (Figure 5.16). The maxima of these peaks were not shifted from the peaks for the (0) and (+2) oxidation states.

The (+1) peaks in this case are best approximated by a simple Gaussian/Lorentzian overlap of two bands present in solution, and not the Bloch-type analysis appropriately used for rapidly exchanging systems. It is concluded that this particular "organic mixed-valence complex" is localised on the infrared timescale.

5.3.5   Ligand Centred Dendrimers

In addition to forming dimeric complexes, these trinuclear ruthenium clusters can be extended into dendrimeric networks by choosing a tridentate central bridging ligand. A 0th-generation ligand-centred dendrimer (LCD) with the central ligand 2,4,6-tri-(4-pyridyl)-*s*-triazine has been synthesised and studied by SEC[36] (Figure 5.17). In the cyclic voltammetry, there are two, three-electron oxidations ($E_{1/2} = +0.14$ V, $+0.92$ V) and a subsequent one-electron reduction ($E_{1/2} = -1.26$ V) followed by a two-electron reduction ($E_{1/2} = -1.40$ V). In the case of this LCD, the reduction of the bridging triazine ligand is observed in the cyclic voltammetry as well ($E_{1/2} = -1.82$ V).

Although the cyclic voltammetry for this complex seems fairly similar to the trinuclear ruthenium cluster dimers, the lower-lying LUMO (as evidenced by the lower reduction potential) of the triazine bridging ligand plays an important role in explaining the details of the spectroelectrochemistry observed for this LCD. The CO stretching frequency of the neutral complex appears at 1940 cm^{-1}. Upon application of -1.050 V (*vs.* Ag pseudoreference, roughly 0 to

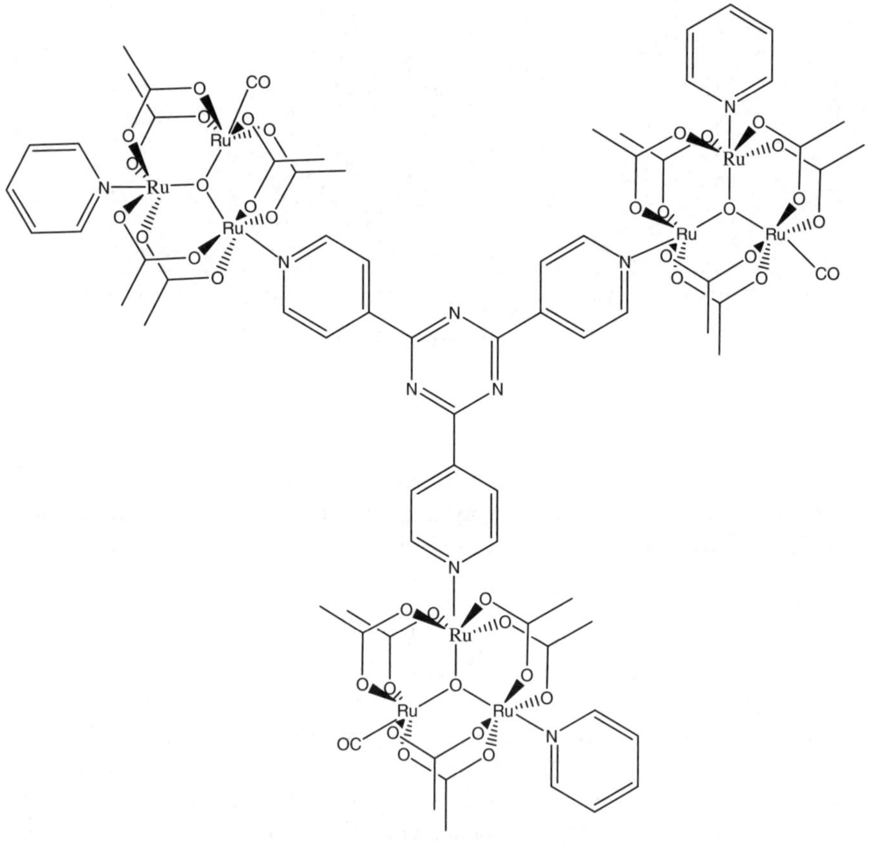

Figure 5.17   Structure of [Ru$_3$(OAc)$_6$O(CO)(py)]$_3$-μ_3-2,4,6-tri-(4-pyridyl)-*s*-triazine.

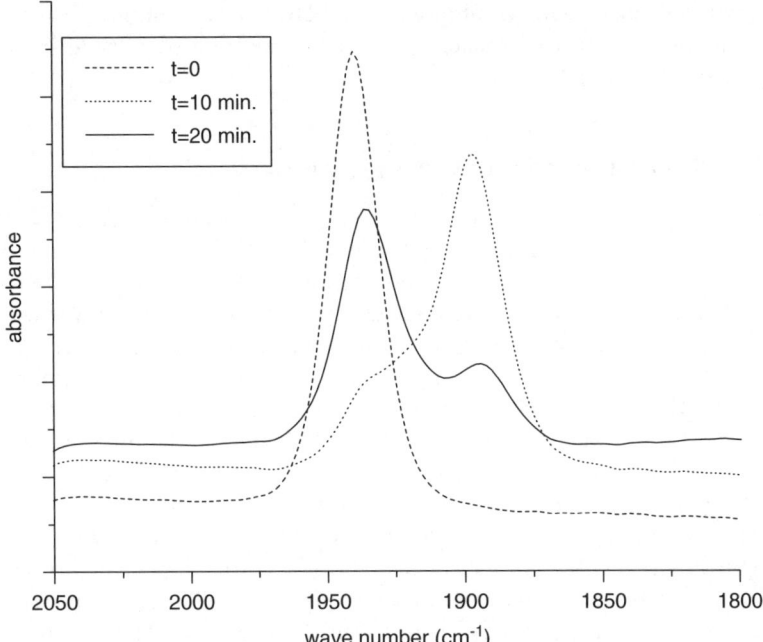

Figure 5.18 IR-SEC spectra of [Ru$_3$(OAc)$_6$O(CO)(py)]$_3$-μ_3-2,4,6-tri-(4-pyridyl)-*s*-triazine in the neutral state ($t=0$, 0 V, dashed line), -3 ($t=10$ min, -1.05 V, dotted line), and -2 ($t=20$ min, -1.05 V, solid line).

-1.42 V) at 25 °C, the v(CO) band at 1940 cm^{-1} diminishes in intensity, while a new band at *ca.* 1900 cm^{-1} grows in until (after 10 min) the bands are present in a 1:2 ratio (1940 cm^{-1}:1900 cm^{-1}) (Figure 5.18). This is a three-electron reduction, with two electrons essentially localised on two of the three ruthenium clusters and the third localised on the triazine ligand (as evidenced by the shift of v_{triazine}(C–C) *ca.* 1550 cm^{-1} upon reduction). Exchange intensity indicates the rapid electron transfer between sites ($k_{\text{et}} = 1.1 \times 10^{12}$ s^{-1}).

After holding at this potential for an additional 10 min, the band for the neutral cluster *increases* in intensity, while the band due to the reduced clusters at 1900 cm^{-1} *diminishes* in intensity. Therefore, an oxidation event occurs *without changing the potential of the cell*. This can be explained by considering that the reduced triazine ligand is a strong electron donor, driving the cluster reduction potentials more negative. Also, these LCDs are highly resonance stabilised, which contributes to $\Delta E_{1/2}$ (c.f. Section 5.3.1).

At -30 °C, reduction of the triazine ligand is not observed. Instead, full reduction of the clusters *via* a three-electron process occurs, and no exchange intensity is apparent. This is likely due to restriction of rotation about the cluster–triazine bonds, resulting in unfavourable orbital overlap at low temperature.

When the triazine ligand is reduced, electron transfer is "on" and when it is not reduced, electron transfer is "off." This is an example of electrochemically "switched on" electronic communication between ruthenium clusters. Our work on the spectroelectrochemistry of larger and dendrimeric cluster assemblies is continuing.

5.3.6   Observation of "Mixed-Valence Isomers"

A particularly elegant example of the use of the spectroelectrochemical cell is in the observation of "mixed-valence isomers." Unlike the other complexes presented in this study, the asymmetrically bridged complex, [Ru₃O(μ-CH₃COO)₆(CO)(py)]₂mpz, undergoes electron transfer in the presence of a driving force. Determination of the equilibrium constant for the charge distribution introduced by this driving force by IR spectroscopy is hindered by overlap of the CO stretching frequencies for each charge-transfer isomer. Therefore, isotopic substitution of the CO ligand is required to spectroscopically differentiate the two sides of the mixed-valence complex. Complexes **5** and **6**, Figure 5.19, were synthesised in a stepwise fashion.[37]

Cyclic voltammetry of **5** and **6**, in a 0.1 M tetrabutylammonium hexafluorophosphate solution in methylene chloride *vs.* the ferrocene/ferrocenium reference, reveals two two-electron oxidations ($E_{1/2} = 200$ mV, 1000 mV); and two one-electron reductions ($E_{1/2} = -1160$ mV, -1500 mV). The splitting in the reduction waves, ΔE, is 340 mV, and corresponds to a comproportionation equilibrium constant of 5.6×10^5. The total electrochemical splitting reflects both the electronic interactions typical of a strongly electronically coupled

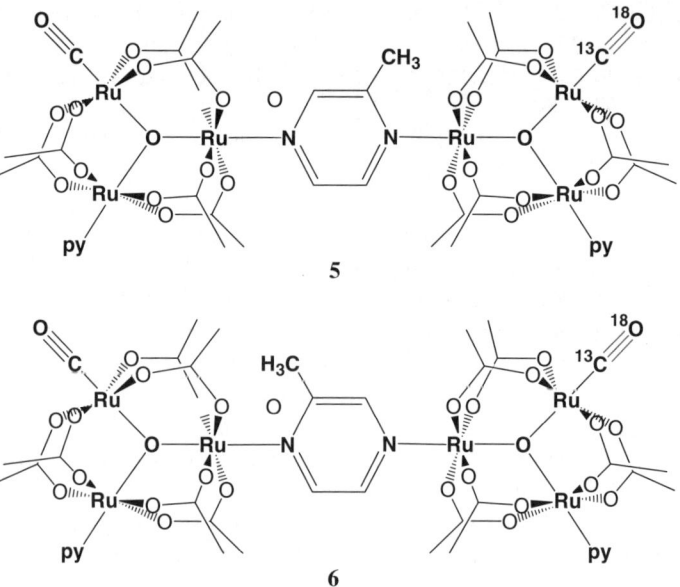

Figure 5.19   The two dimers synthesised for this study, differing only in location of the ¹³C¹⁸O label relative to the methyl group on the bridge.

mixed-valence complex and the intrinsically different reduction potentials of the clusters bound to the different nitrogen atoms of the bridging methyl pyrazine ligand.[14,38] The equilibrium constant between the mixed-valence isomers formed from **5** and **6** is small, and thus the electrochemical splittings observed by cyclic voltammetry are dominated by the electronic interactions. This indicates the presence of a strongly coupled mixed-valence state.

Infrared spectroelectrochemistry of **5** and **6** reveals exchange pairs consistent with mixed-valence isomerism (Figure 5.20). In the case of **5** (Figure 5.20, left), the neutral and doubly reduced (-2) states each show two $\nu(CO)$ bands, separated by *ca.* $90\,\mathrm{cm}^{-1}$, the intrinsic frequency separation due to the isotope substitution, $\nu(^{12}C^{16}O)$ *vs.* $\nu(^{13}C^{18}O)$. In the mixed valence (-1) state of **5**, *four* $\nu(CO)$ bands are observed, and these correspond (from higher to lower energy) to (1) the $\nu(^{12}C^{16}O)$ contribution from the minor (less stable) mixed-valence isomer, (2) the $\nu(^{12}C^{16}O)$ contribution from the major (more stable) mixed-valence isomer, (3) the $\nu(^{13}C^{18}O)$ contribution from the major isomer, and (4) the $\nu(^{13}C^{18}O)$ contribution from the minor isomer. Note that in the mixed-valence

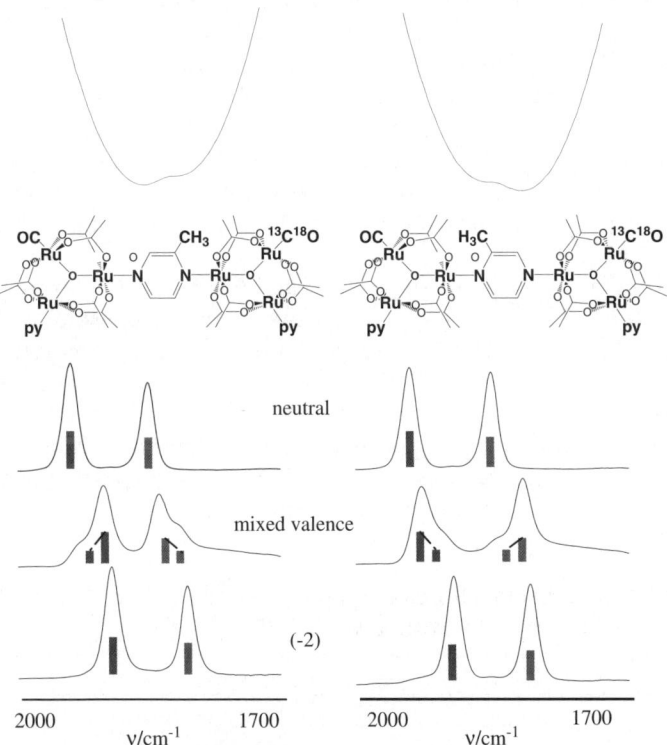

Figure 5.20 Infrared spectroelectrochemistry of **5** (left) and **6** (right) at $-30\,°C$ in a 0.1 M tetra-*n*-butylammoniumhexafluorophospate solution in CH_2Cl_2. Spectra for the neutral (top), charge-transfer (middle), and doubly reduced (bottom) states are shown with a schematic of the exchanging populations. A qualitative potential energy surface is depicted at the top.

state of **5**, the spectral pattern (from higher to lower energy) corresponds to contributions from the minor, major, major, and minor isomers, while in the mixed-valence state of **6**, the spectral pattern is reversed to major, minor, minor, major. This is the expected result of reversing the side of the asymmetric mixed-valence complex that contains the $^{13}C^{18}O$ ligand. On account of the N-atom basicity differences of the mpz ligand, the minor isomer of **5** is expected to have the negative charge mostly on the cluster with the $^{13}C^{18}O$ ligand. The lowest-frequency component of the overall $v(CO)$ spectrum is assigned easily to the cluster of the minor isomer bearing both the negative charge and the $^{13}C^{18}O$ since both the charge and heavier isotopes shift $v(CO)$ to lower frequency. For similar reasons, the highest-frequency part of the overall $v(CO)$ spectrum is assigned to the $^{12}C^{16}O$ substituted cluster of the minor isomer since $v(CO)$ will be unaffected by either charge or heavier isotope substitution. The remaining, more intense bands in the centre of the $v(CO)$ spectrum of the mixed-valence state of **5** are assigned to the major isomer. Reversing the $^{13}C^{18}O$ substituted side of the cluster in going from **5** to **6**, reverses whether each spectral component originates from the major or minor isomer. These data provide the most compelling evidence to date of the existence of mixed-valence isomers as discrete chemical species. It can also be seen that both the high-frequency $v(^{12}C^{16}O)$ and low-frequency $v(^{13}C^{18}O)$ portions of the IR spectra of the mixed-valence states are extensively coalesced exchange lineshapes resulting from the dynamics of the intramolecular electron transfer between the minor and major mixed-valence isomers. Using the same lineshape-simulation methods as for the symmetrically substituted dimers, we can estimate the rate constants for electron exchange, as well as the equilibrium constant between the major and minor mixed-valence isomers.

Analysis of the spectral lineshapes of **5** and **6** gave an uphill rate for charge transfer of 6.5×10^{11} s^{-1} (Table 5.2), and an equilibrium constant of 2.2 for the charge distribution, which compares well with previously reported rate constants for similar mixed-valence dimers of trinuclear ruthenium clusters.[15,38]

[Ru$_3$O(CH$_3$COO)$_6$(CO)(py)]$_2$(μ-clpz), which is also bridged by an unsymmetric bridging ligand (clpz), appears to be at the limit of valence trapping by infrared spectroelectrochemistry (Figure 5.7). The two peaks present in the (−1) state of the dimer are fully resolved and appear at the same frequencies as

Table 5.2 Rate constants for electron transfer. *This is the "downhill" rate of electron transfer with a calculated $K_{eq} = 2.2$.

	L	BL	k_{et} (s^{-1})
1	dmap	pz, d$_4$-pz	2.0×10^{12}
		dmpz	1.5×10^{12}
2	py	pz, d$_4$-pz	1.8×10^{12}
5,6		mpz*	1.2×10^{12}
		dmpz	1.2×10^{12}
3	cpy	pz, d$_4$-pz	1.2×10^{12}
		dmpz	7.0×10^{11}

the single peaks observed in the spectra of the neutral and (−2) species. This suggests that the rate of electron transfer in this dimer can be no faster than $10^{11}\,s^{-1}$. Comparing this value to the electron-transfer rate constant of $1.1 \times 10^{12}\,s^{-1}$ predicted from the relationship of $\Delta E_{1/2}$ to k_{et} (Figure 5.8, which is valid only when the electronic coupling dominates the electrochemical splitting), it can be concluded that it is the intrinsic difference in reduction potentials of each cluster site that dominates $\Delta E_{1/2}$ in this case. Therefore, the observed infrared spectrum is essentially that of one valence isomer, in stark contrast to the highly coupled complexes **5** and **6**.

5.4 Conclusions

The different case studies presented here show how a variable-temperature reflectance spectroelectrochemical cell provides the versatility to carry out studies of UV-Vis electronic spectra, IR spectra, and resonance Raman spectra of inorganic and organic compounds in a multiplicity of different redox states. A key finding is that when rates of intramolecular electron transfer in mixed-valence states approach the "ultrafast" timescale of 10^{11}–$10^{13}\,s^{-1}$, IR line-shapes provide dynamical information and the rate constants for charge transfer. In principle, reflectance SEC can provide detailed dynamical information about charge transfers occurring in larger macromolecular systems including redox-active dendrimers and conducting polymers.

References

1. T. Kuwana, R. K. Darlington and D. W. Leedy, *Anal. Chem.*, 1964, **36**, 2023.
2. J. J. Niu and S. J. Dong, *Rev. Anal. Chem.*, 1996, **15**, 1.
3. P. A. Flowers and S. A. Callender, *Anal. Chem.*, 1996, **68**, 199.
4. P. B. Graham and D. J. Curran, *Anal. Chem.*, 1992, **64**, 2688.
5. W. R. Heineman, J. N. Burnett and R. W. Murray, *Anal. Chem.*, 1968, **40**, 1974.
6. J. P. Bullock, D. C. Boyd and K. R. Mann, *Inorg. Chem.*, 1987, **26**, 3084.
7. S. K. Enger, M. J. Weaver and R. A. Walton, *Inorg. Chim. Acta*, 1987, **129**, L1.
8. K. M. Kadish, X. H. Mu and X. Q. Lin, *Electroanalysis*, 1989, **1**, 35.
9. C. L. Yao, F. J. Capdevielle, K. M. Kadish and J. L. Bear, *Anal. Chem.*, 1989, **61**, 2805.
10. H. O. Finklea, R. K. Boggess, J. W. Trogdon and F. A. Schultz, *Anal. Chem.*, 1983, **55**, 1177.
11. F. Hartl, *Inorg. Chim. Acta*, 1995, **232**, 99.
12. R. Gamage, S. Umapathy and A. J. McQuillan, *J. Electroanal. Chem.*, 1990, **284**, 229.
13. B. J. Brisdon, S. K. Enger, M. J. Weaver and R. A. Walton, *Inorg. Chem.*, 1987, **26**, 3340.

14. T. Ito, H. Hamaguchi, H. Nagino, T. Yamaguchi, H. Kido, I. S. Zavarine, T. Richmond, J. Washington and C. P. Kubiak, *J. Am. Chem. Soc.*, 1999, **121**, 4625.

15. T. Ito, T. Hamaguchi, H. Nagino, T. Yamaguchi, J. Washington and C. P. Kubiak, *Science*, 1997, **277**, 660.

16. H. Kido, H. Nagino and T. Ito, *Chem. Lett.*, 1996, , 745.

17. I. S. Zavarine, Doctoral Thesis, Purdue University, 1998.

18. J. E. Sutton and H. Taube, *Inorg. Chem.*, 1981, **20**, 3125.

19. J. E. Sutton, P. M. Sutton and H. Taube, *Inorg. Chem.*, 1979, **18**, 1017.

20. F. Salaymeh, S. Berhane, R. Yusof, R. Delarosa, E. Y. Fung, R. Matamoros, K. W. Lau, Q. Zheng, E. M. Kober and J. C. Curtis, *Inorg. Chem.*, 1993, **32**, 3895.

21. C. Creutz, *Prog. Inorg. Chem.*, 1983, **30**, 1.

22. Defined as "Ru$_3$O"(0) + "Ru$_3$O"(−2) $\leftrightarrows$ 2"Ru$_3$O"(−1). $K_c = \exp(nF\Delta E)$.

23. J. J. Turner, F. Grevels, S. M. Howdle, J. Jacke, M. T. Haward, W. E. Klotzbucher, 1991, **113**, 8347.

24. C. H. Londergan, J. C. Salsman, S. Ronco, L. M. Dolkas and C. P. Kubiak, *J. Am. Chem. Soc.*, 2002, **124**, 6236.

25. M. L. Horng, J. A. Gardecki, A. Papazyan and M. Maroncelli, *J. Phys. Chem.*, 1995, **99**, 17311.

26. B. J. Lear, S. D. Glover, J. C. Salsman, C. H. Londergan and C. P. Kubiak, *J. Am. Chem. Soc.*, 2007, **129**, 12772.

27. S. D. Glover, B. J. Lear, J. C. Salsman, C. H. Londergan and C. P. Kubiak, *Phil. Trans. Roy. Soc. A.*, 2008, **366**, 177.

28. C. H. Londergan, J. C. Salsman, S. Ronco and C. P. Kubiak, *Inorg. Chem.*, 2003, **42**, 926.

29. K. D. Demadis, C. M. Hartshorn and T. J. Meyer, *Chem. Rev.*, 2001, **101**, 2655.

30. C. H. Londergan and C. P. Kubiak, *J. Phys. Chem. A.*, 2003, **107**, 9301.

31. D. H. Oh, M. Sano and S. G. Boxer, *J. Am. Chem. Soc.*, 1991, **113**, 6880.

32. V. Petrov, J. T. Hupp, C. Mottley and L. C. Mann, *J. Am. Chem. Soc.*, 1994, **116**, 2171.

33. C. H. Londergan, R. C. Rocha, M. G. Brown, A. P. Shreve and C. P. Kubiak, *J. Am. Chem. Soc.*, 2003, **125**, 13912.

34. R. C. Rocha, M. G. Brown, C. H. Londergan, J. C. Salsman, C. P. Kubiak and A. P. Shreve, *J. Phys. Chem. A.*, 2005, **109**, 9006.

35. S. F. Nelsen, G. Q. Li and A. Konradsson, *Org. Lett.*, 2001, **3**, 1583.

36. B. J. Lear and C. P. Kubiak, *J. Phys. Chem. B.*, 2007, **111**, 6766.

37. J. C. Salsman, C. P. Kubiak and T. Ito, *J. Am. Chem. Soc.*, 2005, **127**, 2382.

38. T. Ito, N. Imai, T. Yamaguchi, T. Hamaguchi, C. H. Londergan and C. P. Kubiak, *Angew. Chem. Int. Ed. Engl.*, 2004, **43**, 1376.

CHAPTER 6

Spectroelectrochemical Investigations on Carbon-Rich Organometallic Complexes

RAINER F. WINTER

Institut für Anorganische Chemie der Universität Regensburg,
Universitätsstraßes 31, D-93040 Regensburg

6.1 Introduction

During the past decades spectroelectrochemistry has emerged as a powerful tool for following the course of electrolyte processes and simultaneously collecting the spectroscopic fingerprints of electrogenerated products. Many cell designs with different degrees of sophistication have been reported and put into practical use.[1–5] Ingenious solutions have adjusted spectroelectrochemical cells to various experimental demands such as the rigorous exclusion of moisture and air, more accurate control of the applied potential, the ability to conduct experiments at different and especially at low temperatures or specific cell geometries in order to fit the attached spectrometer. The experimenters' reasons for resorting to spectroelectrochemistry can be just as varied and reach far beyond simply collecting spectroscopic data of the electrogenerated product. They may involve the identification of the actual electron-transfer site within a molecule when two or more redox-active constituents with similar oxidation or reduction potentials but distinct spectroscopic properties of their oxidised or reduced forms are present.

Molecules where identical or similar redox-active moieties are connected by conducting spacers generate mixed-valence systems upon partial oxidation or reduction.[6] Spectroelectrochemistry here serves the purpose of evaluating the strength of the electronic coupling provided by the bridge, which is a central

Spectroelectrochemistry
Edited by Wolfgang Kaim and Axel Klein

issue in this field. Spectroelectrochemistry can also be employed to reversibly alter the electronic properties of a redox-active substituent or ligand in some part of a molecule and to simultaneously monitor the spectroscopic changes. Such experiments allow direct visualisation of electronic substituent effects. Recent work has also disclosed that it is possible to implement or reversibly switch molecular properties such as NLO effects[7] or luminescence by using spectroelectrochemistry.[8] Yet another application of spectroelectrochemical methods is to monitor chemical reactions following electron transfer and to identify the final products. If the subsequent chemical reactions occur on a timescale that well exceeds that of the electrolysis, this approach may also be employed to monitor the kinetics of such follow-up reactions.

A central question in much of the published work is how the uptake or loss of an electron affects the bonding within the analyte molecule. Changes in the electronic or vibrational spectra mirror the altered charge distribution. Electron transfer also changes the spin state of a system. Most electrogenerated products are therefore amenable to ESR spectroscopy and valuable information can be obtained from this method.[4,9] Isotropic g-values and g-tensor anisotropies allow one to differentiate between metal- and organic-centred odd-electron species, while hyperfine splittings, if resolved, provide detailed insight as to the spin distribution within the molecule. If one accepts the approximation that electron transfer mainly involves the HOMO or LUMO frontier orbitals, spectroelectrochemical experiments can be used to map these so-called redox orbitals.[i] Such investigations are particularly rewarding for molecules with highly delocalised bonding. This is certainly true for carbon-rich organometallics where the unsaturated, conjugated or cumulated organic ligand attaches to the metal by a direct metal–carbon single or multiple bond. Spectroelectrochemical investigations on such compounds, including alkynyl, vinyl, vinylidene and higher cumulenylidene complexes, are the subject of this contribution.

There is still the question of what benefits spectroelectrochemistry offers compared to the more traditional approach of electrolytically or chemically generating the electron-transfer product and analysing it separately by various spectroscopic means. While a large variety of chemical oxidants and reductants are available, the success of chemical redox reactions critically depends on the judicious choice of the electron-transfer agent and strict control over the reaction stoichiometry, especially in cases where the analyte may undergo several individual, closely spaced redox events. Intermediate redox states are then accessible by comproportionation of the fully reduced and fully oxidised forms in their appropriate stoichiometric ratios. With chemical redox reactions, one must also keep in mind that the reduced form of a chemical oxidant or the oxidised form of a reductant may engage in chemical follow-up reactions with the oxidised or reduced forms of the analyte or interfere with its

---

[i] In a grossly simplified picture, the redox orbital is often considered as the molecular orbital from where the electron is removed in an oxidation or where it is added to in a reduction process, *i.e.* the HOMO or LUMO orbitals. In reality, electron transfer affects more than just one molecular orbital and may even change the entire ordering of the energy levels in the frontier orbital regime.[18,19]

isolation and spectroscopic detection. A detailed discussion of these issues is available in the literature.[10]

The stepwise electrolytical generation of an electron-transfer product and its separate spectroscopic investigation finds its limitations in the case of highly reactive, unstable oxidised or reduced species. Intimate coupling of bulk electrolysis and spectroscopy, on the other hand, may well allow us to identify and gain structural information on even short-lived intermediates.[1] Moreover, the spectroelectrochemical experiment eliminates the problem of sample transferal to the spectroscopic cell. The greatest advantage of a spectroelectrochemical experiment is, however, that it allows direct correlation of electrochemical and spectroscopic data as a function of potential and time. This may be highly advantageous as compared to acquiring two or more independent data sets. If, for example, the initial electrogenerated product is of only limited stability it may still be monitored at a suitable wavelength and the time course of its concentration established. When combined with the time profile of the total charge passed in the electrochemical process or the decreasing absorbance of the starting material the kinetics of the chemical follow-up reaction can be established. This is again under the premise that the rate of mass transport is fast compared to the kinetic process under study.

There are, nevertheless, some drawbacks and trade-offs.[1,3,11] One relates to the difficulty of realising or maintaining a uniform current–potential distribution within a thin layer cell with often large spatial separations between the working, reference and auxiliary electrodes. This leads to large uncompensated resistance (iR drop) and severely compromises potential control. Redox systems with multiple, closely spaced waves constitute particularly problematic cases. Thin-layer spectroelectrochemical cells with small optical pathlengths may also require high analyte concentrations in order to achieve acceptable signal-to-noise ratios. This is especially true for vibrational spectroscopy where bands with relatively small absorbencies may be involved. High analyte concentrations not only increase the problem of iR drop but may also cause severe adsorption of an electrogenerated product to the working electrode up to the point of electrode passivation. In less unfavourable cases still, exceedingly long electrolysis times may be required to convert all the starting material. The inevitable presence of a solvent and a suitable supporting electrolyte at high concentration levels may also impede with the spectroscopic detection of electrogenerated products. Thus, large sections of the infrared are usually concealed by strong background absorption.[12] Moreover, some spectroelectrochemical cells perform rather poorly at low temperatures. Typical problems are the large increase in supporting electrolyte viscosity and supporting electrolyte precipitation. As a final caveat, spectroelectrochemistry may lead to erroneous spectral assignments in cases where the initial electrogenerated product rapidly converts to a follow-up product and this reverts to the starting material upon backelectrolysis. This corresponds to the square scheme in Scheme 6.1 where the rate constants k_{-1} and k_2 are large and the equilibria are completely displaced toward species A and B^+. Since chemical reversibility is maintained over the whole chemical cycle, the underlying reactivity may completely escape observation under

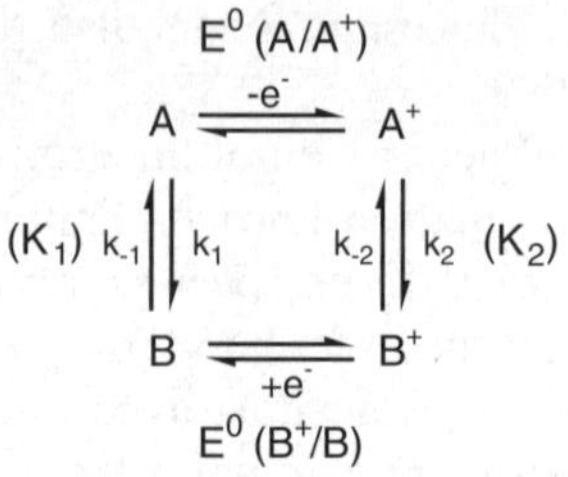

$$E^0 \; (A/A^+)$$
$$A \; \underset{}{\overset{-e^-}{\rightleftharpoons}} \; A^+$$

$$(K_1) \; k_{-1} \Big\Updownarrow k_1 \qquad k_{-2} \Big\Updownarrow k_2 \; (K_2)$$

$$B \; \underset{+e^-}{\rightleftharpoons} \; B^+$$
$$E^0 \; (B^+/B)$$

Scheme 6.1 General square scheme summarising redox and chemical interconversions.

spectroelectrochemical conditions. Electron-transfer-induced isomerisation or dimerisation reactions are possible candidates of such a scenario. In order to avoid such pitfalls, it seems therefore mandatory to supplement spectro-electrochemistry with conventional electrochemistry and bulk electrolysis, where such phenomena are readily recognised.

The desire to combine the advantages of conventional multicompartment electrolysis cells and the simultaneous collection of spectroscopic information has led researchers to the use of bifurcated fibre-optic cables that connect the electrochemical cell to a remote spectrometer. Source radiation is guided into the analyte solution and returned to the detector by the reflective working electrode surface[13] or a mirror with adjustable separation from the source.[14] Setups for optical and IR spectroscopy have been described and successfully employed to address the issue of chemical reactivity coupled to electron transfer.[15–17]

This overview on the use of spectroelectrochemistry within the field of carbon-rich organometallics is organised by the experimenters' main purpose for conducting these experiments rather than by individual compound classes. With that, we wish to emphasise the various applications of spectro-electrochemical methods and to demonstrate their utility. Some complementary work based on chemical redox agents or bulk electrolyses and remote spectroscopic investigations is also cited to provide a more complete picture, though not in a wholly comprehensive manner.

6.2 Uses of Spectroelectrochemistry in the Field of Carbon-rich Organometallics

6.2.1 Spectroelectrochemistry as a Tool to Identify Redox Sites

Several complexes offer two or more redox-active moieties that may undergo electron transfer at rather similar potentials. In such cases voltammetric or potentiometric techniques alone may not suffice to unambiguously assign individual redox processes or to elucidate the order of redox events within such molecules. If, however, the individual redox-active constituents generate oxidised or reduced forms with clearly distinguishable spectroscopic properties, a suitable combination of electrochemical and spectroscopic investigations can

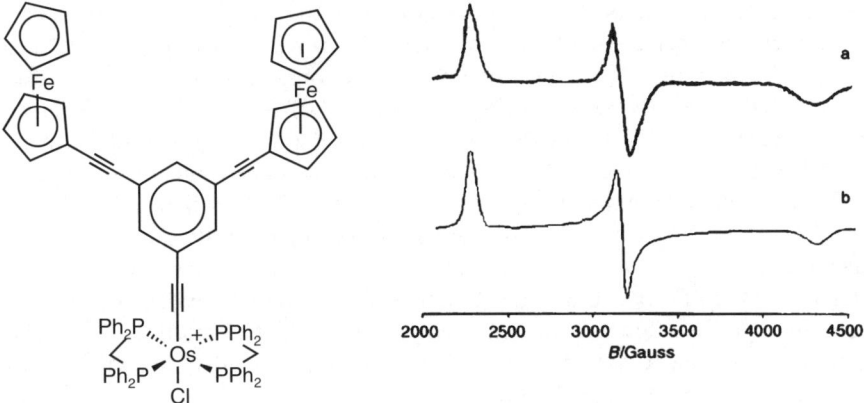

Figure 6.1   Schematic structure of the radical cation of {Cl(dppm)$_2$Os}(−C≡CC$_6$H$_3$−{C≡CFc}$_2$) and its ESR spectrum; a) experimental, b) simulated. Reproduced from ref. 20 with permission.

solve this issue. Illustrative examples for such use of spectroelectrochemistry can also be found within the class of carbon-rich organometallics. Long and co-workers[20] have reported on heterodimetallic complexes of the 1,3,5-triethynylbenzene bridging ligand. Cyclic voltammetry on {Cl(dppm)$_2$Os}(−C≡CC$_6$H$_3$−{C≡CFc}$_2$) (Fc = ferrocenyl, dppm = Ph$_2$PCH$_2$PPh$_2$) reveals one reversible one-electron wave followed by a second reversible feature with about twice the peak current of the first couple. This allowed the authors to assign the first wave to an Os-centred process and the second wave to the simultaneous oxidation of the ferrocenylethynyl subunits. Unequivocal proof came from *in situ* ESR spectroelectrochemistry on the mono-oxidised cation. ESR activity at 100 K and the rhombic pattern are clearly at odds with a ferrocenium based radical (Figure 6.1). In *trans*-(FcC≡C−)$_2$Ru(dppm)$_2$ the situation is just reversed in that the ferrocenyl end groups are oxidised prior to the central ruthenium site.[21,22] IR investigation reveals that the chemically generated dication is a centrosymmetric species, while NMR spectroscopy shows a paramagnetic broadening of the ferrocenyl CH resonances. Ferrocene oxidation has also been corroborated by X-ray photoelectron spectra (XPS) showing a shift of the Fe 2p$_{1/2}$ and 2p$_{3/2}$ peaks to higher binding energy.[22] Complexes *all trans*-(FcC≡C−)$_2$Ru(CO)L(PBu$_3$)$_2$ (L = CO, pyridine, P(OMe)$_3$) show the same behaviour.[23] Particularly revealing are the only slight changes of the CO stretching frequencies in the mono- and dicarbonyl complexes upon dioxidation, which rules out a direct involvement of the central ruthenium atom in the oxidation process. Interestingly, the coligands at the ruthenium atom modulate the degree of electronic interactions between the ferrocenyl substituents (see Section 6.2.3).[23]

Heterodimetallic complexes with fully conjugated oligoynediyl or diethynyl arene bridges constitute another interesting family of compounds where an assignment of individual redox sites may not be trivial. In these systems, strong electron delocalisation of the frontier orbitals onto the bridging ligand even poses the more fundamental question of whether there is such a thing as a defined redox site at all. The oligoynediyl-bridged dinuclear Re/Pd complexes Cp*(NO)(PPh$_3$)Re$-$(C$\equiv$C)$_n$$-$Pd(PEt$_3$)$_2$Cl ($n = 1$, 2) and the related trimetal complex *trans*-{Cp*(NO)(PPh$_3$)Re$-$C$\equiv$C$-$C$\equiv$C$-$}$_2$Pd(PEt$_3$)$_2$ present clear cut cases because of the large differences of the intrinsic redox potentials of the individual metal subunits ($E_{1/2} = 0.52$ V for Cp*(NO)(PPh$_3$)Re$-$C$\equiv$C$-$C$\equiv$ C$-$CH$_3$, $E_{p,a} = 1.48$ V for *trans*-(PEt$_3$)$_2$Pd($-$C$\equiv$CPh)$_2$).[24] ESR spectroscopy on chemically generated radical monocations of the dinuclear complexes and the dication of the Pd-bridged dirhenium complex showed sextet patterns due to Re hyperfine splittings. Since the A_{Re} values are of the same magnitude as for similar mononuclear Re alkynyl complexes, the oxidation was safely assigned as occurring at the Re units (Figure 6.2).

In the butadiynediyl C$_4$-bridged Fe/Re derivative Cp*(dppe)Fe$-$C$\equiv$C$-$C$\equiv$ C$-$Re(NO)(PPh$_3$)Cp* and the *para*-diethynylbenzene-bridged *fac*-{(CO)$_3$(bipy)Re}(μ-C$\equiv$CC$_6$H$_4$C$\equiv$C-1,4) {Fe(dppe)Cp*} the intrinsic redox potentials of the different metal-alkynyl subunits are much closer. Iron-centred oxidation of the diethynyl phenylene complex was established by a combination of cyclic voltammetry, quantum-chemical calculations, X-ray structural studies and by UV–Vis spectroelectrochemistry.[25] Upon oxidation, the Fe$\rightarrow$$\pi$* transition in the visible bleaches, leaving the Re$\rightarrow$bipy and Re$\rightarrow$$\pi$* transitions at somewhat lower energies when compared to the neutral. The observation of oxidation-state-dependent luminescence constitutes another interesting aspect of this work as we will detail in Section 6.2.4. Substantial electron delocalisation in the ground state and the various oxidised forms was observed, and this will be discussed in Section 6.2.3. Of importance here is that spectroscopic and Mössbauer investigations on the chemically generated radical cation argue for a mainly iron-centred process.[26] Indicative are the strongly decreased hyperfine splittings to Re in ESR spectroscopy and the close resemblance of the Mössbauer quadrupole splitting and isomer shift values to those of related di-iron complexes.[27]

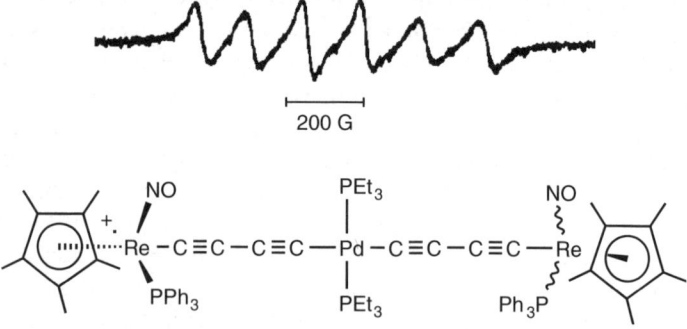

Figure 6.2   The monocation *trans*-Pd(PEt$_3$)$_2${$-$C$\equiv$C$-$C$\equiv$C$-$Re(NO)(PPh$_3$)Cp*}$_2$ and its ESR spectrum. Reproduced with permission from ref. 24.

The question about the identities of the redox sites and the sequence of redox events also applies to complexes built from redox-active metal ions and electroactive substituents or ligands. One such class of compounds are alkynyl complexes derived from ethynyl substituted mono-, bi-, or terthiophenes. Ferrocenylethynyl[28] and ruthenium mono- and bis(alkynyl) derivatives[29] have been studied. In each case, the first oxidations were assigned as metal-based processes. This follows from the insensitivity of the first oxidation potential on the number of linked thiophenyl subunits and its close correspondence to the half-wave potentials of similar ferrocenylethynyl and ruthenium alkynyl derivatives. The number of thiophenyl repeat units on the ligand has, however, a strong bearing on the number and potentials of additional waves at more anodic potentials. The above assignment was further substantiated by comparing the UV–Vis–NIR spectra of the neutrals and their chemically or electrochemically generated mono-oxidised forms. Mono-oxidation produces only slight shifts of the intraligand $\pi \rightarrow \pi^*$ type transitions while intense π(oligothiophenyl) $\rightarrow M^{III}$ LMCT bands appear at low energy. In the ferrocenyl series, the LMCT band maxima and intensities linearly correlate with the differences between the metal- and ligand-based redox potentials, and this has been taken as an indication for an electronic modulation of the donor/acceptor interactions.

Due to its highly developed derivative chemistry and its ideal (or nearly so) electron-transfer behaviour, the ferrocenyl substituent constitutes an especially attractive redox tag. In the field of carbon-rich organometallics, the ferrocenylethynyl ligand is a widely used building block that endows these systems with electroactivity and the possibility to reversibly alter their electronic properties. Once the ferrocenyl entity is connected to another redox-active transition-metal moiety that oxidises at similar potentials, there is ambiguity as to the order of redox events. Examples of such systems can be found in the work of Sato et al.,[30-34] Long and coworkers,[21] Wolf and coworkers[22,23] and Ren and coworkers,[35] but they will be discussed in greater detail in later sections of this chapter since the main purpose of this work was either to investigate the bonding changes resulting from ferrocene oxidation or to establish the degree of the electronic interactions between two ferrocene sites or a ferrocene and another redox site at the mixed-valence oxidation level.

Allenylidene complexes of the type *trans*-$[Cl(L_2)_2Ru = C = C = C(ER_n)R']^+$ (L_2 = chelating diphosphine, ER_n = NR_2, SR, SeR, alkyl, aryl) are reversibly oxidised and possess mainly metal-based HOMO and HOMO-1 orbitals.[36,37] Introducing electroactive ferrocenyl-derived substituents ER_n or R', or the 1,1'-bis(diphenylphosphanyl)ferrocene (dppf) ligand, thus generates some ambiguity as to the sequence of redox events. Since oxidation of the allenylidene chromophore itself drastically changes its spectroscopic properties,[36,37] spectroelectrochemical techniques were successfully employed to resolve this issue. Cyclic voltammetry on *trans*-$[Cl(dppm)_2Ru = C = C = C(NMe_2)(C_2H_4Fc)]^+$ (Chart 6.1) revealed two reversible well-separated one-electron waves. When the first oxidation was followed by means of IR- and UV–Vis spectroelectrochemistry, the intense Ru $\rightarrow$ allenylidene π^* charge-transfer band in the visible region shifted to

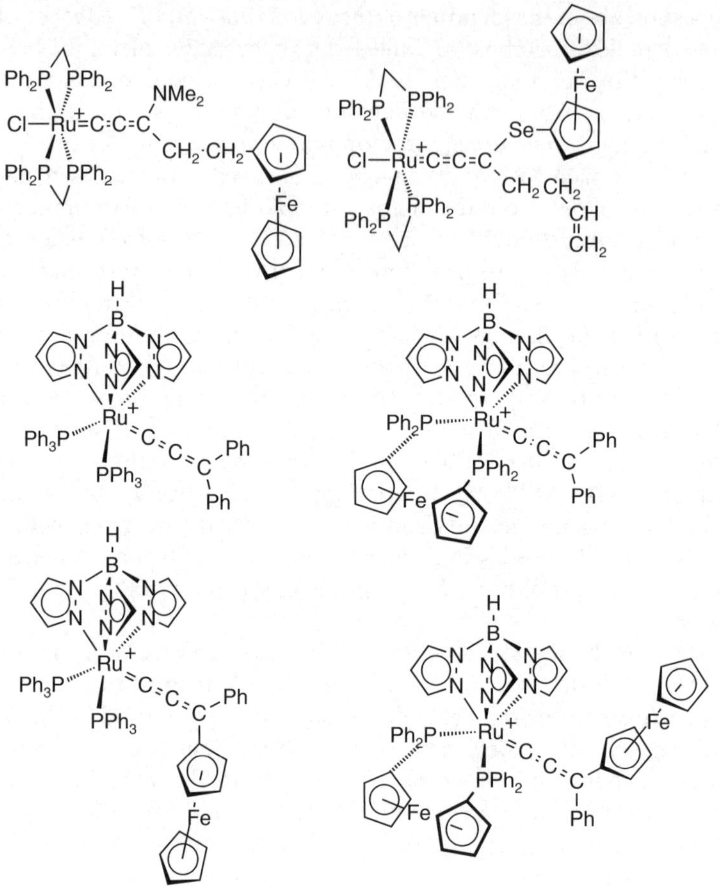

Chart 6.1 Allenylidene complexes with redox-active substituents and ligands.

slightly lower energy. This was accompanied by the growth of a ferrocenium-type absorption and a very slight shift of the allenylidene IR band. These results are as expected for the oxidation of the remote ferrocenyl substituent. Low-temperature ESR spectra of the *in-situ* generated radical cation displayed the characteristic signature of a ferrocenium-based radical.[38] The second oxidation caused considerably larger blueshifts of the MLCT and IR bands in accord with a substantial decrease of the electron density at the ruthenium atom. In ESR spectroscopy, the initial ferrocenium-type signal was gradually replaced by a spectrum that is ascribed to a triplet species. This indicates the presence of uncoupled spins at the ferrocenyl and the ruthenium sites.

In *trans*-$[Cl(dppm)_2Ru = C = C = C(SeFc)(C_2H_4CH = CH_2)]^+$, the redox-active ferrocenyl substituent is directly attached to the allenylidene ligand. Oxidation of the ferrocenylselenyl group caused a redshift of the allenylidene IR band by $10 \, cm^{-1}$.[39] This result fully conforms to what one would expect of the replacement of a substituent ER_n by a weaker donor. Such substitution

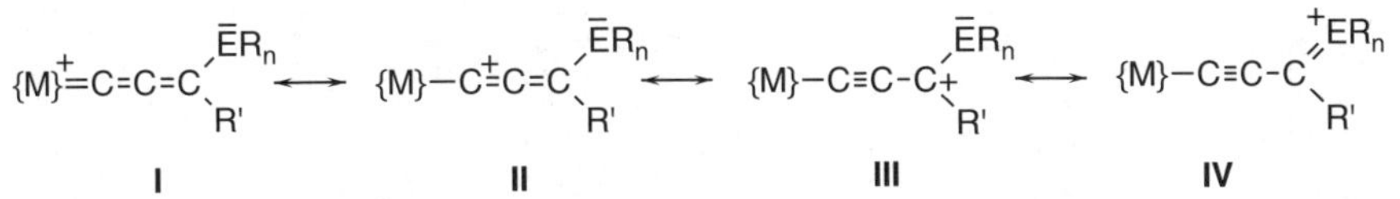

$$\{M\}\overset{+}{=}C=C=C\overset{\overline{E}R_n}{\underset{R'}{\diagup}} \longleftrightarrow \{M\}-C\overset{\pm}{=}C=C\overset{\overline{E}R_n}{\underset{R'}{\diagup}} \longleftrightarrow \{M\}-C\equiv C-C\overset{\overline{E}R_n}{\underset{R'}{\diagup}} \longleftrightarrow \{M\}-C\equiv C-C\overset{\overset{+}{E}R_n}{\underset{R'}{\diagup}}$$

         I   II   III   IV

Scheme 6.2   Resonance forms for cationic allenylidene complexes.

increases the weight of the cumulenic resonance forms in Scheme 6.2 and hence lowers the energy of the allenylidene stretch. The energy of this latter band reflects the bond order between the metal-bonded and the internal carbon atoms.[37,40] Direct attachment of ferrocenyl-based redox tags to either the terminal carbon atom of the allenylidene ligand, the ruthenium atom, or both was achieved in the complexes $[Tp(L_2)Ru=C=C=C(Ph)R]^+$ ($L_2 = 2PPh_3$, dppf; R = Ph, Fc; Tp = hydridotris(pyrazolyl)borate, $HB(pz)_3^-$).[41,42] All complexes with the ferrocenyl substituent attached to the cumulated ligand display an intense absorption band in their visible spectra, and this is located at even lower energies than the Ru→π*(allenylidene) band. This additional absorption has been assigned as an intraligand charge transfer (ILCT) from the electron-donating ferrocenyl group to the electron-accepting allenylidene ligand. One electron oxidation of $[Tp(PPh_3)_2Ru=C=C=C(Ph)(Fc)]^+$ in an OTTLE cell caused this band to bleach (Figure 6.3(b)). IR spectroelectrochemistry of the same process revealed a redshift of the allenylidene band of 20 cm^{-1}, which is about twice that for the ferrocenylselenyl substituted congener (Figure 6.3(a)). The larger effect can be traced to more efficient conjugation between the allenylidene ligand and the redox-active tags. In $[Tp(dppf)Ru=C=C=CPh_2]^+$, oxidation of the dppf ligand had the opposite effect, thus inducing a shift of the allenylidene ligand by an average of 15 cm^{-1} to higher energy (Figure 6.3(c)). This finding is readily accommodated by invoking larger contributions from the alkynyl-type resonance forms III and IV (Scheme 6.2) as the backbonding capability of the {TpRuL_2} unit decreases. The complex $[Tp(dppf)Ru=C=C=C(Ph)(Fc)]^+$ finally poses the question as to which of the two ferrocene containing subunits, the dppf ligand or the ferrocenyl substituent, is oxidised first. IR- and UV–Vis spectroelectrochemistry allowed for an unequivocal assignment. During the first oxidation the ferrocene→π*(allenylidene) ILCT band is bleached (Figure 6.3(d)) while under IR monitoring a redshift of the allenylidene band by 20 cm^{-1} was observed. Both findings are only compatible with the oxidation of the allenylidene-bound ferrocenyl substituent. The second oxidation, now at the dppf site, shifted the allenylidene IR band back to its original position in the parent monocation. The direction of this shift is the expected one for oxidation at the dppf site. For most of these complexes the effects of further oxidation at the ruthenium atom could also be addressed by IR and UV–Vis spectroelectrochemistry. The most notable feature is the decrease of the intensity of the MLCT band along with a slight blueshift as the largely metal-based donor orbitals are depopulated and energetically stabilised.

    Spectroelectrochemistry has also allowed discrimination between metal- or ligand-dominated oxidations in the 1,3-divinylphenylene-bridged diruthenium

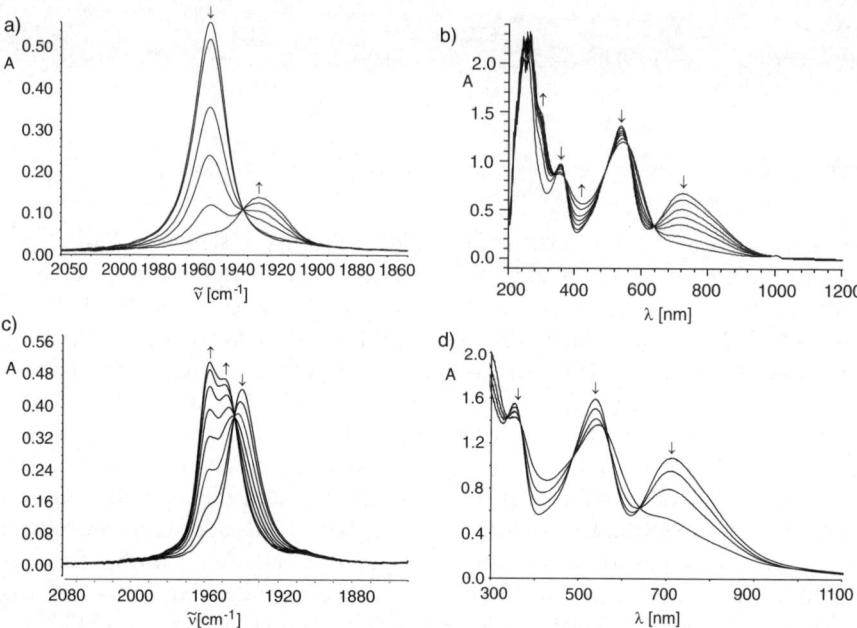

Figure 6.3 Spectroelectrochemical investigations on $[Tp(L_2)Ru=C=C=C(Ph)(R)]^+$ complexes bearing electroactive substituents ($R=Fc$) and/or ligands ($L_2=dppf$). a) IR and b) UV–Vis traces for the first oxidation of $[Tp(PPh_3)_2Ru=C=C=C(Ph)(Fc)]^+$; c) IR traces during the first oxidation of $[Tp(dppf)Ru=C=C=C(Ph)_2]^+$ and d) UV–Vis-spectroelectrochemistry on the first oxidation of $[Tp(dppf)Ru=C=C=C(Ph)(Fc)]^+$.

complex *E,E-trans*-{(PPh$_3$)$_2$(CO)Cl(4-EtOOCpy)Ru}$_2$(μ-CH=CH–C$_6$H$_4$–C=CH-1,3) (4-EtOOCpy = ethylisonicotinate) to be made.[43] This complex undergoes two consecutive one-electron oxidations that are separated by 320 mV. As was shown by IR spectroelectrochemistry, the initial single CO band splits into two resolved features at the mono-oxidised state. These new bands are blue-shifted by 15 and 37 cm^{-1} with respect to the neutral. During the second oxidation both bands collapse to a single feature, which is now at 46 cm^{-1} higher energy as in the neutral starting complex. Much smaller shifts of 3 and 4 cm^{-1}, respectively, of the isonicotinate ester band were noted for the first and second oxidations. Overall, the CO band shifts are much smaller as is expected for a metal-based process, and this suggests a sizable contribution of the bridging ligand to the redox orbital.[ii] UV–Vis–NIR and ESR spectroelectrochemistry revealed that the oxidation is even dominated by the bridge. Structured low-energy absorption bands in the visible and the near-IR grow in upon mono-oxidation (Figure 6.4) and closely resemble those observed for radical cations or anions of diphenylethylene or distyrylbenzene. The *in-situ* electrogenerated

---

[ii] Removal of an electron from a mononuclear carbonyl complex normally causes CO band shifts of 100 cm^{-1} and more.

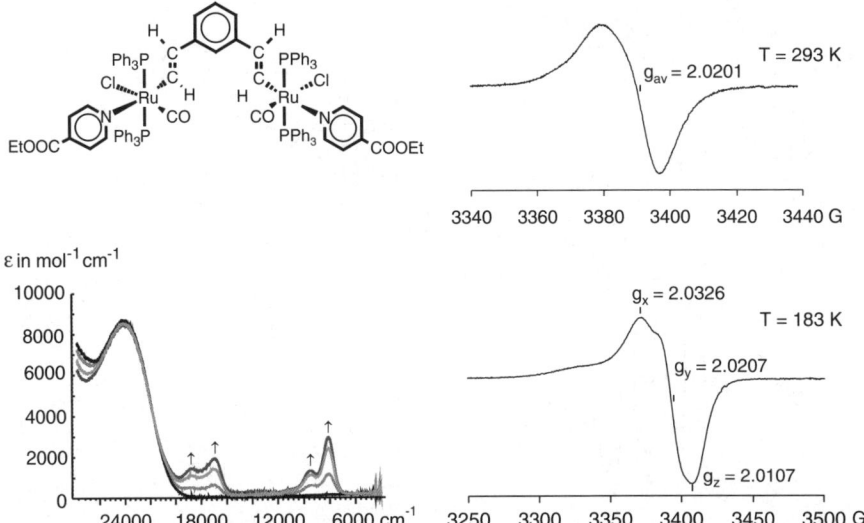

Figure 6.4 UV–Vis–NIR traces recorded during the *in-situ* mono-oxidation of *E,E-trans*-{(PPh$_3$)$_2$(CO)Cl(4-EtOOCpy)Ru}$_2$(μ-1,3-CH = CH–C$_6$H$_4$-CH = CH) (bottom left) and ESR spectra recorded at different temperatures.

monocation displays a strong isotropic ESR spectrum in fluid solution, which clearly argues for a ligand-centred paramagnetic species. Some rhombic splitting of the *g*-tensor in the frozen state and a deviation of the isotropic *g*-value (2.0201) from that of the free-electron (g_{el} = 2.0023) are nevertheless in line with a detectable metal contribution to the SOMO of the mono-oxidised complex. Quantum-chemical calculations agree with this view.

Ambiguities as to the identity of the redox site may also arise in complexes bearing different electroactive ligands. The work of Adams and coworkers[44] on 2,2′-bipyridine ruthenium and platinum complexes with 4-nitrophenylethynyl ligands provides instructive examples of the use of spectroelectrochemistry for deciphering the sequence of redox events. Thus, the ESR spectrum of the *in-situ* generated [(4,4′-^{t}Bu$_2$bipy)Pt(−C≡C-C$_6$H$_4$NO$_2$-4)$_2$]$^{2-}$ strongly deviates from that of the related phenylacetylide congener [(4,4′-^{t}Bu$_2$bipy)Pt(−C≡C–C$_6$H$_5$)$_2$]$^-$ in displaying much larger couplings of the unpaired spin to ^{14}N (10.1 compared to 3.4 G) and smaller coupling to the Pt^{2+} ion (15.8 *versus* 21.0 G). Moreover, the unpaired spin is coupled to just one ^{14}N nucleus in the nitrophenyl-substituted dianion as opposed to two ^{14}N nuclei in the case of the reduced phenylacetylide complex. This suggests reduction at the nitro groups of the alkynyl ligand for the nitrophenyl but at the bipy site for the phenylethynyl complex (Figure 6.5). Differences in the UV–Vis spectroscopic traces recorded during reduction inside an OTTLE cell point in the same direction. Whereas [(4,4′-^{t}Bu$_2$bipy)Pt(−C≡C–C$_6$H$_5$)$_2$]$^-$ displays the typical signatures of an anionic bipy $^·$$^-$-type ligand, such features are completely absent from the spectra of

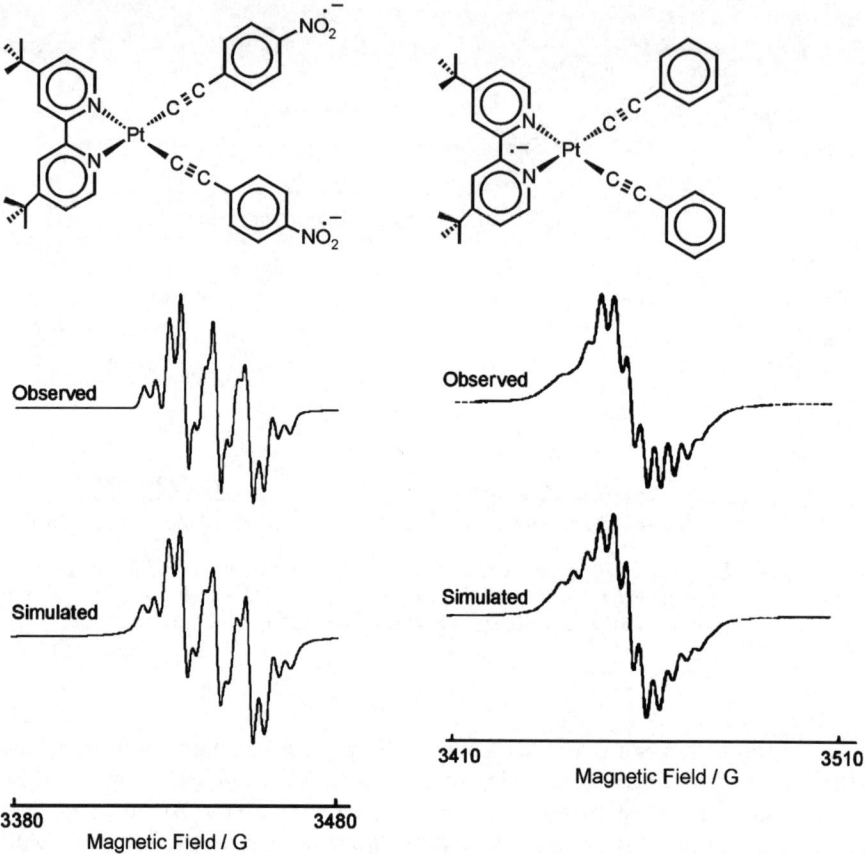

Figure 6.5 Comparison of the ESR spectra of the *in-situ* generated reduced forms of
$(4,4'-{}^tBu_2bipy)Pt(-C\equiv C-C_6H_4NO_2-4)_2$ and of $(4,4'-{}^tBu_2bipy)Pt(-C\equiv C-C_6H_5)_2$. Reproduced with permission from ref. 44.

$[(4,4'-{}^tBu_2bipy)Pt(-C\equiv C-C_6H_4NO_2-4)_2]^{2-}$.[44] The ruthenium complexes $[(4,4'-X_2bipy)(PPh_3)_2RuCl(-C\equiv C-C_6H_4NO_2-4)]$ (X = Me, Br, I) and $[(4,4'-X_2bipy)(PPh_3)_2Ru(-C\equiv C-C_6H_4NO_2-4)_2]$[45] behave in an identical fashion.[45]

6.2.2   Spectroelectrochemistry as a Probe for Metal–Ligand Bonding

6.2.2.1   *The Issue of Backbonding in Alkynyl Complexes*

The nature of the metal–alkynyl bond has long been a matter of debate, especially with regard to the metal–C(sp) bond order. Multiple bonding could in principle arise from interactions between occupied alkynyl π levels and empty d-orbitals at the metal or from backbonding between filled metal d-orbitals and unoccupied π* levels on the alkynyl ligand. Early photoelectron spectroscopy work on

$Cp(CO)_2Fe–C\equiv CR$ ($R = H$, Ph, ^tBu, $C\equiv CH$) and $Cp^*(CO)_2Fe–C\equiv CH$ concluded that alkynyl-to-metal π donation is significant and destabilises the metal d_π levels due to filled–filled interactions, while metal-to-ligand π^* backdonation is very weak.[46,47] The same overall picture obviously applies to bis(alkynyl) complexes *trans*-M($-C\equiv CR$)$_2$(PR$_3$)$_2$ of palladium and platinum.[48] Application of alkynyl complexes in the fields of nonlinear optics and photoluminescence has given new impetus for research into this topic. Countless new derivatives have meanwhile been synthesised, many of them strongly polarised with pronounced donor/acceptor characteristics. It soon became evident that in complexes of electron-rich metal ions and electron-poor acceptor-substituted alkynyl ligands metal-to-alkynyl π^* backdonation may significantly contribute to the overall bonding.[49,50] Clear experimental evidence for backbonding to alkynyl ligands comes from the work of Hopkins and coworkers[51] who compared the $C\equiv C$ stretching frequencies of $Mo_2(PMe_3)_4(-C\equiv CSiMe_3)_4$ (MoMo) and its chemically reduced monoanion (MoMo$^-$) in their ground and electronically excited states (MoMo*, MoMo$^-$ *) (Scheme 6.3). In these Mo–Mo quadruply bonded systems, the δ and δ^* orbitals have appropriate symmetry to interact with the alkynyl π^* orbitals and are involved in the low-energy excitations. With increasing occupancy of the δ^* orbital the energy of the $C\equiv C$ stretch drops from 1991 cm^{-1} in MoMo ($\delta^2\delta^{*0}$) to 1970 cm^{-1} in MoMo* ($\delta^1\delta^{*1}$), and 1954 cm^{-1} in MoMo$^-$ ($\delta^2\delta^{*1}$) to finally 1890 cm^{-1} in MoMo$^-$ * ($\delta^1\delta^{*2}$). This implies a weakening of the alkynyl $C\equiv C$ bond by increasing backdonation from the Mo$_2$ core.

Paul and coworkers have scrutinised a series of differently substituted iron half-sandwich complexes $Cp^*(dppe)Fe(-C\equiv C-C_6H_4X-4)$ and their chemically oxidised cationic Fe(III) congeners bearing *para* substituents X of variable donor or acceptor capacity. They made a convincing case for π backdonation in acceptor-substituted derivatives based on IR, Raman, Mössbauer,[52,53] and quantum-chemical studies.[54] In the oxidised Fe(III) forms, on the other hand, alkynyl to metal donation dominates the Fe-alkynyl π bonding, especially when conjugated electron-rich substituents are present at the phenyl *para* position.[55] These studies, in essence, demonstrate that the issue of the nature of the metal–alkynyl bond is not a trivial one. Every parameter, such as the metal ion, its

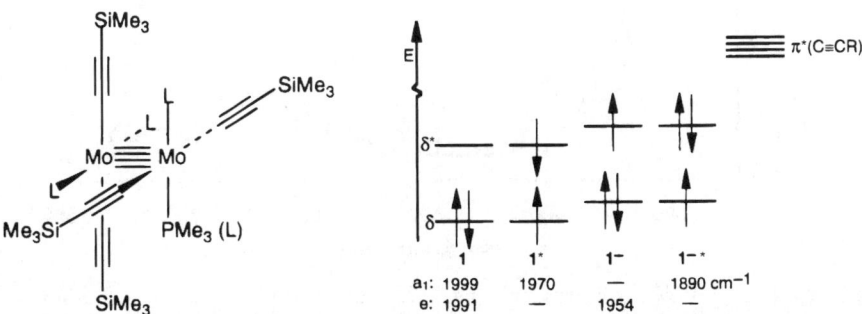

Scheme 6.3   Orbital occupation and $C\equiv C$ IR frequencies in quadruply bonded {Mo($-C\equiv C$-TMS)$_2$L$_2$}$_2$ dimers. Adapted from ref. 51 with permission.

oxidation state and the substituent on the alkynyl ligand exert their influence on the metal–alkynyl π-bond.[56] There is a subtle interplay of σ donor, π donor and π acceptor metal–alkynyl interactions with sometimes congruent, sometimes opposing contributions to the overall bonding and on the structural and spectroscopic properties.

The above examples emphasise that redox reactions have a direct bearing on the nature of the metal–alkynyl bond through changes in the occupancy of the metal d_π levels. It is therefore only natural that several alkynyl complexes have been subjected to spectroelectrochemical investigations with the aim to monitor bonding changes as a function of oxidation state. Ruthenium bis(alkynyl) complexes of the macrocyclic 1,5,9,13-tetramethyltetraazacyclohexadecane (16-TMC) ligand with differently substituted arylethynyl ligands are interesting examples.[57] These complexes undergo two consecutive oxidations at easily accessible potentials that were assigned as the Ru(II/III) and the Ru(III/IV) couples. Electrochemical or chemical oxidation of the Ru(II) congener was followed by UV–Vis–NIR spectroscopy (Figure 6.6). Upon oxidation, the initial metal→alkynyl π* MLCT band near 390 nm bleaches while an intense vibrationally structured π(alkynyl)→d_π(Ru) LMCT absorption at ca. 720 nm grows in. Vibrational progressions are due to v(C≡C) of the alkynyl ligands and range from 1730 to 1830 cm^{-1}. The strong decrease from the $\tilde{\nu}$ (C≡C) ground state values of 2002 to 2012 cm^{-1} has been taken as an indication for a

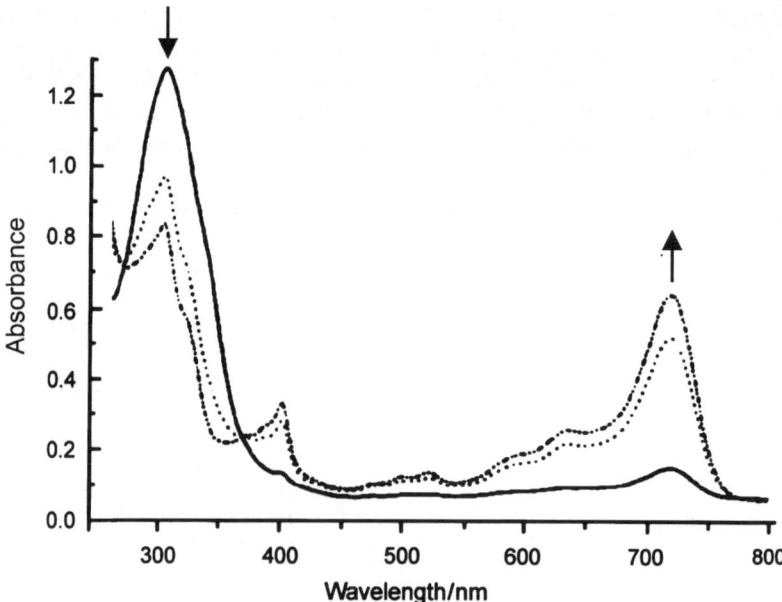

Figure 6.6 UV–Vis spectroscopic changes during electrochemical oxidation of (16-TMC)Ru(C≡C–C$_6$H$_4$Cl-4)$_2$ in an optically transparent thin-layer electrolysis cell (0.1 M ^{n}Bu$_4$PF$_6$/ CH$_2$Cl$_2$ solution). Reproduced with permission from ref. 57.

distortion of the alkynyl ligand, most probably toward a vinylidene structure, as a consequence of increased metal–carbon π-bonding in the LMCT excited state.

In the same vein, oxidation of the complexes *trans*-(dppe)$_2$RuX($-$C≡CR) (dppe = 1,2-bis(diphenylphosphino)ethane, X = Cl or C≡CPh, R = Ph or C≡CPh) inside an optically transparent thin-layer electrolysis (OTTLE) cell gives rise to an intense, structured low-energy band. Based on quantum-chemical calculations, this absorption was assigned as a mixed LLCT/LMCT transition involving charge transfer from out-of-plane π orbitals of the chloride or phenylacetylide ligands to a SOMO that is heavily delocalised across the Ru$-$C≡CR units.[7] Here, the vibrational spacings are only slightly reduced from the $v_{C≡C}$ values of the Ru(II) congeners. No attempts have, however, been made to establish the energies of the IR stretch for the ground states of the oxidised forms of any of these complexes.

A more comprehensive collection of $v_{C≡C}$ values was reported for [(4,4'-Me$_2$ bipy)(PPh$_3$)$_2$RuCl(C≡CR)]$^{n+}$ ($n = 0$, 1; R = *para*-tolyl, Ph, ^{t}Bu; 4,4'-Me$_2$ bipy = 4,4'-dimethyl-2,2'-bipyridine).[58] Oxidation of the Ru(II) congeners to their Ru(III) forms removes an electron from the plane containing the Cl, bipyridine and alkynyl ligands and thus directly influences the Ru–alkynyl bonding *via* the in-plane alkynyl π orbitals. ESR spectroscopy on chemically oxidised samples identified the metal centre as the oxidation site. Oxidised samples display their C≡C stretch at up to 100 cm^{-1} lower energies when compared to their Ru(II) forms, and this is in line with increased alkyne to ruthenium π donation in the oxidised state (*viz.* resonance forms II and III in Scheme 6.4). The crystallographically observed contraction of the Ru$-$C(sp) and the slight elongation of the alkynyl C≡C bond in the oxidised state further support this view. While chemical oxidation served the purpose of IR and ESR characterisation of the Ru(III) forms well, UV–Vis–NIR spectroelectrochemistry in an OTTLE cell was the method of choice for establishing the optical data since it avoids overlapping bands of the ferrocenium/ferrocene or any other chemical redox system. In addition to the expected bleaching of the Ru→bipy* MLCT bands near 330 to 350 cm^{-1} and at *ca.* 500 cm^{-1} the growth of a new low–energy LMCT absorption was observed (Figure 6.7). Its position strongly depends on the nature of the alkynyl substituent with a shift to lower energy as the conjugation length increases. Vibrational subbands at spacings of 1676 or 1223 cm^{-1} were observed for the aryl-substituted congeners and an even more substantial reduction of C≡C bond order in the LMCT excited state was thus proposed.

6.2.2.2 Reorganisation of π Electron Density within Conjugated or Cumulated M–L Entities

Redox reactions can induce significant redistribution of π electron density along the metal–carbon backbone when conjugated or cumulated carbon-rich ligands are present and may thus lead to major bonding changes. Particularly common are interconversions between alkynyl and cumulenic structures.

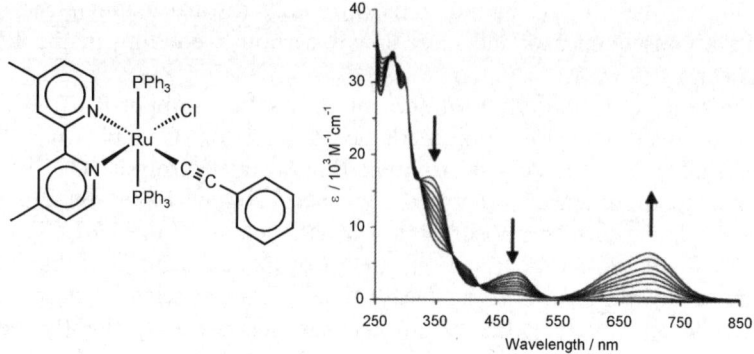

Figure 6.7 Changes in the UV–Vis spectrum of Cl(PPh$_3$)$_2$(4,4′-Me$_2$bipy)Ru–C upon oxidation in an OTTLE cell showing the bleach of the Ru→bipy MLCT bands and the growth of a structured low-energy LMCT band. Reproduced with permission from ref. 22.

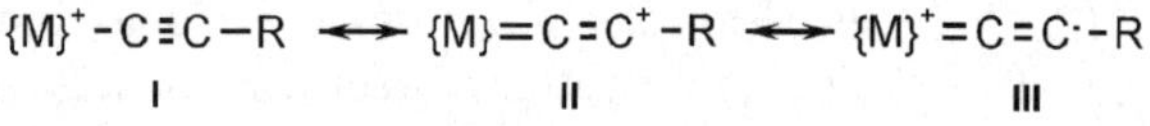

$$\{M\}^+\text{-C} \equiv \text{C-R} \longleftrightarrow \{M\} = \text{C} = \text{C}^+\text{-R} \longleftrightarrow \{M\}^+ = \text{C} = \text{C}^-\text{-R}$$

I   II   III

Scheme 6.4 Resonance forms for the oxidised forms of transition metal–alkynyl complexes.

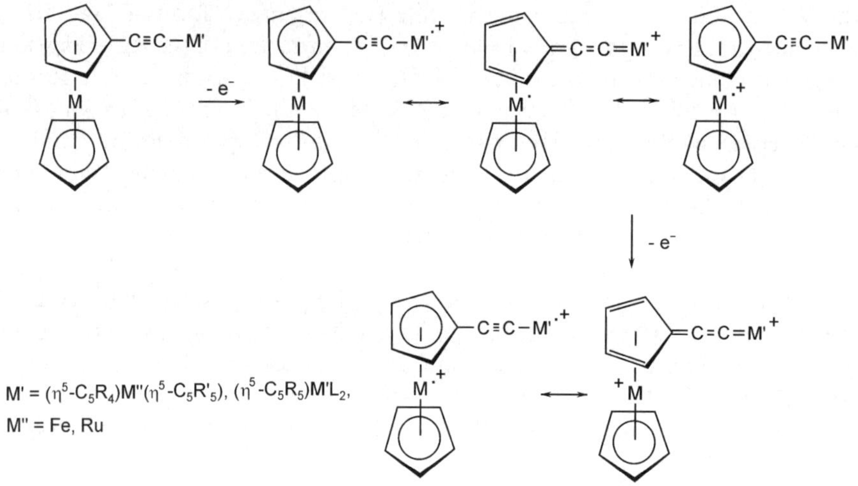

M′ = (η^5-C$_5$R$_4$)M″(η^5-C$_5$R′$_5$), (η^5-C$_5$R$_5$)M′L$_2$,
M″ = Fe, Ru

Scheme 6.5 Stepwise oxidation of complexes bearing metallocenylethynyl ligands.

Just as vinylidene-type resonance forms constitute one extreme description for the oxidised forms of alkynyl complexes (Scheme 6.4), cumulenic structures may result from the oxidation of oligoynediyl-bridged dimetal complexes (Schemes 6.5 to 6.8). Interesting examples come from the work of Sato, who

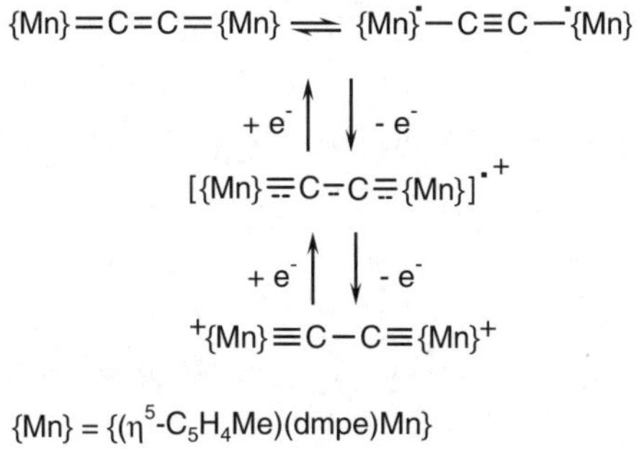

$$\{Mn\}=C=C=\{Mn\} \rightleftharpoons \{Mn\}\overset{\bullet}{-}C\equiv C-\overset{\bullet}{\{Mn\}}$$

$$+ e^- \uparrow \quad \downarrow - e^-$$

$$[\{Mn\}\equiv C=C\equiv\{Mn\}]^{\bullet+}$$

$$+ e^- \uparrow \quad \downarrow - e^-$$

$$^+\{Mn\}\equiv C-C\equiv\{Mn\}^+$$

$$\{Mn\} = \{(\eta^5\text{-}C_5H_4Me)(dmpe)Mn\}$$

Scheme 6.6   Redox-induced bonding changes in a dimanganese acetylide complex.

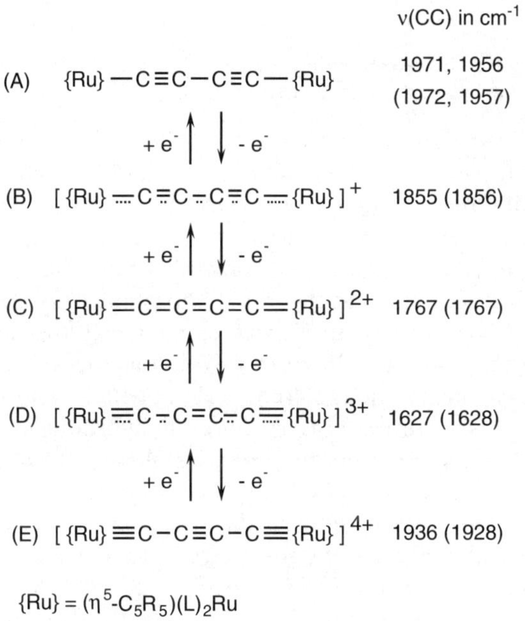

$\nu(CC)$ in cm^{-1}

(A) $\{Ru\}-C\equiv C-C\equiv C-\{Ru\}$ 1971, 1956 (1972, 1957)

$$+ e^- \uparrow \quad \downarrow - e^-$$

(B) $[\,\{Ru\}\cdots C\equiv C\cdots C\equiv C\cdots\{Ru\}\,]^+$ 1855 (1856)

$$+ e^- \uparrow \quad \downarrow - e^-$$

(C) $[\,\{Ru\}=C=C=C=C=\{Ru\}\,]^{2+}$ 1767 (1767)

$$+ e^- \uparrow \quad \downarrow - e^-$$

(D) $[\,\{Ru\}\equiv C\cdots C=C\cdots C\equiv\{Ru\}\,]^{3+}$ 1627 (1628)

$$+ e^- \uparrow \quad \downarrow - e^-$$

(E) $[\,\{Ru\}\equiv C-C\equiv C-C\equiv\{Ru\}\,]^{4+}$ 1936 (1928)

$$\{Ru\} = (\eta^5\text{-}C_5R_5)(L)_2Ru$$

Scheme 6.7   Resonance forms for complexes $\{Ru\}_2(\mu\text{-}C_4)]^{n+}$ in different oxidation states. IR data apply to $\{Ru\} = (\eta^5\text{-}C_5H_5)(PPh_3)_2Ru$ or $(\eta^5\text{-}C_5H_5)(PPh_3)$ $(PMe_3)Ru$ (data in brackets).

investigated a series of ruthenium alkynyl complexes featuring ferrocenyl- or ruthenocenylethynyl ligands in different oxidation states. Chemical oxidation of $(\eta^5\text{-}C_5H_5)(L_2)Ru-C\equiv C-Fc$ ($L_2 = 2$ PPh$_3$, dppf), $(\eta^5\text{-}C_5Me_5)$ $(PPh_3)_2Ru-C\equiv C-Fc$ and of $(\eta^5\text{-}C_5R_5)(CO)_2Fe-C\equiv C-Fc$ ($R = H$, Me) shifts the $\nu_{C\equiv C}$ band by about 90 cm^{-1} to lower energy into a position intermediate between those typical of alkynyl and allenylidene structures.[30,31] X-Ray crystallography on $[(\eta^5\text{-}C_5H_5)(PPh_3)_2Ru-C\equiv C-Fc]^+$ revealed a

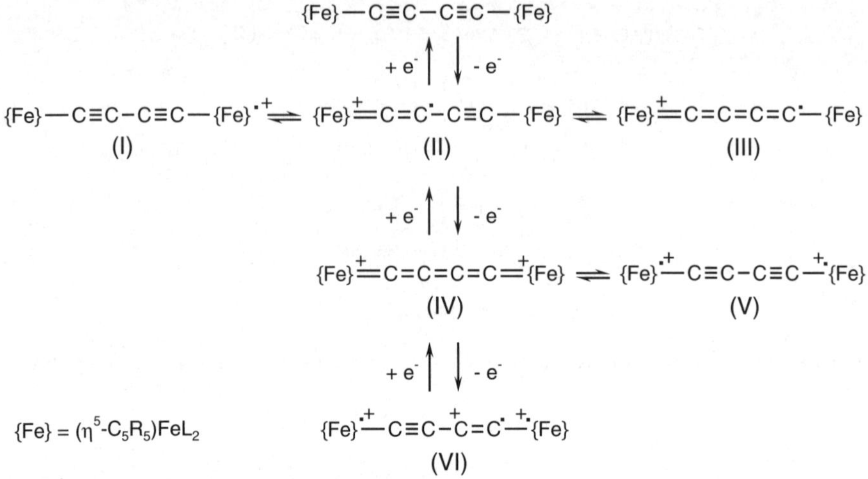

$$\{Fe\}-C\equiv C-C\equiv C-\{Fe\}$$

$$+ e^- \uparrow \downarrow -e^-$$

$$\{Fe\}-C\equiv C-C\equiv C-\{Fe\}^{\cdot+} \rightleftharpoons \{Fe\}\overset{+}{=}C=C^{\cdot}-C\equiv C-\{Fe\} \rightleftharpoons \{Fe\}\overset{+}{=}C=C=C=C^{\cdot}-\{Fe\}$$

$$\text{(I)} \qquad\qquad\qquad\qquad \text{(II)} \qquad\qquad\qquad\qquad \text{(III)}$$

$$+ e^- \uparrow \downarrow -e^-$$

$$\{Fe\}\overset{+}{=}C=C=C=C\overset{+}{=}\{Fe\} \rightleftharpoons \{Fe\}\overset{\cdot+}{-}C\equiv C-C\equiv C\overset{\cdot+}{-}\{Fe\}$$

$$\text{(IV)} \qquad\qquad\qquad\qquad \text{(V)}$$

$$+ e^- \uparrow \downarrow -e^-$$

$$\{Fe\} = (\eta^5\text{-}C_5R_5)FeL_2 \qquad\qquad \{Fe\}\overset{\cdot+}{-}C\equiv C-\overset{+}{C}=C^{\cdot}\overset{+}{-}\{Fe\}$$

$$\text{(VI)}$$

Scheme 6.8   Redox-induced bonding ranges and resonance structures for buta-diynediyl-bridged di-iron complexes.

fulvalenyl-type distortion of the substituted ferrocenyl ring (Scheme 6.5) and a shortening of the Fc−C and Ru−C bonds, while Mössbauer spectroscopy served to identify the ferrocenyl entity as the primary oxidation site. In the related ferrocenylacetylide diphosphine iron complexes $(\eta^5\text{-}C_5R_5)$ $(L)_2Fe-C\equiv C-Fc$ ($R = H$, Me; L_2 = chelating diphosphine) the order of redox events is inverted such that the iron diphosphine half-sandwich moiety is oxidised first.[34] Chemical oxidation with either the ferrocenium ion or 2,3-dichloro-5,6-dicyano-p-benzoquinone (DDQ) caused the ν (C≡C) band to shift to lower energies by again about $100\,cm^{-1}$ along with a considerable intensity increase. Similar cases are the bis(ferrocenylethynyl)ruthenium complexes *all trans*-$(FcC\equiv C-)_2RuL_4$ reported by Long, Wolf and their coworkers.[21–23] Oxidation of the outer ferrocenyl substituents again redshifts the C≡C stretch by some 50 to $90\,cm^{-1}$, depending on the coligands at the ruthenium atom. Particularly instructive are the mono-oxidised mixed bis(alkynyl) complexes *trans*-$[(dppm)_2Ru(-C\equiv CFc)(-C\equiv CC_6H_4R\text{-}4)]^+$ ($R = H$, Me, Ph, NO_2). These show one C–C stretch at around $1985\,cm^{-1}$ for the FcCC and one at *ca.* $2070\,cm^{-1}$ for the CCAryl groups, the latter being almost unshifted with respect to the neutrals.[21] Oxidising the ferrocenylethynyl entity of complexes $(PPh_3)_2(Aryl)Pt(-C\equiv CFc)$ has a considerably smaller effect and this can be traced to the reluctance of the platinum centre to engage in Pt–carbon multiple bonding.[33]

Oxidation of both terminal sites was finally achieved in the alkynyl-bridged bis(ruthenocenes) $\{(\eta^5\text{-}C_5R^1{}_5)Ru(\eta^5\text{-}C_5R^2{}_4)\}_2(\mu\text{-}C\equiv C)$[58] and the ruthenocenylethynyl half-sandwich ruthenium complexes $(\eta^5\text{-}C_5R^1{}_5)Ru(\eta^5\text{-}C_5R^2{}_4)-C\equiv C-Ru(P_2)L$ ($R^1 = H$, $R^2 = Me$, $P_2 = 2$ Ph$_3$, dppe, L = Cp, Cp*, indenyl; $R^1 = R^2 = H$, Me, $P_2 = $ dppe, 2 PPh$_3$, L = indenyl).[59,60] Multiple pieces of evidence support cumulenic structures in the dioxidised states (Scheme 6.5). Characteristic attributes are the extreme low field shift of the cyclopentadienylidene carbon atoms of more than 300 ppm and a low-energy shift of the C≡C stretch of more than $200 \, cm^{-1}$ when compared to the neutrals. Structural comparison between the neutral and the dioxidised forms of $\{(\eta^5\text{-}C_5Me_5)Ru(\eta^5\text{-}C_5H_4)\}_2(\mu\text{-}C\equiv C)$ and of $[(\eta^5\text{-}C_5H_5)(dppe)Ru\text{-}C\equiv C\text{-}(\eta^5\text{-}C_5Me_4)Ru(\eta^5\text{-}C_5H_5]$ point in the same direction. In an intuitive and simple manner, the above results are readily explicable by the preference of 17 valence-electron $\{ML_n\}$ entities toward monoanionic ligands and of 16 valence-electron fragments toward neutral two electron donors. The unsubstituted C_2, and, by analogy, the homologous C_{2n} ligands can easily accommodate these preferences of the attached metal entities by either assuming dianionic (oligo)ynediide ($|C\equiv C|^{2-}$ and $C\equiv(C\text{-}C\equiv)_{n-1}C^{2-}$) or neutral cumulenic $=C=C=$ and $=C=(C=C=)_{n-1}C=$ structures, or hybrids between both, in the odd-electron, mixed-valence states.

Electron removal from metal-capped cumulenylidene chains causes further bonding changes to bis(carbyne) structures. Berke and coworkers[61] have recently reported rare examples of such redox conversions for the C_2-bridged dimanganese complexes $[\{(\eta^5\text{-}C_5H_4Me)(dmpe)Mn\}_2(\mu\text{-}C_2)]^{n+}$ (dmpe = $Me_2PC_2H_4PMe_2$, $n = 0$, 1, 2). The resonance forms in Scheme 6.6 serve to visualise the redox-induced bonding changes within the Mn_2C_2 core. The mixed-valence monocation reveals an intense Raman band at $1564 \, cm^{-1}$ that has been assigned as the ν_{CC} vibration of the C_2 ligand. Ferrocenium oxidation to the dication causes this band to shift to $1314 \, cm^{-1}$, which is indicative of a C–C stretch of a Mn≡C–C≡Mn array. This view also agrees with the diamagnetism of the dication. Neutral $\{(\eta^5\text{-}C_5H_4Me)(dmpe)Mn\}_2(\mu\text{-}C_2)$ was obtained by Na/Hg reduction and showed no Raman CC bands. Of note is the observed spin equilibrium. Magnetic measurements reveal that $\{(\eta^5\text{-}C_5H_4Me)(dmpe)Mn\}_2(\mu\text{-}C_2)$ possesses a singlet ground state with an excited triplet state at just slightly higher energy. The latter is thermally populated even at 2 K in the solid state but probably only above 183 K in solution. Crystallographic data are available for every member of this redox series. The continuous shortening of the MnC and a concomitant lengthening of the internal CC bonds with an increase in the overall oxidation state support the proposed bonding changes.

The unsubstituted ("naked") C_4 bridge offers an even larger electronic and structural flexibility when spanning two metal centres and several examples for redox-induced conversions between butadiynediyl, butatrienylidene and ethynylbis(carbyne) structures have been reported. Particularly instructive examples come from the work of Bruce *et al.*[62,63] on C_4-bridged diruthenium complexes $[\{(\eta^5\text{-}C_5R_5)(L_2)Ru\}_2(\mu\text{-}C_4)]^{n+}$ (R = H, $L_2 = 2PPh_3$ or one PPh_3 and one PMe_3; R = Me, $L_2 = $ dppm, dppe). Each complex undergoes four consecutive, well-separated one-electron oxidations. The first three waves are chemically reversible on the voltammetric timescale. All partially oxidised forms up to the trication level could be generated inside a modified infrared reflection absorption spectroscopy (IRRAS) or OTTLE cell. Each of these oxidations caused a shift of $\nu_{C\equiv C}$ by about $100 \, cm^{-1}$ to lower energies as is shown in Figure 6.8 for the $[\{(\eta^5\text{-}C_5H_5)(PPh_3)_2Ru\}_2(\mu\text{-}C_4)]^{n/(n+1)+}$ ($n/(n+1) = 0 \rightarrow 1$, $1 \rightarrow 2$ and $2 \rightarrow 3$) series. Of note is the large increase in band intensity upon the first oxidation. Taken together with the shift, this

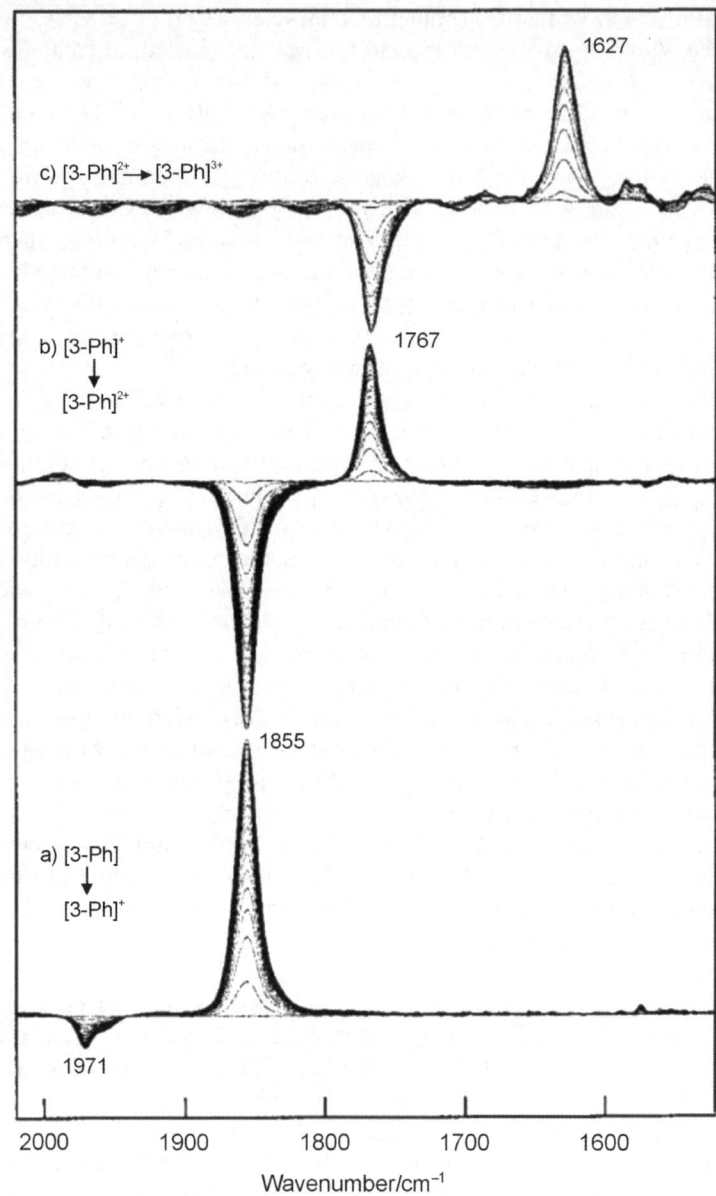

Figure 6.8   IR spectroscopic changes upon the stepwise oxidation of {Cp(PPh$_3$)$_2$ Ru}$_2$(μ-C$_4$) (3-Ph) from a) the neutral to the monocation, b) the mono-cation to the dication, and c) the dication to the trication in an IRRAS cell. Negative peaks indicate consumption of the starting material while positive peaks apply to the electrogenerated products. Reproduced with permission from ref. 62.

absorptivity increase provides clear evidence for large contributions of a cumulenic structure (Scheme 6.7). Weakening of the internal CC bonds continues up to the trication level. At $-40°C$ it was even possible to electrogenerate and IR spectroscopically characterise the tetracations $[\{(\eta^5\text{-}C_5H_5)(PPh_3)_2Ru\}_2(\mu\text{-}C_4)]^{4+}$ and $[\{(\eta^5\text{-}C_5H_5)(PPh_3)(PMe_3)Ru\}_2(\mu\text{-}C_4)]^{4+}$.[62] This last oxidation step is accompanied by a marked increase of $v_{C\equiv C}$ by about $300\,cm^{-1}$, from 1627 or 1628 to 1936 or 1928 cm^{-1}, respectively. The tetracations are thus best described by the ethynylbiscarbyne structure (structure E in Scheme 6.7). Comparisons between the crystallographically determined structures of $[\{(\eta^5\text{-}C_5Me_5)(dppe)Ru\}_2(\mu\text{-}C_4)]$ and its mono- and dioxidised forms confirm the conclusions from IR spectroelectrochemistry in showing successive shortenings of the Ru$-$C and the internal $C^2$$-$$C^3$ bonds along with a lengthening of the outer $C^1$$-$$C^2$ and $C^3$$-$$C^4$ bonds. All CC bonds become virtually identical at the dication state.[63] Quantum-chemical calculations on various models nicely reproduce these experimental findings, especially with regard to the bond sequence along the Ru$-$$C_4$$-$Ru chain. They also indicate an increasing contribution of the $\{CpL_2Ru\}$ entities to the frontier orbitals as more electrons are removed. In accordance with the proposed cumulenic and ethynylbis(carbyne) structures, every even-electron system possesses a diamagnetic ground state. Electronic spectroscopy also revealed the presence of intense $\pi \rightarrow \pi^*$-type absorption bands for the di- and trications. The mixed-valence mono- and trioxidised forms also exhibit sharp and intense intervalence charge-transfer bands that are typical of strongly coupled Class III systems.[64] These bands are of pivotal importance for assessing the degree of the electronic coupling between the metal centres as we will discuss in Section 6.2.3.

Very similar findings have been made for C_4-bridged complexes of rhenium, iron, or manganese. For the latter metal, rigid-rod like $[\{(X)(dmpe)_2Mn\}_2(\mu\text{-}C_4)]^{n+}$ have been prepared in three different oxidation levels ($n = 0, 1, 2$; dmpe $= Me_2PC_2H_4PMe_2$, X $=$ I, CCR, (R $=$ H, SiMe$_3$).[65–67] Interconversions between the different redox states were achieved by chemical oxidation/reduction with the ferrocenium ion or cobaltocene or by comproportionation reactions between the extreme redox states. The iodo complexes have been structurally characterised in all three oxidation levels while the structures of the mono- and dications are available for the (trimethylsilyl)ethynyl-substituted complex. Structural variations within the central $\{Mn\}_2(\mu\text{-}C_4)$ core closely resemble those observed for the above diruthenium complexes, that is a diynediyl-type structure in the neutral and a cumulenic butatrienylidene one in the dioxidised state, while the monocations assume an intermediate position.[65] The rhenium complex $\{(\eta^5\text{-}C_5Me_5)(NO)(PPh_3)Re\}_2(\mu\text{-}C_4)$ has likewise been converted to the dication by chemical oxidation with Ag$^+$ while the intermediate redox state was generated by comproportionation of the neutral and the dication. Crystallographic and quantum-mechanical investigations are again indicative of successive interconversions between conjugated $\{Re\}-C\equiv C-C\equiv C-\{Re\}$ and cumulated $^+\{Re\}=C=C=C=C=\{Re\}^+$ forms. Raman spectroscopy revealed a stepwise low-energy shift of the CC stretch from $2056\,cm^{-1}$ in the neutral to $1990\,cm^{-1}$ in the monocation and to $1883\,cm^{-1}$ in

the dication.[68-70] Oxidatively induced transformations between butatrienylidene ($\{M\} = C = C = C = C = \{M\}^{2-}$ and ethynylbis(carbyne) structures $\{M\}\equiv$
$C-C\equiv C-C\equiv\{M\}$ have also been reported for the $\{Tp'(CO)_2M\}$ (M = Mo, W)
complexes (Tp' = hydridotris(3,5-dimethylpyrazolyl)borate).[71] Again, these
redox transformations were effected by chemical redox agents.

The C_4-bridged di-iron complexes $\{(\eta^5\text{-}C_5Me_5)(L_2)Fe\}_2(\mu\text{-}C_4)$ (L_2 = dppe,
dippe, dippe = $^iPr_2PC_2H_4P^iPr_2$) behave, however, differently in that they form
butadiynediyl-bridged dications upon dioxidation with even stronger $C\equiv C$
bonds as in the neutral. Evidence comes from the blueshift of $v_{C\equiv C}$ upon
dioxidation.[72,73] This is a direct consequence of a distinctly higher metal
character of the occupied frontier levels as is expressed by the dominant contributions of resonance forms I and V in Scheme 6.8. In keeping with a
description as a largely metal-based diradical, the diamagnetic ground and
the paramagnetic triplet states are separated by only a small energy gap.
Replacement of the dppe by the stronger donor ligand dippe even allowed the
highly oxidised trication to be accessed. ESR investigations indicate an unusual
quartet ground state with two unpaired spins on the iron centres and one on the
all-carbon bridge (resonance form VI in Scheme 6.8).[74]

Considerably less is known about redox-induced bonding changes in complexes with even longer all-carbon chains. Chemically oxidised $[\{(\eta^5\text{-}$
$C_5Me_5)(dppe)Fe\}_2(\mu\text{-}C\equiv C-C\equiv C-C\equiv C-C\equiv C)]^+$ displays an average redshift
of the $v_{C\equiv C}$ bands of $200\,cm^{-1}$ when compared to the neutral. Based on this
observation a similar reorganisation of π-electron density as in the C_4-bridged
analogues and a structure intermediate between a polyacetylenic and a cumulenic form have been proposed.[75]

Extending the butadiynediyl ligand by inserting aromatic spacers such as
phenylene-1,4-diyl or thiophene-2,5-diyl leads to similar phenomena with
bis(alkynyl) structures in the reduced state and increasing contributions of
cumulenic/quinoidal structures in the oxidised ones. Evidence comes from IR
spectroscopic investigations as well as quantum-chemical calculations on series
of different redox congeners, extending from the neutral to the mono-oxidised[76]
or di-oxidised states.[77-79] Just as is the case for the C_4-bridged systems, half-
sandwich iron end groups have a high tendency toward spin localisation on the
metal. This reduces the contribution of the cumulenic resonance forms and
renders the dications ESR active.[77-79] *Meta*-connections at the interspersed
phenylene ring lead to a high-spin ground state with a thermally populated
excited low-spin configuration.[80-82] None of the oxidised forms were, however,
generated by spectroelectrochemical techniques.

Reductively induced alterations from cumulenic to alkynyl resonance structures have been observed for mononuclear and dinuclear ruthenium allenylidene
complexes. The half-wave potentials for the one-electron reduction of allenylidene complexes $[\{Ru\} = C = C = C(ER_n)(R')]^+$ ($\{Ru\}$ = *trans*-Cl$(L_2)_2$Ru; L_2 =
chelating diphosphine; ER_n = NR_2, SR, SeR, aryl, alkyl; R' = aryl, alkyl)
strongly depends on the nature of the ER_n substituent. Amino- and aryl-
substituted congeners with reduction potentials of *ca.* $-2.2\,V$ and $-1.0\,V$, respectively, constitute the two extremes within this series.[36,37] These sizable potential

variations indicate that the allenylidene ligand is strongly involved in the reduction process. Quantum-chemical calculations agree with this view.[37,83] Experimental confirmation comes from IR-, UV–Vis-, and ESR spectro-electrochemistry. Vibrational spectroscopy has revealed that reduction leads to distinct bonding changes within the metallacumulene chromophore. Irrespective of the substituents attached to the terminal carbon atom, reduction causes the collapse of the strong and intense allenylidene band and the growth of a considerably weaker absorption at 2065 to 2045 cm^{-1}, as is typical of the alkynyl C≡C stretch in ruthenium(II) complexes (Figure 6.9(a)). An exception is the diphenylallenylidene complex *trans*-Cl(dppm)$_2$Ru$=$C$=$C$=$CPh$_2$, which upon reduction displays a much smaller IR shift from 1930 to 1968 cm^{-1} along with an only moderate decrease in band intensity (Figure 6.9(b)).[36] This may be due to a larger confinement of the unpaired spin onto the phenyl substituents. *In-situ* ESR spectroscopy on the electrogenerated neutral radicals has yielded intense

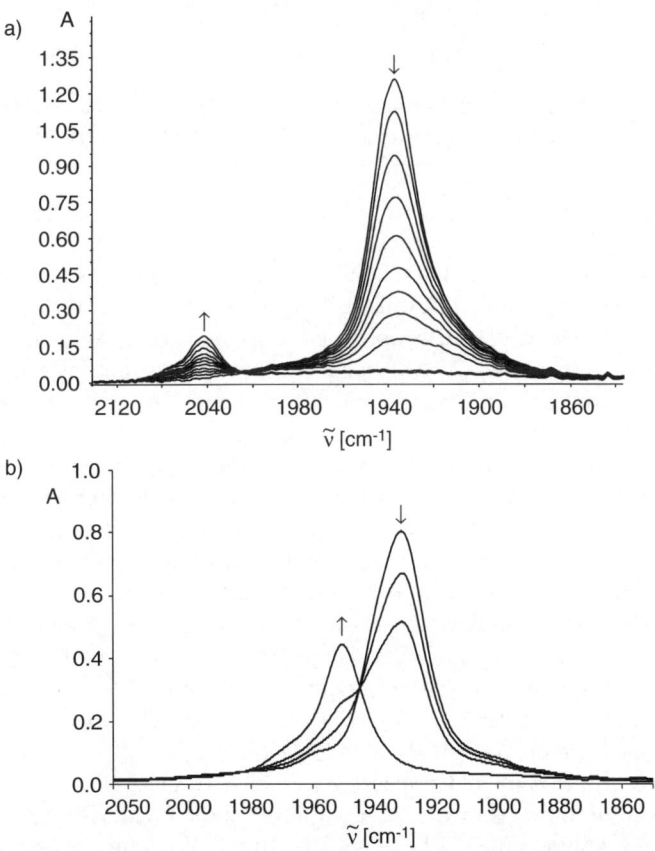

Figure 6.9 IR spectroelectrochemical reduction of a) *trans*-[Cl(dppm)$_2$Ru$=$C$=$C$=$C(SeFc)(C$_4$H$_7$)]$^+$ (reproduced with permission from ref. 39 and b) *trans*-[Cl(dppm)$_2$Ru$=$C$=$C$=$CPh$_2$]$^+$.

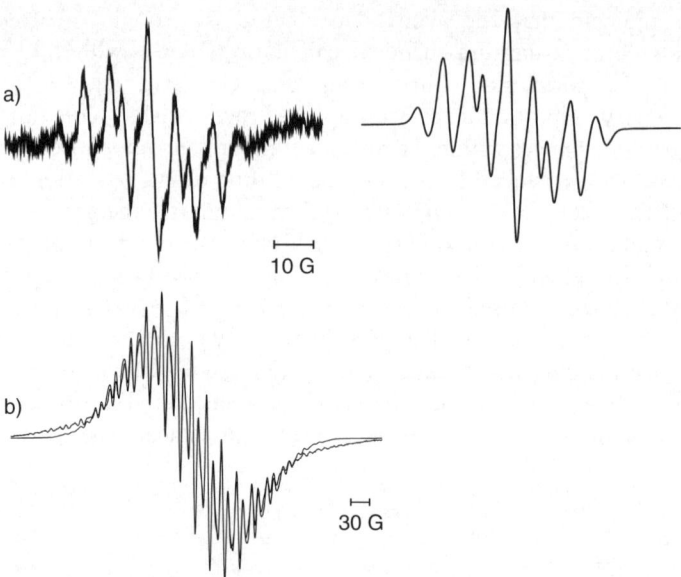

a)

$\vdash\!\!-\!\!\dashv$
10 G

b)

$\vdash\!\!\dashv$
30 G

Figure 6.10   ESR spectra of electrogenerated $Cl(dppm)_2Ru=C=C=C(SeFc)(C_4H_7)$·
(left: experimental, right: simulated) and of $Cl(dppm)_2Ru=C=C=CPh_2$·
(overlay of experimental and simulated spectra).

and richly structured spectra from fluid solutions. This finding strongly supports reduction on the unsaturated ligands and argues against ruthenium-centred odd-electron species. The observed coupling patterns show, however, interesting differences. When $ER_n = NR_2$,[84] SR[85] or SeR,[39] hyperfine splittings to the phosphorus nuclei amount to some 7 G. Once a phenyl group is attached to the terminal carbon atom, the ^{31}P hyperfine splitting decreases to 2.39 G while large couplings to the *ortho* and *para* hydrogens on the phenyl groups are observed (Figure 6.10).[36,86,87] When the reduction of *trans*-[$Cl(dppm)_2Ru=C=C=C(Ph)(R)$][+] was monitored by UV–Vis spectroelectrochemistry, two strong and sharp bands at around 330 and 350 nm were observed (Figure 6.11), and these strongly parallel those found in aryl methyl radicals. In the absence of phenyl substituents no such bands were present. It was also noted that the electronic spectra of the reduced forms of such allenylidene complexes and those of alkynyl complexes of ruthenium(II) closely resemble each other.[36,39,85] The complexes [$TpL_2Ru=C=C=C(Ph)(R)$][+] ($L_2 = 2\ PPh_3$ or dppf, R = Ph or ferrocenyl)[42] and dinuclear derivatives [$\{Ru\}=C=C=C(R')–X–(R')C=C=C=\{Ru\}$][2+] with a connecting conjugated π-spacer give similar results.[87,88]

  Reduction of diaryl-substituted complexes also gives rise to interesting chemistry. Trapping the radical formed from cobaltocene reduction of [$Cl(dppe)_2Ru=C=C=CR_2$][+] with Ph_3SnH gives the alkynyl complexes $Cl(dppe)_2Ru–C\equiv C–CR_2H$ (R = Me, Ph). When the same sequence was applied to the "cumulogous" pentatetraenylidene, regioselective formation of the

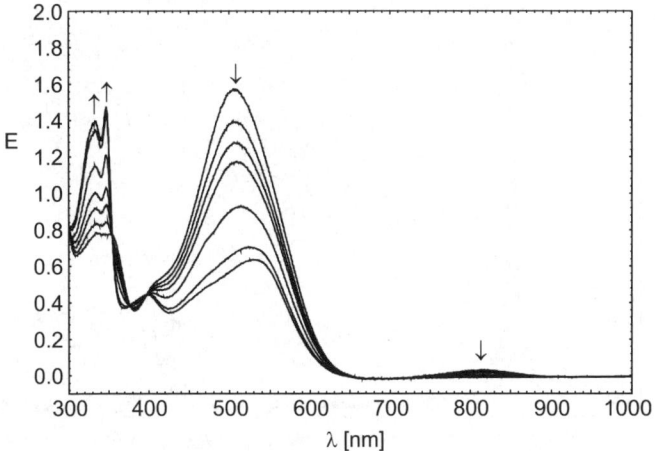

Figure 6.11 Spectroscopic traces during the reduction of *trans*-[Cl(dppm)$_2$Ru=C= C=CPh$_2$]$^+$ in an OTTLE cell.

corresponding substituted butadiynyl derivative Cl(dppe)$_2$Ru–(C≡C)$_2$–CPh$_2$H was observed.[86]

Redistribution of π-electron density as a consequence of redox processes is not restricted to unsubstituted all-carbon chains and their elongated arylene-bridged congeners. In essence, any conjugated C$_n$R$_m$ ligand may give rise to such phenomena. Well-documented examples are the dimetallapolyenes M–CR(=CH– CH)$_n$=CR–M. These are commonly oxidised in two consecutive one-electron steps to finally give bis(carbenes) $^+$M=CR–(CH=CH)$_n$–CR=M$^+$ where the metal centres are bridged by an olefinic ligand with one C=C bond less than in the starting material. Evidence for such bond reorganisation mainly rests on X-ray crystallography of different redox congeners and quantum-chemical calculations.[89–96] Both indicate a shortening of the metal–carbon bonds from typical single to double bond values as well as a shortening of the inner and lengthening of the outer CC bonds of the polyenic ligands.[89,90,93,95,96] Spin distributions in the intermediate redox states have been probed by ESR spectroscopy and revealed interesting differences depending on the attached metal entities. In [{(η^5-C$_5$H$_4$Me) (CO)$_2$Mn}$_2${μ-C(OEt)–CH=CH–C(OEt)}]$^{·−}$,[89,90] [{(η^5-C$_5$H$_5$)(CO)(PMe$_3$)Fe}$_2$ (μ-C$_4$H$_4$)]$^{·+}$ and the two diastereoisomers of [{(η^5-C$_5$H$_5$)(CO)(PPh$_3$)Fe}$_2$ (μ-C$_4$H$_4$)]$^{·+}$[92] hyperfine coupling to the metal (Mn) and the phosphine P nuclei exceeds that to the hydrogen atoms on the bridge. The opposite is true for the radical cation [{(PPh$_3$)$_2$(CO)(Cl)Ru}$_2$(μ-C$_4$H$_4$)]$^{·+}$, where ^{31}P and ^{99,101}Ru hyperfine splittings are smaller than those to the bridge protons.[96] This is suggestive of a larger metal contribution to the SOMO in the 3d metal complexes. Additional support comes from IR spectroscopy. Owing to the synergistic nature of the metal–carbonyl bond, CO band shifts provide a highly sensitive measure of the electron density at a metal atom. In the ruthenium complexes {(PPh$_3$)$_2$(CO)(Cl)(L)Ru}$_2$(μ-C$_4$H$_4$) the overall CO band shift upon dioxidation

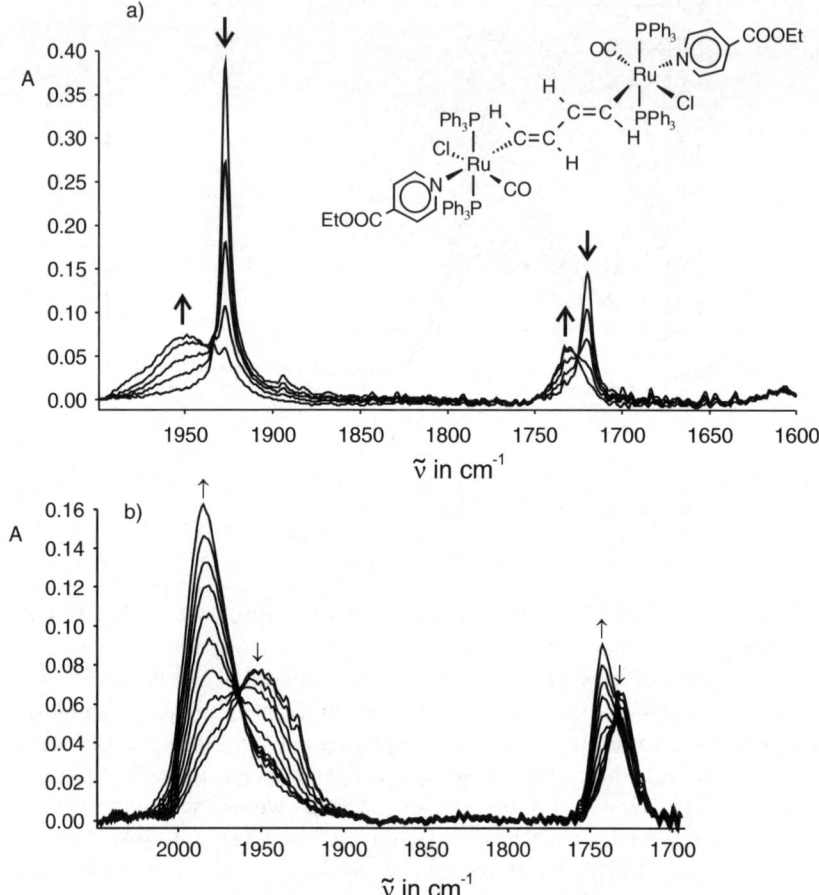

Figure 6.12   IR spectroelectrochemistry on {(PPh$_3$)$_2$(CO)(Cl)(4-EtCOOpy)Ru}$_2$(μ-C$_4$H$_4$); a) first oxidation and b) second oxidation.

(*ca.* 55 cm^{-1})[96] is considerably smaller than those for the C$_4$R$_4$-bridged di-iron (*ca.* 100 cm^{-1}) and dimanganese (*ca.* 90 cm^{-1}) derivatives. Only the ruthenium complexes were investigated by means of spectroelectrochemistry. IR traces acquired during the stepwise oxidation of {(PPh$_3$)$_2$(CO)(Cl)(4-EtCOOpy) Ru}$_2$(μ-C$_4$H$_4$) to the mono- and dication are displayed in Figure 6.12. In that specific example, the isonicotinate ester band provides an additional spectroscopic label for monitoring the oxidation-induced electron density changes on the metal. The di-iron complex [{(C$_5$Me$_5$)(dppe)Fe}$_2$(μ-C(OEt)CHCH (OEt)C) is unique among this series in that it displays a spin equilibrium between the antiferromagnetically coupled singlet and thermally excited, ferromagnetically coupled triplet forms with a larger energy gap between these states in solution.[91] This state ordering is opposite that for the C$_4$-bridged derivatives.

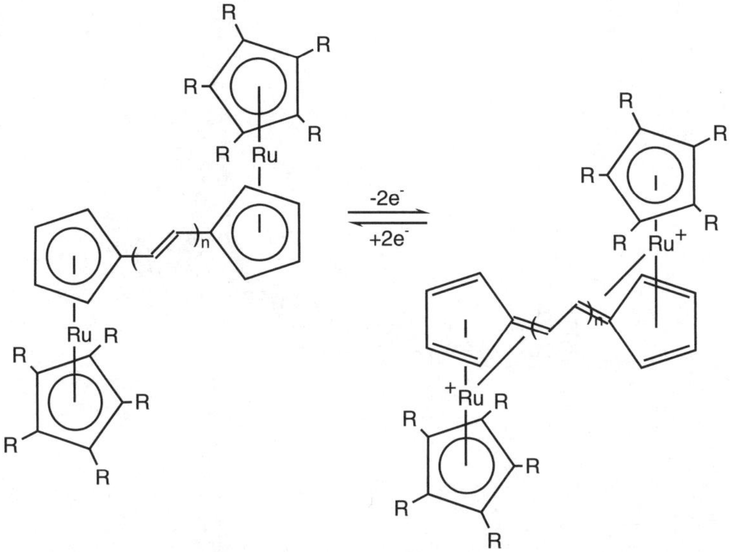

Scheme 6.9 Redox-induced bonding range in a vinyl bridges bis(ruthenocene).

The terminal olefinic carbon atoms may also be part of a metallocenyl cyclopentadienyl ring, and redox-induced interconversions between neutral bis(ruthenocenyl)oligoenes and dicationic (cyclopentadienyl)pentafulva-oligoene diruthenium complexes have been reported (see Scheme 6.9).[97-99] X-Ray crystallography on the oxidised dications clearly shows a bending of the fulvalenic carbon atoms toward the metal, thus allowing for significant contacts to the ruthenium atom. There is also substantial shortening of the C_5H_4–C and the former C–C bonds as well as elongation of the former olefinic C=C bonds upon oxidation. Further support for fulvalenepentadiene-type substructures in the oxidised forms comes from NMR spectroscopy and from quantum-chemical calculations.[99] A combination of bulk electrolysis and Raman spectroscopy provided yet another convincing piece of evidence for reorganisation of the π-electron density within the polyolefin ligand. Upon oxidation, the original C=C band at $1641 \, cm^{-1}$ is gradually replaced by a new, stronger band at $1540 \, cm^{-1}$. These changes are completely reversed during reduction (see Figure 6.13).[99] Bis(ferrocenyl)polyenes[100,101] behave quite differently. The largely metal-centred nature of the oxidation process renders these systems diradicals with 17 valence electron ferrocenium sites such that oxidation does not involve reorganisation of the π-system.[102]

Redox-induced changes of the bonding sequence may also affect conjugation between individual subunits. Examples are dyads of ferrocenylethynyl (FcC≡C) and acridine, acridone or anthraquinone units as were described by Robinson and coworkers.[103] Due to a symmetry mismatch between the FcC≡C substituent and the polycyclic arene there is little conjugation between them in the parent state. Optical spectra are thus simply the sum of the absorptions characteristic of their individual FcC≡C and arene constituents without any

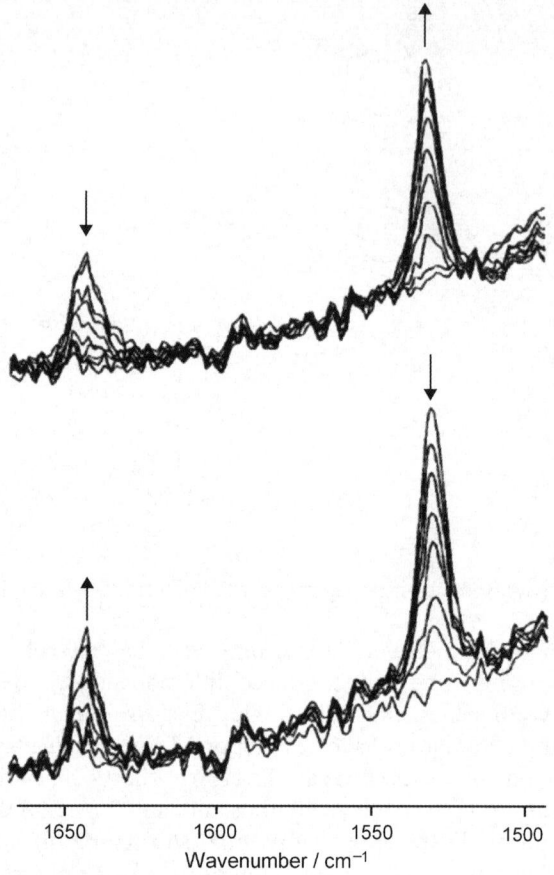

Figure 6.13 Evolution of Raman spectra during the bulk oxidation of *E*- $(C_5Me_5)Ru$ $(C_5H_4–CH = CH–C_5H_4)Ru(C_5Me_5)$ (upper trace) and upon back-reduction (lower trace). Reproduced with permission from ref. 98.

additional charge-transfer band. Likewise, $\nu_{C\equiv C}$ of the $FcC\equiv C$ entity is essentially unperturbed by the attached fluorophore. Oxidation was performed in an OTTLE cell and established the appearance of low-energy bands that were assigned as involving a charge transfer from the organic dye to the oxidised $FcC\equiv C^+$ site. A redshift of this band with an increasing donor capacity of the attached arene and a large negative solvatochromism are consistent with this view. Increased conjugation in the oxidised state as expressed by cumulenic resonance forms also has a strong bearing on the luminescence of these dyads, and this aspect will be discussed in Section 6.2.4. Despite irreversible reduction waves in cyclic voltammetry, weak ESR signals of the reduced forms could be recorded on samples that were generated by *in-situ* electrolysis at low temperature. The *g* values and hyperfine splittings are virtually identical to those of the reduced forms of the respective fluorophore, confirming localisation of the LUMO on the polyarene.

6.2.3 Spectroelectrochemistry in the Assessment of the Electronic Coupling Between Redox Sites

6.2.3.1 General Comments

Much of the work on di- or polymetallic complexes with unsaturated, conjugated carbon-rich ligands was motivated by these ligands' ability to provide efficient pathways for electronic interactions (often referred to as "communication") between the redox-active end groups. This makes them interesting candidates for molecular wires[6,104–106] and, in a broader context, as constituents of miniaturised devices in the emerging field of molecular electronics. Shorter-chain oligoynediyl bridges are particularly popular since they combine, in a highly favourable manner, convenient access routes, efficient electron delocalisation *via* two orthogonal π-systems regardless of the torsional angle between the individual subunits and effective control of the electron-transfer distance. Oligoenediyl bridges, though less common, may even give superior performance owing to a better energy match between the metal-based and bridge orbitals.[6,94,102] In such highly conjugated systems the differentiation between redox-active end groups and the bridge becomes somewhat arbitrary, if not impossible. The underlying reason is that the spacer contributes significantly to the frontier orbitals. The spacer thus constitutes an integral part of the redox orbital that is directly involved in the electron-transfer process, and there are even cases where the redox processes are dominated by the nominal bridge.[43]

The ability of the spacer to transmit electronic information between the terminal redox sites is often probed by electrochemistry. The parameter of interest is the splitting of half-wave potentials between consecutive redox events, $\Delta E_{1/2}$. According to common belief, $\Delta E_{1/2}$ increases as electronic interactions across the bridge get stronger. Some caveats, however, apply. Redox potentials and their differences depend on a variety of factors other than an electronic stabilisation of the mixed-valence state and these contributions often dominate the observed potential splittings.[107–109] A more direct and unambiguous measure of the extent of electron delocalisation within these systems is clearly warranted. The rate at which the odd electron is transferred from the formally reduced to the formally oxidised site in the mixed-valence state is one such. The intramolecular electron-transfer rate has been probed by various experimental techniques such as ESR, IR or, if applicable, Mössbauer spectroscopy. Each of these techniques is associated with its own intrinsic time constant,[110] and there are cases where molecules appear completely delocalised by a slower method (*e.g.* Mössbauer or ESR spectroscopy) but valence localised on the faster vibrational timescale.[111,112] Systems where even the fastest spectroscopic probe cannot distinguish chemically equivalent redox-sites in the mixed-valence state are termed intrinsically delocalised and assigned as Class III compounds according to the Robin and Day classification scheme.[64] The other extreme, Class I, is characterised by essentially noninteracting redox sites. Each individual subunit behaves as it would in a similar mononuclear analogue in the respective redox state. In the intermediate regime of Class II the formally oxidised and reduced forms already acquire some of the properties of the other redox state. The intramolecular electron-transfer rate is, however, still too low to render them truly equivalent. Another asset of electronically coupled mixed-valence compounds is the appearance of a characteristic intervalence charge-transfer (IVCT) band in the low-energy region of the visible regime or the near-infrared. The energy, shape, intensity and solvent dependence of this band allow for a determination of the electronic coupling V_{AB} and of the interaction parameter α.[113] Both are quantitative measures of the degree of electronic interactions between the bridged sites. All of these methods require the generation and spectroscopic characterisation of the intermediate mixed-valence state. Given the main purpose of this contribution we will largely restrict ourselves to work that actually employed spectroelectrochemistry for that purpose and less comprehensively cover related work where the mixed-valence state was assessed by chemical or redox agents or bulk oxidation in a conventional electrolysis cell.

6.2.3.2 Electron Transfer across All-carbon Bridges

Dimetal complexes with unsubstituted all-carbon bridges have been realised for chain lengths of up to ten acetylenic repeat units.[114–116] Within the extensive series of complexes with redox-active $\{(\eta^5\text{-}C_5Me_5)(NO)(PPh_3)Mn\}$ termini, electrochemical wave splittings prevail right up to the C_{16} homologue that spans a length of about 23 Å. Although electrochemical wave splittings by themselves do not necessarily prove the existence of electronic interactions between the terminal sites, their mere existence over such large distances is nevertheless remarkable.[114,115,117] To date, $[\{Cp^*(dppe)Fe\}_2(\mu\text{-}C_8)]^+$ seems to be the most extended all-carbon-bridged complex to be characterised in the mixed-valence state.[75,79,105] While completely localised on the vibrational timescale in solution, temperature-dependent valence delocalisation is seen already on the slower Mössbauer timescale in the solid state. Any shorter oligoynediyl linker produces intrinsically delocalised mixed-valence systems and exhibits wire-like behaviour.[61–63,65,67,68,70,72,73,118–121]

The C_4-bridged bis(diruthenium) complex $\{Ru_2(ap)_4\}_2(\mu\text{-}C_4)$ (ap = 2-anilinopyridinate) is quite remarkable in that it generates strongly coupled mixed-valence species upon oxidation *and* reduction.[120] Both redox processes were effected inside an OTTLE cell using spectroelectrochemistry. Figure 6.14 displays the optical and near-IR spectra for both processes along with the crystallographically determined structure of the neutral. The new absorption bands in the near-infrared have no counterparts in the oxidised or reduced forms of related complexes $Ru_2(ap)_4\text{-}(C\equiv C)_nR$ containing just one redox-active diruthenium end group. Class III behaviour was deduced from the low half-widths of these IVCT bands.

The diruthenium complexes $[\{Cp(PPh_3)(L)Ru\}_2(\mu\text{-}C_4)]^{n+}$ (L = PPh_3, PMe_3) also form two different mixed-valence states at the mono- (formally Ru^{II}/Ru^{III}) and trication (formally Ru^{III}/Ru^{IV}) levels.[62] Spectroelectrochemistry was again

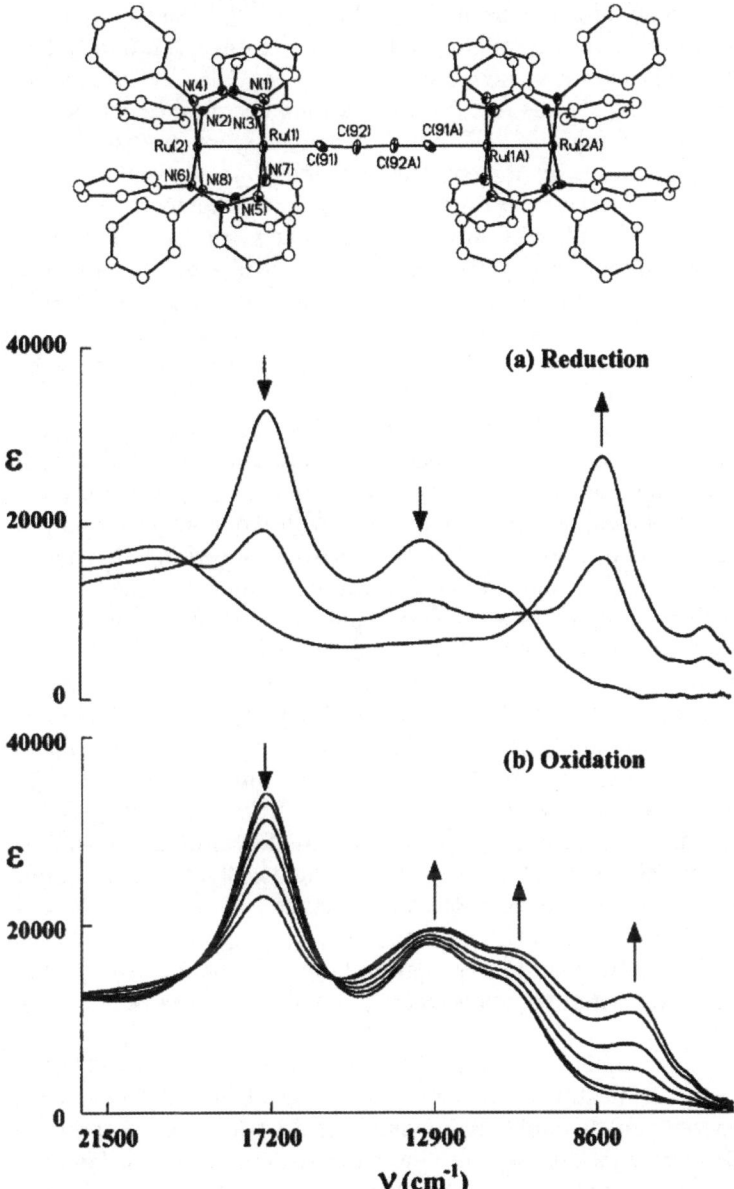

Figure 6.14   Spectroelectrochemical traces upon the reduction (a) and oxidation (b) of
{Ru₂ap₄}₂(μ-C₄). The crystallographically determined structure of the
neutral is also shown (top). Reproduced with permission from ref. 120.

the method of choice to generate and characterise any of the oxidised forms.
Complete valence delocalisation follows from the IR pattern of the all-carbon
bridge (see also Section 6.2.2.2 and Figure 6.8) and the appearance of IVCT
bands with half-widths that are appreciably narrower than those expected for

Class II systems applying the model of Hush.[113,122,123] Data analysis points to a slightly larger intermetal coupling at the trication stage. It is quite interesting to note that this parallels a contraction along the Ru–C_4–Ru axis upon progressive oxidation as was revealed by quantum-chemical calculations[62] and by X-ray crystallography.[63, iii] UV–Vis–NIR spectroscopic traces recorded during stepwise oxidation of the bis(triphenylphosphane) diruthenium complex (L = PPh_3) are displayed in Figure 6.15. Cyclic voltammetry also points to the existence of two different mixed-valence states for $[\{Cp^*(L_2)Ru\}_2(\mu\text{-}C_4)]^{n+}$ (L_2 = dppm, dppe) ($n = 1$, Ru^{II}/Ru^{III}; $n = 3$, Ru^{III}/Ru^{IV}),[63] $[\{CpRu(dppf)\}_2(\mu\text{-}C_4)]^{n+}$ ($n = 1$, Ru^{II}/Ru^{III}; $n = 3$, Ru^{III}/Ru^{IV})[121] and $[\{(\eta^5\text{-}C_5Me_4H)dmpe)Mn\}_2(\mu\text{-}C_2)]^n$ (dmpe = $Me_2PC_2H_4PMe_2$) ($n = +1$, Mn^{III}/Mn^{II}, $n = -1$, Mn^{II}/Mn^{I}).[61] Only the monocations were, however, investigated in these cases and they were all found to be intrinsically delocalised.

As stated above, spectroscopic investigations on the mixed-valence congeners provide the most conclusive information on the strength of the electronic interactions across a bridging ligand or a spacer. Given the large variety of metal-capped all-carbon chains with redox-active termini that have been generated over the years, it is somewhat surprising that most studies pertaining to their mixed-valence congeners are limited to the C_4-bridged systems. While the ability of all-carbon chains to function as a molecular wire is undebatable, it is still difficult to quantify their attenuation factor, *i.e.* the distance dependence of the strength of the electronic interactions conveyed by these ligands[iv, 6,108,124,125] This is because for most series of C_n-bridged complexes differing in bridge lengths, V_{AB} has been determined for a maximum of just two members. From the literature data, a moderate decrease of V_{AB} from 0.47 in $[\{Cp^*(dppe)Fe\}_2(\mu\text{-}C_4)]^{+}$[73] to 0.32 eV in $[\{Cp^*dppe)Fe\}_2(\mu\text{-}C_8)]^{+}$[75,79] can be deduced. This translates into an attenuation factor of about 0.032 Å^{-1}. The most extensive series of mixed-valence oligoynediyl-bridged complexes reported to date consists of $[\{Ru_2(ap)_4\}_2\mu\text{-}(C\equiv C)_n]^{-}$ where $n = 2$, 3, 4 and 6 (ap = 2-anilinopyridinate). Based on the crystallographically determined or estimated Ru–Ru separations across the polyynediyl bridge and the energies of the IVCT band an attenuation factor γ of 0.064 Å^{-1} has been estimated (see also Figure 6.14).[120]

Electronic coupling across oligoynediyl bridges is not restricted to systems comprising identical redox systems. Due to the inherent differences in ground-state energies of dissimilar redox sites there is, however, a larger tendency towards valence localisation onto the easier reduced or oxidised one. Since this area has just been reviewed,[126] we just mention here recent work on unsymmetrically substituted C_4-bridged di-iron or iron/rhenium complexes. Upon oxidation of $[Cp^*(CO)_2Fe-(\mu\text{-}C_4)-Fe(dppe)Cp^*]$ to its monocation the IR

---

iii According to Hush analysis the intermetal coupling is reciprocal to the intramolecular electron-transfer distance between the donor and the acceptor sites, r_{AD}.

iv The efficiency of long-distance energy transfer and its gradual decrease with intercomponent distance is usually described by an exponential decay law of the type $V_{AB} = V_0 \exp(-\gamma R_{AB})$, where γ is denoted the attenuation factor and R_{AB} is the spatial separation between the redox sites. In the case of ligand-bridged complexes R_{AB} is set as the distance between the metal centres.

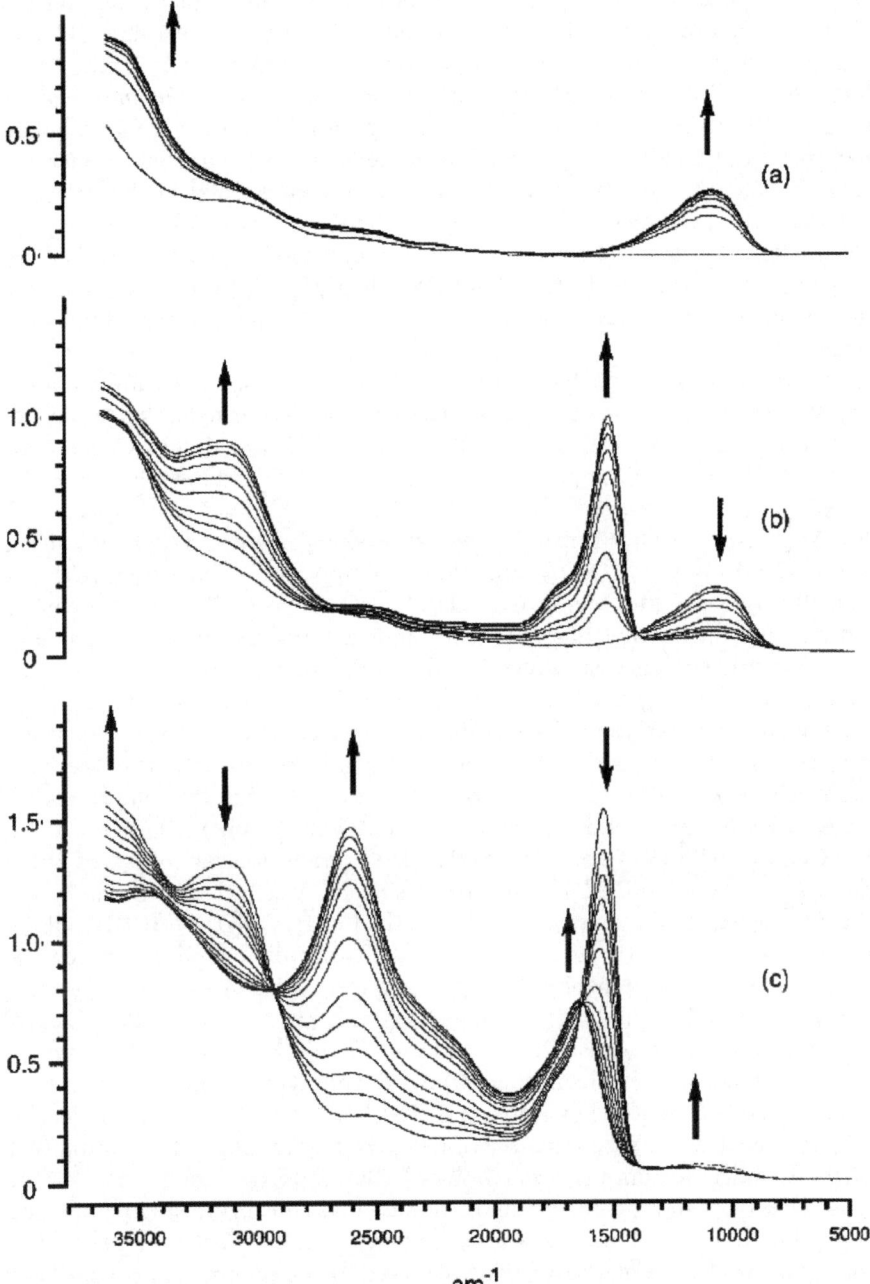

Figure 6.15 UV–Vis–NIR spectroscopic traces recorded upon the first (a), second (b) and third (c) oxidation of $\{(\eta^5\text{-}C_5H_5)(PPh_3)_2Ru\}_2(\mu\text{-}C_4)$. Reproduced with permission from ref. 62.

carbonyl bands shift on average by merely $28\,cm^{-1}$, and this shows that the charge is predominantly localised on the electron-rich {Cp*(dppe)Fe} end group. Further support comes from the observation that ^{31}P hyperfine couplings in the ESR spectrum are of a similar size as in related *mononuclear* $CpL_2Fe(III)$ alkynyl complexes.[119] In keeping with this, the electronic coupling parameter of $0.21\,eV$ as was derived from the NIR IVCT band is considerably smaller than that of the symmetrically substituted $[\{Cp^*(CO)_2Fe\}_2(\mu\text{-}C_4)]^+$ counterpart $(0.32\,eV)$. Even stronger charge and spin localisation $(V_{AB} = 0.19\,eV)$ was reported for the butadiynediyl-bridged iron rhenium complex $[Cp^*(dppe)Fe-(\mu\text{-}C_4)-Re(NO)(PPh_3)Cp^*]^+$.[26] In all these cases the mixed-valence forms were generated by chemical oxidation of the homovalent precursors.

Electron delocalisation has also been observed in the chemically oxidised, mixed-valence forms of complexes bearing the ferrocenylacetylide ligand, {M}$-C\equiv C$-Fc, where {M} is $(\eta^5\text{-}C_5R_5)L_2Fe$ (R = H, Me, $L_2 = 2$ CO; R = H, $L_2 = 2$ $P(OMe)_3$, 2 PPh_3, dppe, dppm, dmpe or one CO and one PPh_3; R = Me, $L_2 = 2$ PPh_3, dmpe, dppe),[32,34] $(\eta^5\text{-}C_5R_5)L_2Ru$ (R = H, $L_2 = 2$ PPh_3, dppe, dppf; R = Me, $L_2 = 2$ PPh_3, dppe),[30] *trans*-$Ru(dppm)_2(-C\equiv CC_6H_4R)$ (R = 4-Ph, 4'Ph-C_6H_4-4, 4-Me, 4-NO_2) and *trans*-$Ru(dppm)_2Cl$,[21] or *trans*-$(C_6H_4R$-4)$Pt(PPh_3)_2$ (R = OMe, Me, H, Cl, COOMe, COMe).[33] Similar ruthenocenyl or tetramethylruthenocenyl acetylide half-sandwich ruthenium complexes have also been studied.[60] It was demonstrated that the tuning of the coligands at the iron half-sandwich unit allows determination of the primary redox site (ferrocenyl *vs.* the iron half-sandwich one) in the di-iron complexes. This has a strong bearing on the coupling between them, and primary oxidation at the half-sandwich site seems to enhance intermetal interactions.[34] In the complexes (Fc-$C\equiv C-)_2Ru(dppm)_2$ and all-*trans* (Fc-$C\equiv C-)_2M(PPh_3)_2(L)(CO)$, where L = CO, py or $P(OMe)_3$, Ru(II)$\rightarrow$Fe(III) IVCT transitions are observed at the mono- *and* the dioxidised levels.[23] Increasing the electron density at the ruthenium decreases the energy gap between the Fe(III)-Ru(II)-Fe(II/III) and the Fe(II)-Ru(III)-Fe(II/III) states such that the electronic coupling between the ferrocenium and the ruthenium sites gets stronger. Some hints as to long-range IVCT between the remote ferrocene and ferrocenium sites across the bis(ethynyl) ruthenium linker were found for mono-oxidised $[(Fc\text{-}C\equiv C-)_2Ru(dppm)_2]^+$ only, which constitutes the most electron-rich congener within this series.

For (Fc-$C\equiv C-)_2Ru_2(DMBA^R)_4$ ($DMBA^R$ = 3-substituted dimethylbenzimidate, see Chart 6.2) the sequence of redox events seemingly differs from that of the monoruthenium-bridged relatives.[35] Based on the results of Vis–NIR spectroelectrochemistry, the first oxidation involves the diruthenium amidinate core while the second and third anodic processes occur at the peripheral ferrocenyl sites. The single NIR bands observed at the monocation stages have thus been ascribed to a Fc $\rightarrow$ Ru_2(III/IV) IVCT. For any of these complexes the second oxidation produces another NIR band at higher energy while leaving the first one essentially unaltered. This latter band has thus been ascribed as an Fc $\rightarrow$ Fc^+ IVCT across the bis(acetylide) diruthenium connector. Both NIR bands vanish upon oxidation of the second ferrocenyl group.

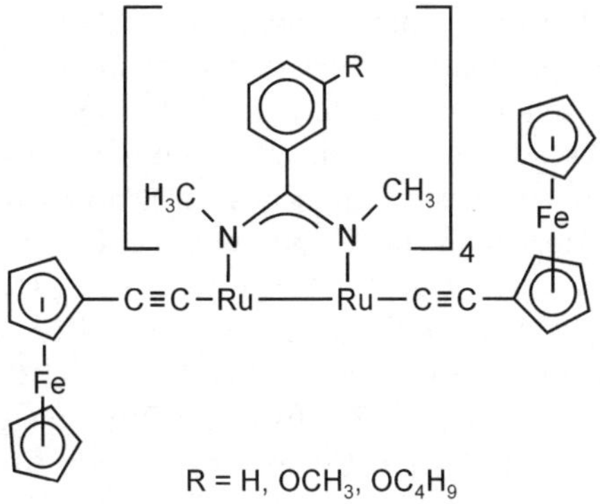

Chart 6.2   FcC≡C-substituted DMBA diruthenium complexes.

Yip and coworkers[127] have reported interesting cases of interactions between remote ferrocenyl sites across a triangular diphosphine-bridged array of copper atoms in $[Cu_3(\mu\text{-dppm})_3(\mu_3\text{-}\eta^1\text{-C}\equiv\text{CFc})_2]^{2+}$ or a Pt–Pt σ-bond in $[Pt_2(\mu\text{-dppm})_2(-C\equiv\text{CFc})_2]^+$.[128] Since no IVCT bands are observed in similar complexes with only one ferrocenylacetylide unit, the ones observed upon chemical oxidation of the bis(ferrocenyl)acetylide complexes were ascribed as arising from an intramolecular Fe(II)→Fe(III) electron transfer. The Pt–Pt σ-bond was found to be a particularly effective mediator. Cluster expansion to a Pt_2Au array such as in $Pt_2(\mu\text{-AuX})(\mu\text{-dppm})_2(-C\equiv\text{CFc})_2$ (X = Cl, Br) forces the ferrocenylethynyl entities away from a collinear arrangement with each other and the Pt–Pt bond and shuts down any observable interaction between them.[128] A dramatic difference in the splitting of the individual redox waves of the FcC≡C subunits has also been observed for *trans*-Ru(dppm)$_2$(C≡CFc)$_2$ and *cis*-Ru(dppm)$_2$($\mu\text{-}\eta^1\text{:}\eta^2$-C≡CFc)$_2$CuI.[23] It is, however, difficult to disseminate between the effects of comparing two different isomers and copper coordination to the alkynyl ligands. The latter event withdraws electron density from the π-system and may thus additionally contribute to lessening the electronic interactions between the two C≡C-Fc sites. There is also an interesting case of a *trans*-bis(acetylide) metal unit serving as an electronic coupler between terminal organic (triaryl)amine-based redox systems.[129] In this instance, the *trans*-Pt(PEt$_3$)$_2$(-C≡C-)$_2$ moiety was found to be slightly less efficient as a coupling unit than the 1,4-phenylene spacer.

6.2.3.3   Oligoenediyl-bridged Complexes

Though less numerous than the oligoynediyl-bridged systems, several examples of oligoenediyl-bridged dinuclear complexes have been reported, most notably for the group 8 metals. The most extensive series derives from {Ru(PR$_3$)$_2$(CO)(Cl)(L)} or {Ru(PR$_3$)$_2$(CO)(L$_2$)} end groups, where L denotes a neutral two-electron donor (mostly a substituted pyridine) and L$_2$ a monoanionic chelate ligand. Within this series, spacer lengths of four, six, and eight CH-units have been realised. Electrochemical studies show that the splitting of the individual redox waves gradually decreases with increasing chain length with a still sizable splitting even for the octatetraenediyl congener. This has been taken as an indication for strong intermetal coupling across the unsaturated bridge.[130–132] In depth studies to support such a notion have, however, been restricted to the shorter-chain butadienediyl-bridged di-iron and diruthenium congeners [{(η^5-C_5R_5)(LL′)Fe}$_2$(μ-CH = CH−CH = CH)]   (R = H,   LL′ = dppm, dppe, L = CO, L′ = PPh$_3$, PMe$_3$; R = Me, LL′ = dppm),[92–93,95] [Cp*(dppm)Fe}$_2${μ-C(OMe) = CH−CH = C(OMe)}],[92] [{Ru(PR$_3$)$_2$(L)(CO)Cl}$_2$ (μ-C_4H_4)]   (R = Ph,   L = none,   R = Et,   L = PEt$_3$)   and   [{Ru(PPh$_3$)$_2$(CO) Cl(NC$_5$H$_4$COOEt-4)}$_2$(μ-C_4H_4)].[96] IR spectra of the mixed-valence monocations   of   the   carbonyl   containing   [{Cp(CO)(PPh$_3$)Fe}$_2$(μ-CH = CH− CH = CH)]$^+$ and [*mer*-{(PEt$_3$)$_3$Ru(CO)(Cl)}$_2$(μ-CH = CH−CH = CH)]$^+$ complexes provide clear evidence for complete valence delocalisation. There is just one CO band as in the neutrals, and the CO band energy in the mixed-valence state resembles the average of v_{CO} observed for the neutrals and the dications.[92] This   is   also   true   for   the   isonicotinate   ester   bands   of   [{Ru(PPh$_3$)$_2$(- CO)Cl(NC$_5$H$_4$COOEt-4)}$_2$(μ-C_4H_4)] and shifts of 11 cm^{-1} were observed for each consecutive oxidation step.[96] For the ruthenium complexes the stepwise oxidations were followed by spectroelectrochemical techniques, and this allowed even for the spectroscopic detection of the rather labile dications. Of note is the significant increase of the CO bandwidths for [{Ru(PPh$_3$)$_2$ (CO)Cl}$_2$(μ-C_4H_4)]$^+$ and   [{Ru(PPh$_3$)$_2$(CO)Cl(NC$_5$H$_4$COOEt-4)}$_2$(μ-C_4H_4)]$^+$ (see Figure 6.12). The latter possibly results from exchange broadening, *i.e.* intramolecular electron transfer occurring at a comparable timescale as the molecular vibration.[133,134] Several of the mono-oxidised forms of the butadienediyl-bridged dimetal complexes gave well-resolved ESR spectra with couplings to two sets of two identical CH protons each and the appropriate number of equivalent phosphorus nuclei on the metals. All butadienediyl-bridged dimetal complexes examined thus far produce mono-oxidised forms that   are   fully   delocalised   on   the   ESR   timescale.   This   makes   the −CH = CH−CH = CH− an as efficient electronic coupler as is its more unsaturated butadiynediyl counterpart, despite the larger conformational requirements for efficient coupling such as a coplanar arrangement of the individual subunits.[94] Large V_{AB} values in the range of 0.5 to about 0.85 were derived from the positions and intensities of the IVCT bands in the low-energy region of the visible or the near-IR and support this notion.[95,96] We have already pointed out that the iron- and ruthenium-based butadienediyl complexes differ insofar that in the ruthenium systems the bridge contributes considerably more to the redox orbitals and even dominates the HOMO and HOMO-1 levels. Consistent with this view is the finding that hyperfine splittings of the unpaired spin of the phosphine ligands' ^{31}P nuclei are twice as large for the iron complexes as for the ruthenium ones. It remains, however, to be explored how the more extended oligoenediyl ligands compare to their oligoynediyl counterparts   as   far   as   charge   and   spin   delocalisation   in   odd-electron   states   are concerned.

Reference should also be paid to Launay's pivoting series of oligoene-bridged diferrocenes Fc$-$(CH$=$CH)$_n$$-$Fc, where n ranges from 1 to 6.[100] Careful redox titrations coupled with periodical spectra sampling provided the optical/NIR traces of all mixed-valence congeners within this series. This is despite only small comproportionation constants that even reach the statistical limit for the more extended analogues. Adherence to an exponential decay law with a comparatively small attenuation factor of 0.087 Å^{-1} (or rather of 0.054 Å^{-1} for the shorter congeners with up to three conjugated double bonds and of 0.113 Å^{-1} for the higher homologues) was observed. This finding places polyenediyl bridges amongst the most efficient electronic couplers.

6.2.3.4 *Arylene–Ethynylene and Arylene–Vinylene Spacers*

A popular method in order to access "rigid-rod" complexes with more extended bridges spanning the metal sites is to insert arylene units into the bridging oligoynediyl or, less commonly, the oligoenediyl ligand. Several arylene-ethynylene-bridged complexes have been synthesised and investigated with respect to the degree of the intermetal electronic coupling conveyed by these linkers. Even more extensive are the series of complexes with arylene-ethynylene or arylene-vinylene spacers inserted into the backbone of ligands such as pyridine, bipyridine, terpyridine or phthalocyanine, but they are not considered here because of the lack of a direct metal–carbon bond. The reader who is interested in that topic may refer to authoritative reviews available in the literature.[6,108] In the case of arylene-ethynylene- or arylene-vinylene-type spacers additional complexity arises from the different topologies that are possible with these architectures. From simple considerations one may infer that the coupling is more efficient with *ortho* and *para* linkages than with the *meta* one, and there are numerous examples to support such a notion, particularly from Lapinte's group.[25,77–82] Theoretical analysis has led to topological quantum-interference rules that provide a rationale for these effects.[135] Thus, the mixed-valence cations of the *meta*-diethynylbenzene-bridged [{Cp*(dppe)Fe}$_2${(μ-C$\equiv$C)$_2$C$_6$H$_4$-1,3}]$^+$ and [[{Cp*Fe(dppe)}$_3${(μ-C$\equiv$C)$_3$C$_6$H$_3$-1,3,5}]]$^{n+}$ ($n = 1$, 2) display two different C$\equiv$C stretches in the infrared. The higher-energy absorption is typical of a largely localised Fe(II)-arylethynyl site while the one at lower energy corresponds to a localised Fe(III) unit. Slight deviations of the band energies from those of the corresponding homovalent forms have been ascribed to weak interactions between the iron sites and some degree of charge and spin delocalisation onto the central phenylene ring. Further studies have shown that localised valencies still exist on the slower Mössbauer and ESR timescales. Analysis of the broad, weak IVCT bands observed in the near-infrared provided coupling parameters in the range of 71 to 161 cm^{-1} (0.009 to 0.02 eV) that are typical of weakly coupled class II mixed-valence systems.[82] In contrast, the mono-oxidised form of the *para*-diethynylphenylene isomer, [{Cp*(dppe)-Fe}$_2${(μ-C$\equiv$C)$_2$C$_6$H$_4$-1,4}]$^+$, is seemingly situated at the borderline of a completely delocalised (class III) and strongly coupled, yet valence-trapped class II system according to Mössbauer and ESR studies.[78] Curiously enough, the simultaneous existence of valence-trapped and valence-detrapped forms is seen for crystalline samples on the Mössbauer timescale with some bias toward the valence-trapped form as the temperature decreases. This parallels the behaviour of the corresponding octatetraynediyl congener.[79] A strong bias to valence delocalisation in solution seems to be a more general phenomenon.[61]

Replacing the phenylene by a thiophene-2,5-diyl entity enhances the electronic coupling such that the mixed-valence radical cation $[\{Cp^*(dppe)Fe\}_2 \{(\mu-C\equiv C)_2C_4H_2S-2,5\}]^+$ appears fully delocalised even at 4 K in Mössbauer spectroscopy. IR data are, however, less conclusive. The radical cation has a $C\equiv C$ band at an energy close to the average of the homovalent bis-Fe(II) and bis-Fe(III) forms, just as expected for complete valence detrapping. In addition, there is another absorption at an even lower energy than that for the homvalent bis-Fe(II) form, and this was proposed to arise from the coexistence of a second form that is represented by the cumulenic structure $[Cp^*(dppe)Fe^+ = C=C^{\cdot}-C_4H_2S-C\equiv C-Fe(dppe)Cp^*]^+$ where the oxidation has occurred on the bridge. This complex thus provides another example of a system where discrimination between metal- and ligand-based redox events is probably inadequate.[79] All variously oxidised forms of these bis(iron) complexes were prepared by means of chemical oxidation, though.

Bis(ethynyl)phenylene (1,3- and 1,4- isomers)-, 2,5-dimethyl-1,4-bis(ethynyl) phenylene-, 2,5-bis(ethynyl)pyridinylene- and 2,5-bis(ethynyl)thiophenylene-bridged diruthenium and diosmium complexes containing *trans*-Cl(dppm)M end groups provide another interesting series of complexes where the influence of the conjugated spacer and of the transition metal on the degree of metal–metal interactions becomes apparent.[76] Judged by the IVCT bands observed for the electrogenerated monocations, a pattern grossly consistent with the results on the di-iron complexes from above appears. The thiophenyl spacer allows for more efficient coupling than the 1,4-phenylene, and the pyridinylene group occupies a position intermediate between the methyl-substituted and the unsubstituted 1,4-phenylene. No IVCT band at all was observed for the *meta* phenylene linker. Couplings for the diruthenium complexes are consistently larger than those for their osmium counterparts. This has been explained by a more favourable interaction of the bis(acetylide) donor with a stronger electron acceptor, *i.e.* Ru(III) as compared to Os(III). Even the stronger coupled diruthenium complexes are, however, more localised than their Cp*(dppe)Fe based counterparts. Comparison of the voltammetric data of $[\{Cp(dppf) Ru\}_2(\mu-C\equiv C-arene-C\equiv C-)]$ (arene = 1,4-phenylene, 1,4-naphthalenylene, or 9,10-anthracenylene) and of the NIR absorption bands of their monocations have revealed an enhancement of the bridge-mediated interactions between the metal sites with increasing size of the arene π-system.[121] This may be due to a better energy match of the bridge and the metal-based orbitals as more bridge π-levels become available. The half-widths of the IVCT bands are appreciably lower than those expected for a class II systems based on the Hush model. This led the authors to describe them as situated on the borderline between class II and class III systems.

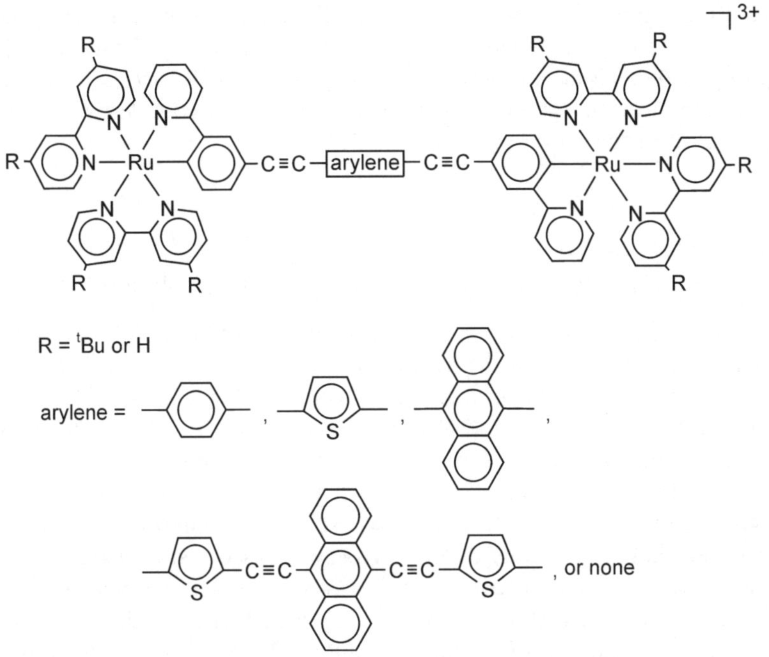

Chart 6.3 Ru tris(bipy)-complexes with arylene-ethynylene bridged bipy-ligands.

In a similar vein, Launay and coworkers[136] have compared metal-metal interactions between two {Ru(bpy)$_2$(pp-)} groups, where pp- denote cyclometalated 2-phenylpyridine ligands that connect to a $-C\equiv C$-X-$C\equiv C-$ spacer *via* their 4-positions (see Chart 6.3). Within the series X = none, 1,4-phenylene, 2,5-thiophenylene and 9,10-anthracenylene, insertion of the first two aromatic spacers was found to decrease the intermetal coupling significantly below that of the butadiynediyl complex. Here, the thiophenylene and phenylene bridges gave almost identical results. The remarkably enhanced interaction *via* the anthracenylene unit was attributed to a more favourable energy match between the lower-lying bridge and the higher-lying metal orbitals. In this instance, electrochemical generation of oxidised samples and spectra acquisition were performed in a discontinuous manner, that is by removing small aliquots for spectra sampling from the electrolysis cell at certain stages and returning them into the cell after recording the spectra.

As we have already pointed out in Section 6.2.1, divinylphenylene-bridged diruthenium complexes [{Ru(CO)Cl(PR$_3$)$_3$L}(μ-CH$=$CH-C$_6$H$_4$-CH$=$CH)] (R=Ph, L=4-EtOOCC$_5$H$_4$N, 4-MeC$_5$H$_4$N, 4-OMeC$_5$H$_4$N, L=^{i}Pr, L=none) present unique cases of a situation where the anodic redox processes are not only influenced but dominated by the bridge.[43] Despite a splitting of the individual oxidation waves by some 250 to 350 mV, none of the PPh$_3$-substituted complexes displayed any observable IVCT band in the near-infrared or the infrared. Instead, a low-energy structured transition due to the oxidised form of the bridge occurred which closely resembles a characteristic absorption reported for the purely organic analogues. It remains to be explored whether similar 1,3,5-trivinylphenylene triruthenium- or 1-ethynyl-3,5-divinylpheny-lene-bridged diosmium complexes[137] display similar behaviour. This also applies to phenylene-bridged diruthenium cyclopropenyl complexes where each ethylene group is part of a cyclopropenyl ring.[138]

6.2.3.5 Dimetal Complexes Bridged by Highly Unsaturated C_nR_m (m<n) Cumulenic or Cumulene Arylene Linkers

As we have noted in a previous section of this chapter, cumulenic resonance forms present one extreme description for dimetal complexes linked by all carbon bridges. They are frequently encountered in some oxidation state (mostly the dioxidised one) and also contribute to intermediate ones between the purely cumulenic and oligoynediyl or the cumulenic- and the alkynyl-bridged dicarbyne forms (see Schemes 6.6–6.8). Purely cumulenic wires are encountered in α,ω-di- or -tetraferrocenyl substituted cumulenes, but no evidence other than a splitting of the ferrocenyl-based redox waves has been presented to support the presence of electronic interactions between the ferrocenyl end groups across the cumulenic ligands. Based on the results of experimental[139,140] and quantum-chemical studies[141] a splitting of the individual redox waves prevails up to the 1,6-diferrocenyl-1,6-diphenyl[5]cumulene where the redox-active end groups are separated by a chain of six cumulated carbon atoms. Quantum-chemical studies pose a conjugation length limit of about 16 Å before the breakdown of electronic interactions occurs. This compares favourably to other classes of organic conjugated linkers such as oligoynediyl or oligoenediyl chains.[6,108]

Dimetal complexes bridged by an odd-numbered all-carbon chain present another interesting class of compounds. Depending on the identity of the metal units, the all-carbon ligand preferably assumes oligoynylcarbyne (*i.e.* $\{M\} \equiv C-(C \equiv C)_n-\{M'\}$) or cumulenic (*i.e.* $\{M\} = C = (C = C)_n = \{M'\}$) resonance structures. There is ample precedence for both forms. Dominant contributions of the cumulenic structures were proposed for the C_3- or C_5-bridged manganese/rhenium complexes $[Cp*(NO)(PPh_3)Re = C = (C = C)_n = Mn(CO)_2$ $(\eta^5\text{-}C_5Cl_{5-c}H_c)]$ ($n = 1$, $c = 1$, 4 or 5; $n = 2$, $c = 0$),[142] the iron/rhenium complex $[Cp*(NO)(PPh_3)Re = C = C = C = Fe(CO)_4]^+$,[142] and the molybdenum/tungsten derivative $[Tp'(CO)_2Mo = C = C = C = W(CO)_2Tp']^{2-}$ (Tp' = hydridotris (3,5-dimethylpyrazolyl)borate).[143] In contrast, the tricarbido complexes $[Tp'$ $(CO)_2M \equiv C-C \equiv C-M'(O)_2Tp']$ (M = Mo, M' = W or M = W, M' = Mo),[143] $[Cp*(NO)(PPh_3)Re-C \equiv C-C \equiv W(O^tBu)_3]$[144] and *trans*-$[Tp'(CO)_2W \equiv C-C \equiv$ $C-M(CO)(PPh_3)_2]$ (M = Ir, Rh), all-*trans*-$[\{Tp'(CO)_2W \equiv C-C \equiv C-\}_2Ir(PPh_3)_2$ $H(CO)]$, as well as heterodimetal ruthenium/molybdenum or ruthenium/tungsten complexes of the type *trans*-$\{(CO)(H)(PPh_3)_2(L')Ru\}-C \equiv C-C \equiv \{M'\}$ (L' = 3,5-dimethylpyrazole, CO, PPh_3, 2,4,6-Me_3C_6H_2NC) adopt ethynyl carbyne structures.[145–147] Quantum-chemical calculations have aided in rationalising the experimental findings. They predict symmetrical cumulenic structures when {M} and {M'} are electronically similar, but polarised, unsymmetrical ones when {M} and {M'} differ from each other with an increasing bias toward the alkynyl carbyne resonance forms as the electronic disparity increases.[148,149] While calculations and experimental data for the ground-state forms of electronically more balanced representatives suggest sizable metal–metal interactions,[149] there seemingly have been no attempts to directly probe for such effects in their partially oxidised mixed-valence states.

The same holds for the vast majority of dinuclear complexes with various types of linear[150–155] or cyclic[156–163] C_nR_m ligands ($m < n$), where the unsaturated ligands constitute partially hydrogenated forms of all-carbon chains. Notable exceptions are oligonuclear allenylidene complexes where the individual $M = C = C = CR_2$ subunits radiate from a common central arylene unit[87,88,164] or are interconnected by an ethynediyl spacer.[87] Owing to the mainly ligand-centred reductions but metal-centred oxidations of allenylidene complexes,[37,83–88] two different kinds of mixed-valence systems may be expected to arise from such architectures. In fact, large redox splitting has been observed for the stepwise reduction[87,88,164] and oxidation[88] processes. This has been substantiated by recent spectroelectrochemical studies on the monooxidised and monoreduced forms of a bianthracenylidene-bridged diruthenium bis(allenylidene) complex. UV–Vis–NIR spectroscopic traces acquired during the gradual reduction and oxidation of this complex are depicted in Figure 6.16, along with the basic structure of this complex. Given the fact that the half-widths of the IVCT bands are much lower than those predicted by Hush theory for moderately coupled Class II systems, both mixed-valence forms appear to be highly delocalised.[88] ESR spectra recorded on frozen glasses of the singly reduced monocation show a distinct rhombic splitting of the *g*-tensor and thus a detectable contribution of the metal to the spin-bearing orbital. The mono-oxidised trication is, however, ESR silent under these conditions in accord with an appreciably higher Ru(III) character. Sizable redox splittings for the individual ligand-centred reduction waves have also been observed for similar arylene-bridged dinuclear allenylidene complexes (arylene = 1,4- or 1,3-phenylene, 2,5-thiophenylene, 2,2'-bithiophenyl-5,5'-ene or 2,2',2''-terthiophenyl-5,5''-ene) (Chart 6.4).[87] The authors also report on large anodic shifts of both reduction potentials with respect to the corresponding monoruthenium complexes in all cases except for the largely insulating 1,3-phenylene linker. Such potential shifts seem to constitute another indication of the presence of electronic interactions across the carbon-rich bridging ligand. Spectroscopy on the chemically generated monoreduced form of the 1,4-phenylene-bridged congener provides additional evidence for electron delocalisation over both ruthenium allenylidene constituents.

Mixed allenylidene ethynyl and allenylidene butenynyl ligands are present in the C_5H-bridged diruthenium, diosmium and diiron complexes $[\{(\eta^5\text{-}C_5R_5) L_2M\}_2(\mu\text{-}C = C = CH-C\equiv C-)]^+$ ($R = Me$, $L_2 = dppe$, $M = Ru$; $R = H$, $L_2 = 2 PPh_3$, $M = Ru$, Os; $R = H$, $L_2 = dppe$, $M = Fe$)[155,156] and the vinylogous C_7R_3-bridged diruthenium complexes *trans*-$[\{Cl(dppe)_2Ru\}_2(\mu\text{-}C = C = C(CH_3)-CR = CR'-C\equiv C-)]^+$ ($R = H$, $R' = Me$, Ph; $R = Me$, $R' = Ph$).[165] The splitting

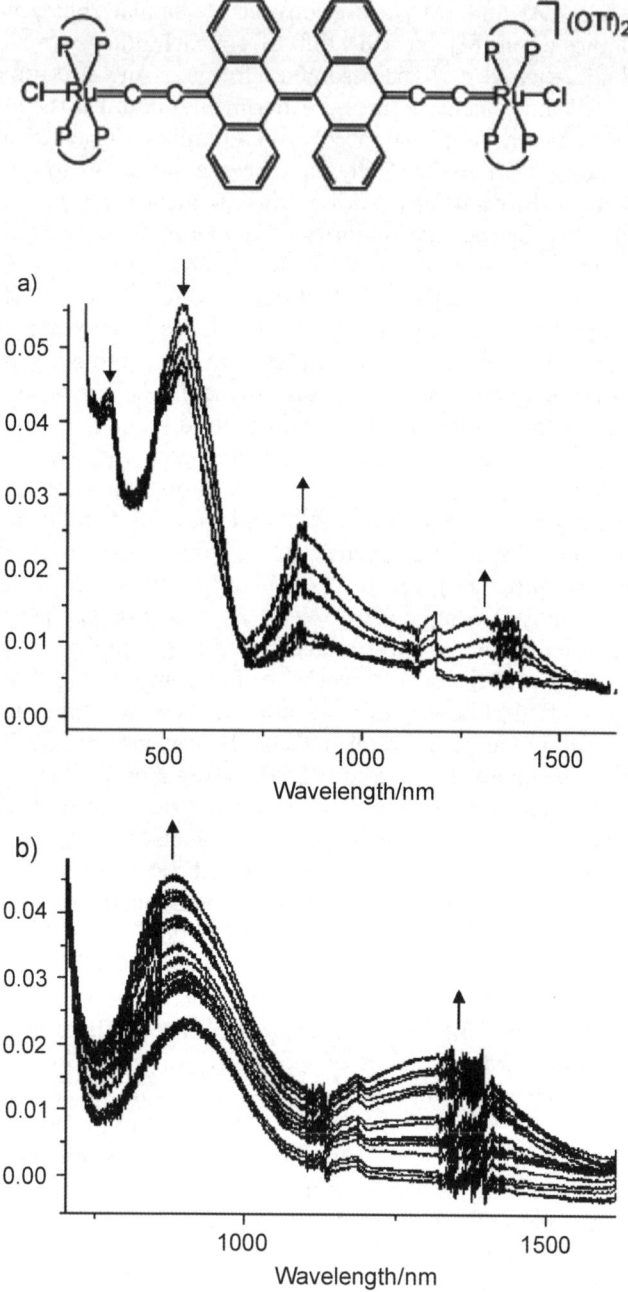

Figure 6.16 UV–Vis–NIR spectroscopic traces recorded during stepwise reduction, a) and oxidation, b) of *trans*-[{Cl(dppe)$_2$Ru$=$C$=$C$=$C}$_2$(bianthracenylene)]$^{2+}$. Adapted with permission from ref. 88.

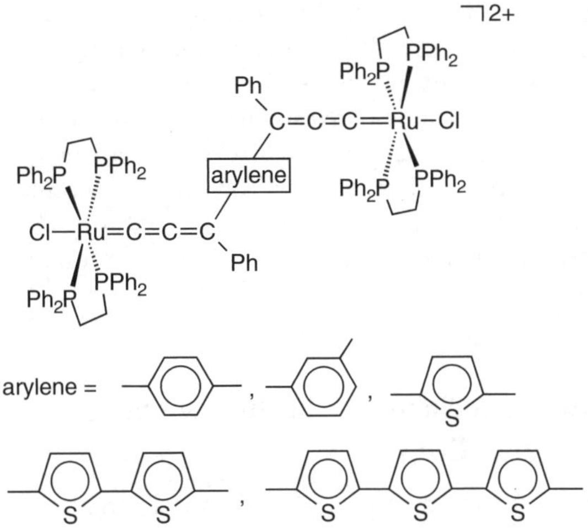

Chart 6.4   Arylene-bridged bis(allenylidene)ruthenium complexes.

of the oxidation waves exceeds 650 mV in each case and points to complete electron delocalisation at the mono-oxidised mixed-valence state. No experimental data on the mono-oxidised forms are, however, presently available.

An interesting case of charge and spin delocalisation over two allenylidene chains has been reported for the monoreduced form of the unique mono-ruthenium bis(allenylidene) complex $[(Ph_2C=C=C=)_2Ru(dppe)_2]^{2+}$.[166] Spin delocalisation follows from the observation of an ESR quintet that arises from coupling of the unpaired spin to the four equivalent phosphorus nuclei of the diphosphine ligands. Charge delocalisation is indicated by the presence of just one single broad absorption for the unsaturated carbon-rich ligands in the IR. Quantum-chemical calculations support the view that the unpaired spin mainly resides on the odd-numbered carbon atoms with an almost equal spin distribution over both allenylidene chains and a sizable lowering of the IR frequency following reduction.

Comparatively little work has been done on complexes with highly unsaturated linear C_nH_m bridges composed of alternating ethynyl/ethenyl or carbyne/ethenyl bridges. Examples are hex-3-ene-1,5-diyne-1,6-diyl-bridged complexes $\{Ru_2(ap)_4\}_2$ $(\mu\text{-}C{\equiv}C\text{-}C(CH_2OSi^tBuMe_2)=C(CH_2OSi^tBuMe_2)\text{-}C{\equiv}C\text{-})$ (ap = 2-anilinopyridinate)[167] and $Tp'(CO)_2M{\equiv}C\text{-}CH=CH\text{-}C{\equiv}M(CO)_2Tp'$.[168] In both cases the presence of some interactions between the metal-based redox sites was concluded from a splitting of the individual redox waves but nothing more has been done to test this assumption by investigating their partially oxidised and reduced forms.

The work summarised in this chapter gives an overview of the various forms of highly unsaturated linear or cyclic hydrocarbon bridges C_nH_m ($n < m$) that have been utilised in connecting redox-active transition-metal moieties. We wish to emphasise how important it is to not just rely on electrochemical data for evaluating the degree of electronic interactions across these bridges and how misleading such a notion can be. In fact, there are cases of complete delocalisation despite small redox splitting, and *vice versa*,[169] while another study demonstrated the impressive effects on ion pairing on the redox potentials.[109] Any conclusion with regard to the actual strength of the intervalence coupling thus calls for detailed investigations of the mixed-valence forms. To these ends, spectroelectrochemistry provides a powerful tool in addition to the more traditional chemical or stepwise coulometric/spectroscopic approach. More work towards these ends is encouraged in order to fully explore the potential of highly unsaturated hydrocarbons as constituents of conducting "wires".

6.2.4 Spectroelectrochemistry in the Switching of Molecular Properties

One major impetus for the synthesis of carbon-rich organometallics with ever-higher degrees of sophistication arises from their fascinating molecular properties, and these have been met with considerable interest from materials sciences.[170] This pertains, for example, to the fields of nonlinear optics,[171–174] luminescence,[170,175–179] and, as we discussed in the previous section, the emerging field of molecular electronics. In this context it is important that the actual performance of an organometallic compound for any of these purposes may dramatically depend on its oxidation state. The ability to reversibly switch between "ON" and "OFF" states and to exert control over the magnitude of such a molecular property[180] may constitute a decisive advantage of redox-active organometallic systems over their frequently easier-to-synthesise organic competitors. To demonstrate this switching effect has thus been the focus of some recent work. Again, spectroelectrochemistry has been of great utility toward these ends.

Some accounts on the redox-induced switching of hyperpolarisabilities utilised penta-ammine ruthenium complexes bearing substituted 4,4'-bipyridinium coligands. The intense bipyridinium → Ru LMCT band at 580 to 636 nm bleaches upon reduction of the metal centre, thus switching off their quadratic NLO response.[180,181] The oxidation of the ferrocene donor in an ethenyl-linked ferrocene–nitrothiophene dyad likewise results in a decrease of the quadratic NLO response by about one order in magnitude.[182] Oxidation-state-dependent quadratic NLO performance has also been noted by Lapinte and coworkers[185] who compared the complexes $Cp^*(dppe)Fe-C{\equiv}CPh$, $\{Cp^*(dppe)Fe-C{\equiv}C-\}_2(\mu-C_6H_4-1,4)$, $\{Cp^*(dppe)Fe-C{\equiv}C-\}_2(\mu-C_6H_4-1,3)$, and $\{Cp^*(dppe)Fe-C{\equiv}C-\}_3(\mu-C_6H_4-1,3,5)$ in their various oxidation states. Redox interconversions were initially brought about by means of chemical redox agents that severely limit practical applicability. Subsequent work, however, led to reversible electrochemical switching of the quadratic NLO effect in an OTTLE cell for the latter dyad.[183] Other examples of redox switching of NLO effects inside an OTTLE cell originate from the groups of Humphrey, Heath and coworkers.[7] Their work is based on the finding that ruthenium alkynyl complexes are optically transparent in their Ru(II) reduced states but display significant absorption in the near-IR in their oxidised Ru(III) states.

The new NIR absorption arises from the promotion of an electron from a lower-lying filled orbital centred on the terminal donor ligand(s) (Cl⁻, alkynyl) to the newly generated hole in the SOMO, which is of a mixed metal/alkynyl character. Such LMCT/LLCT transitions are thus only operative when the oxidised ruthenium moiety acts as an electron acceptor. Mononuclear and triangular-shaped octopolar alkynyl complexes like *trans*-[X−Ru(−C≡CR)] (X = Cl, R = Ph or C≡C−C$_6$H$_4$-4; X = C≡CPh, R = C≡C−C$_6$H$_4$-4) or *trans*-{ClRu(dppe)$_2$−C≡C−C$_6$H$_4$-4}$_3$(C$_6$H$_3$-1,3,5) display significant two-photon absorption at the 800 nm laser wavelength in their reduced state. Strong linear absorption in the oxidised state overrides this effect and causes a change of the magnitude *and* of the sign of the cubic hyperpolarisability. Likewise, cubic NLO is simply turned on when there is no two-photon absorption in the reduced state (see Figure 6.17).[7,184] Later work demonstrated that the NLO switching itself occurs on a exceedingly rapid timescale (within femtoseconds).[181] In practise, this rate is, however, limited by the time required to effect the macroscopic redox transformation.

Other work has demonstrated that it is possible to switch "ON" and "OFF" luminescence by reduction/oxidation,[186,187] and it has been demonstrated that such switching is possible inside an OTTLE cell.[8] Many alkynyl complexes, especially those of rhenium,[176,177,188,189] platinum,[190–195] copper,[178] silver[196,197] or gold[198–204] are highly luminescent from their excited ³MLCT or metal perturbed π* states. This opens up the possibility to significantly influence their emissive properties by redox processes. An interesting example is found in recent work of Wong *et al.*[25] Unlike other rhenium(I)-alkynyl complexes, heterobimetallic

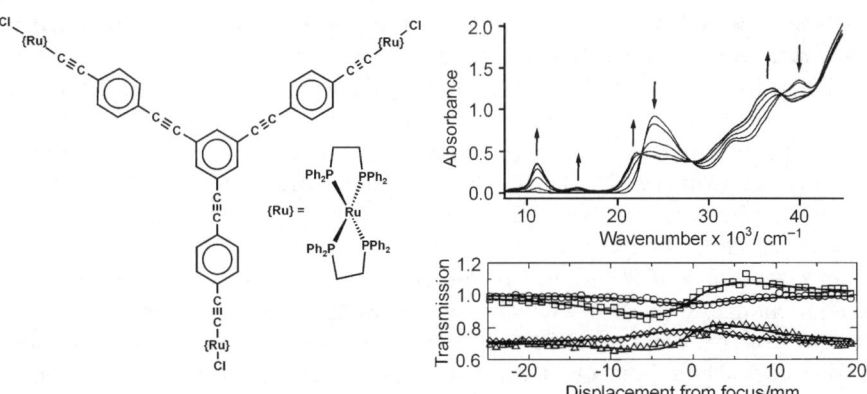

Figure 6.17 UV–Vis–NIR spectroelectrochemical traces (top right) upon oxidation of [{Cl(dppe)$_2$Ru–C≡C–C$_6$H$_4$-4–}$_3$C$_6$H$_3$-1,3,5] (left); closed and open aperture Z-scan traces for the complex and its trioxidised form at 800 nm using 100-fs pulses. Neutral: Closed aperture (squares), open aperture (circles); trication: closed aperture (triangles), open aperture (diamonds). Drawn out theoretical curves assume a thin sample approximation. Adapted with permission from ref. 184.

bis(alkynyl)-bridged *para*-[(bpy)(CO)$_3$Re$-$C$\equiv$C$-$C$_6$H$_4$$-C\equiv$C$-$Fe(dppe)Cp*] is found to be nonemissive due to the intramolecular quenching of the emissive d(Re)$\rightarrow\pi$ *(bpy) ^{3}MLCT state by the low-lying MLCT and LF excited states of the iron moiety. Oxidation of the Fe site removes this quenching pathway and renders the radical cation of this complex emissive. Similar effects were observed upon protonation to the corresponding vinylidene [(bpy)(CO)$_3$Re$-$C$\equiv$C$-$C$_6$H$_4$$-$CH $=$ C $=$ Fe(dppe)Cp*]$^+$.

Ethynyl- or ethenyl-linked ferrocenyl polyarene dyads provide another interesting series of compounds that underpin the viability of the above concept. Depending on the nature of the unsaturated linker and that of the arene, ferrocene oxidation may either result in an enhancement or quenching of the arene-based emission.[103,205] In ferrocenylethynyl derivatives of 2-acridine, 2, 7-acridine, 2-anthraquinone and 9-*N*-acridone the organometallic and the arene entities are electronically decoupled because of a symmetry mismatch between the ferrocenylethynyl and the arene subunits and small contributions of the unsaturated linker to the HOMO and the LUMO orbitals. Multiple pieces of evidence including the results from quantum-chemical studies and from ESR spectroscopy on the radical anions generated by *in-situ* electrolysis suggest, that the LUMOs of these systems reside on the polyarene (see also Section 6.2.1.2). Emission thus occurs from an excited state that is largely dominated by the organic fluorophore. Just like in the above rhenium/iron complex this emission is partially quenched by intramolecular charge transfer from the ferrocenyl donor. Spectroelectrochemistry with monitoring in the Vis–NIR range revealed the growth of a new low-energy charge-transfer band accompanying oxidation of the electroactive ferrocenyl group (Figure 6.18(a)). The oxidised forms of 2-(ferrocenylethynyl)acridine, (Figure 6.18(b)), ferrocenylethynyl-9-*N*-acridone, and of 2-(ferrocenylethynyl)anthraquinone were electrogenerated inside an OTTLE cell and display redshifted and more intense emissions compared to their neutral precursors. Rather unexpectedly, ferrocene oxidation decreased the emission from 2,7-bis(ferrocenylethynyl)acridine (Figure 6.18(c)) and *trans*- 2-(ferrocenylethenyl)anthraquinone. Similar lowering of the emission intensity following ferrocene oxidation was noted for naphthalimides bearing ferrocenylalkyl substituents.[103] This effect probably arises from intramolecular quenching by enhanced association between the charged ferrocinium headgroup and the imide carboxyl group in the excited charge-transfer state.[205]

Electrochromism is another potentially useful property, and examples of colour changes or electroswitchable NIR dyes have been presented in previous sections of this account. Other examples of the redox switching of molecular properties include host–guest chemistry and reversible ion binding, especially by ferrocenes with appended multidonor macrocycles. Some relevant work can be found in references 206,207. The ability of these systems to act as reversible ion scavengers mostly relies of a modulation of electrostatic attraction or repulsion by electron transfer. Although recent work has utilised alkynyl complexes for pH[208] or ion sensing,[199,209,210] the switching ability by electron transfer and its visualisation under the conditions of spectroelectrochemistry remain to be explored.

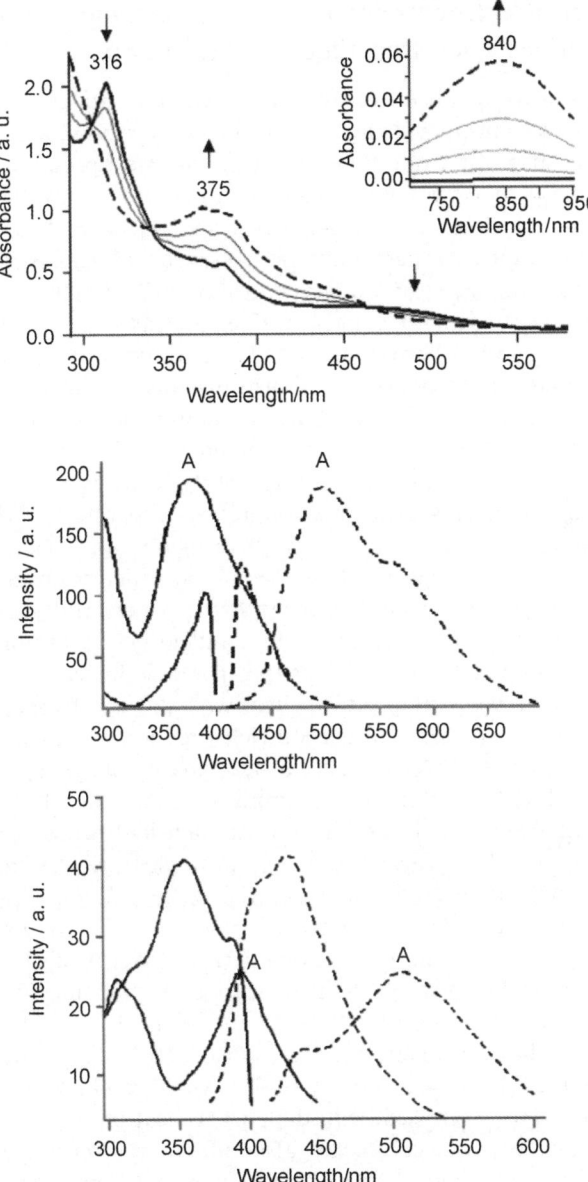

Figure 6.18   a) Effect of ferrocene oxidation inside an OTTLE cell on the Vis–NIR spectrum of a ferroceneethynyl-acridine dyad; b), c) Comparison of the emissions from the oxidised (A) and reduced forms of the ferrocenylethinyl-acridine (b) and the bis(ferrocenylethinyl)acridine dyads. Reproduced with permission from ref. 103.

6.2.5 Spectroelectrochemistry in the Monitoring of Chemical Reactions Following Electron Transfer

Electron transfer from substrates frequently generates highly reactive species, and these may be subject to further chemical reactions. So-called EC processes may well occur at a faster rate than the electrochemical process itself. In such cases the primary redox congener may just appear as a transient or totally escape spectroscopic detection, the only new responses being those of the follow product(s). We emphasise here that the presence of clean isosbestic points that is commonly quoted as a *bona fide* confirmation of the chemical stability of the primarily generated redox congeners does not necessarily preclude such a scenario. Seemingly ideal behaviour can also result when the chemical follow-up step is very fast and produces a uniform product. We therefore encourage probing for overall chemical reversibility by performing a re-electrolysis and comparing the spectroscopic traces at the beginning of the experiment and after performing the full redox cycle and to verify conclusions from spectro-electrochemistry by means of a bulk electrolysis experiment. While spectro-electrochemistry is mostly used to investigate and characterise the electrogenerated redox congener, its scope is even wider when the above situation applies. In fact, spectroelectrochemistry is a powerful tool to monitor reactions following electron transfer and to identify the product(s) formed during such processes by virtue of their spectroscopic fingerprints.

The potential of combining electrochemistry and spectroscopic detection has yet to be fully explored and carbon-rich organometallics may provide a rewarding testing ground. As a matter of fact, many interesting cases of electron-transfer-induced coupling or isomerisation reactions of organometallic radicals derived from complexes with unsaturated hydrocarbon ligands have been reported over the years. Prominent examples are the homocoupling of oxidised metal–alkynyl complexes to the corresponding bis(vinylidenes) owing to delocalisation of the unpaired spin onto the remote β carbon atom.[63,72,73,120,211,217] The radical cations derived from neutral alkynyl complexes may also suffer cleavage of the metal–alkynyl bond. The titanocene complexes $Cp'_2Ti(-C{\equiv}C-C{\equiv}CFc)_2$ ($Cp' = \eta^5\text{-}C_5H_4R$, $R = H$, $SiMe_3$), for example, are oxidised in an irreversible two electron step. Oxidation is accompanied by the release of $Fc-(C{\equiv}C)_4-Fc$, which results from the coupling of two alkynyl radicals released from the parent dication.[212]

Vinylidene and carbyne complexes also offer a rich chemistry following electron transfer, and a comprehensive review on that topic is available in the literature.[213] As an example, oxidation of vinylidene complexes generates highly reactive radical cations. These may undergo a host of different follow-up reactions and they are summarised in Scheme 6.10. Possible follow reactions include dimerisation by direct C_β–C_β homocoupling to dinuclear butanediylidyne complexes $\{M\}^+{\equiv}C-CRH-CRH-C{\equiv}\{M\}^+$,[71,214] deprotonation to 17 valence-electron alkynyl radicals $\{M\}^{\cdot}-C{\equiv}CR$, which subsequently dimerise to the corresponding bis(vinylidenes) (= 1,3-butadiene-1,4-diylidene derivatives) $\{M\}=C=CR-CR=C=C\{M\}$[215,216] and CH-bond homolysis. The latter

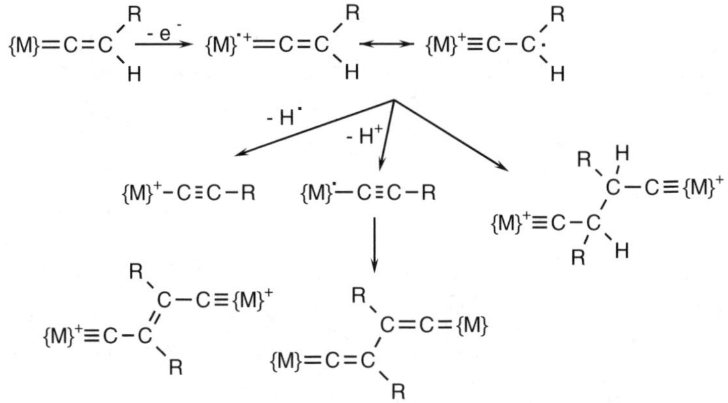

Scheme 6.10 Possible reactions following oxidation of vinylidene complexes.

event generates cationic mononuclear 16 valence-electron alkynyl intermediates $\{M\}^+-C{\equiv}CR$ which in turn may dimerise to dicationic 2-butene-1,4-diylidyne complexes $^+\{M\}{\equiv}C-CR=CR-C{\equiv}\{M\}^{+}$.[217] Bis(vinylidenes) of the type $\{M\}=C=CR-RC=C=\{M\}$ may also be oxidised by two electrons to give electrophilic ethenyl-bridged dicarbyne complexes $\{M\}^+{\equiv}C-CR=CR-C{\equiv}\{M\}^{+}$.[218] Berke and coworkers[67] have reported on the deprotonation of $[(RC{\equiv}C)(dmpe)_2Mn(C{\equiv}CH)]^+$ to neutral alkynyl radicals and their dimerisation to the C_4-bridged $\{(RC{\equiv}C)(dmpe)_2Mn\}_2(\mu\text{-}C{\equiv}C-C{\equiv}C)$. A sequence consisting of one-electron oxidation of alkynyl complexes bearing a propargylic hydrogen atom, deprotonation and further oxidation of the resulting radical by one more electron has been established as a synthetical route to cationic allenylidene complexes[219] and ultimately given access to *trans*-[(dppe)$_2$Ru(=C=C=CPh$_2$)]$^{2+}$, a unique example of a bis(allenylidene) derivative.[166] An intriguing case of the redox-induced coupling of butadiynyl complexes *trans*-[Cl(dppe)$_2$Ru$-$C${\equiv}$C$-$C${\equiv}$CR (R = H, SiMe$_3$) to the highly delocalised alkynyl/allenylidene substituted cyclobutenylidene complex *trans*-[Cl(dppe)$_2$Ru$-$C${\equiv}$C$-$\{CH=CHCCH$_2$)=C=C=C={Ru(dppe)$_2$Cl}]$^+$ (Scheme 6.11) has also been reported.[163] The oxidation of the parent alkynyl complex is thermodynamically unfavourable but becomes feasible through the shifting of the equilibrium by irreversible subsequent processes, here desilylation in the case of the Me$_3$Si-protected alkynyl complex, formal [2 + 2] cycloaddition with another equivalent of the starting complex *via* the terminal $C_\gamma = C_\delta R$ bonds and finally abstraction of a hydrogen atom from the reaction medium.

Compared to this plethora of redox-initiated reactions of carbon-rich complexes, the ones that have been studied by *in situ* techniques constitute a vanishing minority with not even a handful of examples in the literature. The most remarkable study deals with the shifting of the equilibrium between alkyne and vinylidene isomers by electron transfer.[15] Thus, oxidation of the vinylidene complex $[(\eta^6\text{-}C_6Me_6)(CO)_2Cr=C=C(SiMe_3)_2]$ generates a

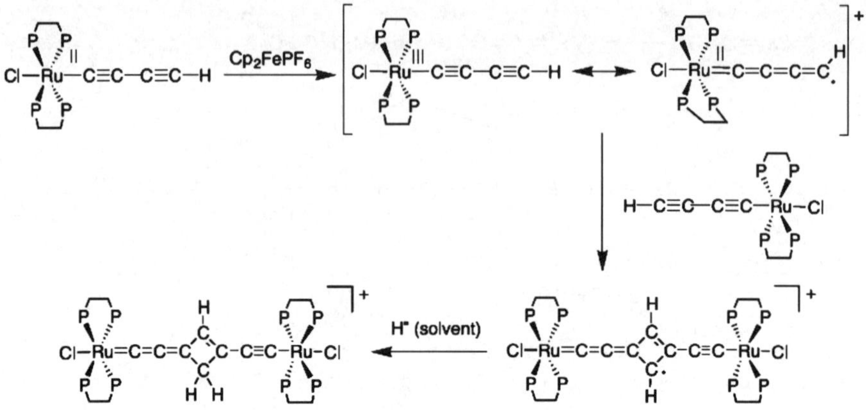

Scheme 6.11   Oxidation induced coupling of a butadiynyl ruthenium complex.

short-lived vinylidene cation that rapidly converts to the cationic alkyne com-
plex $[(\eta^6\text{-}C_6Me_6)(CO)_2Cr(\eta^2\text{-}Me_3SiC\equiv CSiMe_3]^+$ $(k_1 \gg k_{-1}$ in Scheme 6.12).
Reduction of the cationic alkyne isomer initially produces its neutral congener
which in turn backisomerises to the thermodynamically more stable neutral
vinylidene starting complex $(k_2 \gg k_{-2}$ in Scheme 6.12). The rates for the
redox-induced isomerisations in Scheme 6.12 were established by cyclic voltammetry at variable temperature. IR spectroelectrochemistry at low temperature was employed to spectroscopically characterise the neutral alkyne
complex and to determine the rate of the alkyne-to-vinylidene isomerisation at
temperatures where the reaction rate is too small to be determined by electrochemical techniques (Figure 6.19). IR spectroelectrochemistry also served to
probe for the extent and rate of the crossreaction between the neutral alkyne
and the cationic vinylidene complexes to give the cationic alkyne and neutral
vinylidene derivatives $(k_3$ and k_{-3} in Scheme 6.12). The cationic vinylidene
isomer is only present in small equilibrium quantities besides its alkyne isomer,
but appreciably affects the overall isomerisation rate. IR spectroelectrochemistry, along with kinetic CV analysis, thus provided a comprehensive data
set from which the thermodynamics of all the reactions involved in this ECEC
square scheme could be established.

Shaw and coworkers[16] have studied the anodically induced formation of the
σ,π complex $[\{CpFe(CO)_2\}_2(\mu\text{-}\eta^1{:}\eta^2\text{-}C\equiv CPh)]^+$ from $Cp(CO)_2Fe\text{-}C\equiv CPh$ by
infrared reflectance spectroelectrochemistry (IRRAS). This process goes along
with the cleavage of the Fe–alkynyl bond and attack of the $CpFe(CO)_2{}^+$
electrophile thus formed onto the parent alkynyl complex. Such processes are
commonly dubbed as mother–daughter (or father–son) reactions. The identity
of the electrogenerated product was established by comparing its IR absor-
bances to those of the independently prepared product.

In our laboratories we have studied the reduction of *trans*-[Cl(dppm)$_2$
Ru$=$C$=$C$=$CMePh]$^+$ by voltammetric and *in-situ* techniques.[36] The initially

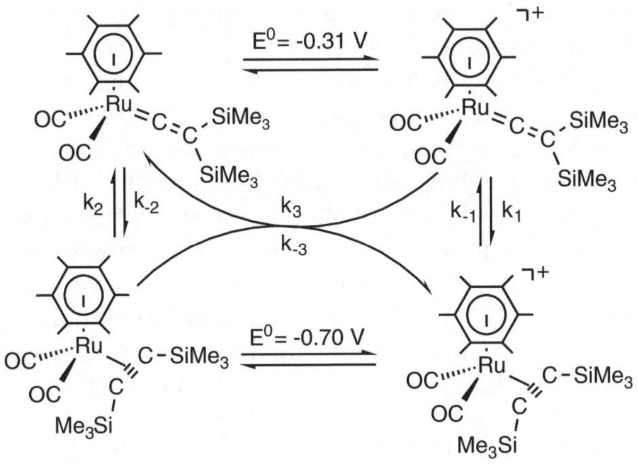

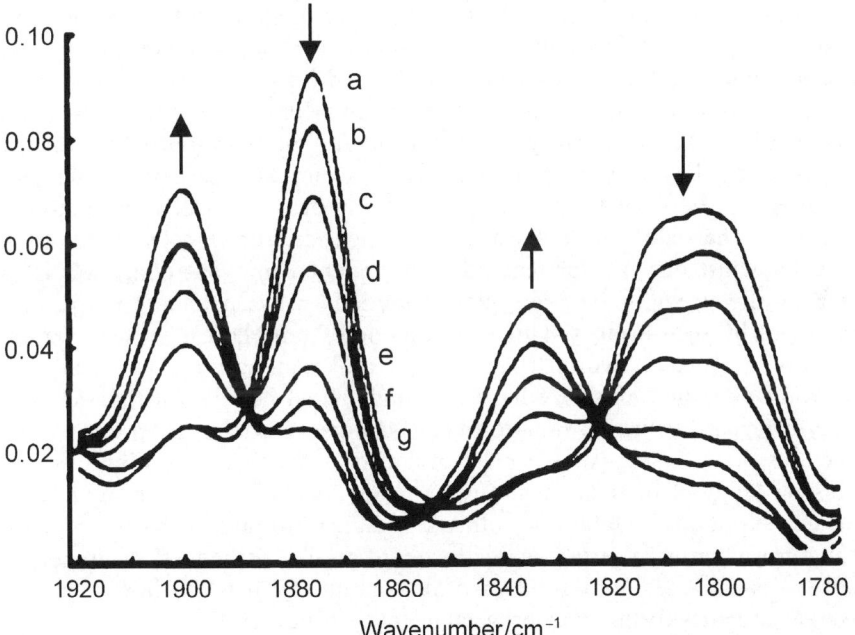

Scheme 6.12 Redox-induced isomerization of a ruthenium vinylidene and alkyne complex.

Figure 6.19 Monitoring of the conversion of electrogenerated $[(\eta^6\text{-}C_6Me_6)(CO)_2Cr(\eta^2\text{-}Me_3SiC\equiv CSiMe_3)]$ to the vinylidene isomer $[(\eta^6\text{-}C_6Me_6)(CO)_2Cr=C=C(SiMe_3)_2]$ inside an OTTLE cell. Spectra were recorded immediately after the electrochemical production at 228 K a) and after b) 2 min, c) 7 min, d) 11 min, e) 19 min, f) 22 min and g) 25 min at 234 K (reproduced with permission from ref. 15.

formed neutral radical gives rise to a weak alkynyl band near $2036\,cm^{-1}$ in agreement with a large contribution of the alkynyl-type resonance structure *trans*-[Cl(dppm)$_2$Ru−C≡C−C·(Me)(Ph)]. Such resonance structures are frequently proposed and have been experimentally verified for the reduced forms of cationic allenylidene complexes.[37,86] This band is gradually replaced by a new alkynyl absorption at a slightly higher energy of $2066\,cm^{-1}$. Chemical reduction of the allenylidene complex by cobaltocene produces the same species. This product has been identified as the butenynyl complex *trans*-[Cl(dppm)$_2$Ru−C≡C−CPh=CH$_2$] resulting from CH-bond homolysis of the primary radical. Chemical reduction of more concentrated solutions under NMR monitoring revealed the simultaneous formation of a chiral complex that was identified as *trans*-[Cl(dppm)$_2$Ru−C≡C−CHMePh], albeit in low quantities. This second product likely arises from disproportionation of the parent [Cl(dppm)$_2$Ru−C≡C−C·(Me)(Ph)]. It thus constitutes the product of a second-order process that requires fairly high analyte concentrations to be operative.

6.3   Concluding Remarks

With this account it was our aim to demonstrate the utility of spectro-electrochemistry within the field of carbon-rich organometallics. Such complexes are particularly amenable to such studies since they present characteristic spectroscopic labels that allow one to monitor the progress of the electrolysis. Simultaneously, these labels are highly revealing with respect to the charge and spin distribution in the electrogenerated forms and the alterations of the former brought about by the electron-transfer event. A spin-off of these characteristics is that spectroelectrochemistry allows one to address such questions as to the identity of the redox site associated with each electron-transfer event when more than just one possible candidate is present. Another application is to quantify the strengths of electronic interactions between remote redox sites connected by such bridges. This is an especially important issue since conjugated or cumulated hydrocarbon bridges rank amongst the most efficient electronic coupling units. We have also highlighted how useful this methodology is when it comes to monitoring the changes of molecular properties or functions or the nature or the rates of chemical reactions following electron transfer. Many of these issues are right at the heart of the current interest in carbon-rich organometallics. It is thus not difficult to predict that more work along these lines will appear, especially as spectroelectrochemical equipment is further disseminated across the scientific community. It is our hope that this account provides further encouragement toward these ends.

Abbreviations

ap	2-anilinopyridinate
bipy	2,2′-bipyridine
Fc	ferrocenyl, (η^5-C$_5$H$_5$)(η^5-C$_5$H$_4$)Fe
DDQ	2,3-dichloro-5,6-dicyano-*p*-benzoquinone

dippe	1,2-bis(diisopropylphosphanyl)ethane
DMBA	dimethylbenzimidate
dmpe	1,2-bis(dimethylphosphanyl)ethane
dppe	bis(diphenylphosphanyl)ethane
dppf	1,1'-bis(diphenylphosphanyl)ferrocene
dppm	bis(diphenylphosphanyl)methane
IRRAS	infrared reflection absorption spectroscopy
4-EtCOOpy	isonicotinate, NC_4H_4C-COOEt-4
4,4'-Me$_2$bipy	4,4'-dimethyl-2,2'-bipyridine
OTTLE	optically transparent thin-layer electrolysis
16-TMC	1,5,9,13-tetramethyltetra-azacyclohexadecane
Tp	hydridotris(pyrazolyl)borate, $HB(pz)_3^-$
Tp'	hydridotris(3,5-dimethylpyrazolyl)borate

References

1. McCreery in "Electrochemical Methods", ed. B. W. Rossiter, J. F. Hamilton, John Wiley & Sons, New York, 1986, Vol. **II**, pp. 591–692.
2. "Spectroelectrochemistry, Theory and Practice", ed. R. J. Gale, Plenum, New York, 1988.
3. F. M. Hawkridge, in *"Laboratory Techniques in Electroanalytical Chemistry"*, ed. P. T. Kissinger, W. R. Heinemann, Marcel Dekker Inc., New York, 1996, pp. 267–292.
4. F. Battaglini, E. J. Calvo and F. Doctorovich, *J. Organomet. Chem.*, 1997, **547**, 1.
5. A. Neudeck, F. Marken and R. G. Compton, in *"Electroanalytical Methods"*, ed. F. Scholz, Springer, Berlin, 2002, pp. 167–189.
6. J.-P. Launay, *Chem. Soc. Rev.*, 2001, **30**, 386.
7. C. E. Powell, M. P. Cifuentes, J. P. Morrall, R. Stranger, M. G. Humphrey, M. Samoc, B. Luther-Davies and G. A. Heath, *J. Am. Chem. Soc.*, 2003, **125**, 602.
8. M. Staffilani, P. Belser, L. De Cola and F. Hartl, *Eur. J. Inorg. Chem.*, 2003, 335.
9. A. M. Waller and R. G. Compton, in *"Comprehensive Chemical Kinetics"*, ed. R. G. Compton, Elsevier, Amsterdam, 1989, Vol. **29**.
10. N. G. Connelly and W. E. Geiger, *Chem. Rev.*, 1996, **96**, 877.
11. P. B. Graham and D. J. Curran, *Anal. Chem.*, 1992, **64**, 2688.
12. Southampton Electrochemistry Group in "Instrumental Methods in Electrochemistry", Horwood Publishing, Chichester, 2001.
13. C.-H. Pyun and S. M. Park, *Anal. Chem.*, 1986, **58**, 251.
14. M. J. Shaw and W. E. Geiger, *Organometallics*, 1996, **15**, 13.
15. N. G. Connelly, W. E. Geiger, M. C. Lagunas, B. Metz, A. L. Rieger, P. H. Rieger and M. J. Shaw, *J. Am. Chem. Soc.*, 1995, **117**, 12202.

16. M. J. Shaw, R. L. Henson, S. E. Houk, J. W. Westhoff, M. W. Jones and G. B. Richter-Addo, *J. Electroanal. Chem.*, 2002, **534**, 47.
17. M. E. Stoll and W. E. Geiger, *Organometallics*, 2004, **23**, 5818.
18. R. Gleiter, I. Hyla-Kryspin, P. Binger and M. Regitz, *Organometallics*, 1992, **11**, 177.
19. M.-H. Baik, T. Ziegler and C. K. Schauer, *J. Am. Chem. Soc.*, 2000, **122**, 9143.
20. N. J. Long, A. J. Martin, A. J. P. White, D. J. Williams, M. Fontani, F. Laschi and P. Zanello, *J. Chem. Soc., Dalton Trans.*, 2000, **3387**.
21. M. C. B. Colbert, J. Lewis, N. J. Long, P. R. Raithby, A. J. P. White and D. J. Williams, *J. Chem. Soc., Dalton Trans.*, 1997, 99.
22. N. D. Jones, M. O. Wolf and D. M. Giaquinta, *Organometallics*, 1997, **16**, 1352.
23. Y. Zhu, O. Clot, M. O. Wolf and G. P. A. Yap, *J. Am. Chem. Soc.*, 1998, **120**, 1812.
24. W. Weng, T. Bartik, M. Brady, B. Bartik, J. A. Ramsden, A. M. Arif and J. A. Gladysz, *J. Am. Chem. Soc.*, 1995, **117**, 11922.
25. K. M.-C. Wong, S. C.-F. Lam, C.-C. Ko, N. Zhu, V.W.-W. Yam, S. Roué, C. Lapinte, S. Fathallah, K. Costuas, S. Kahlal and J.-F. Halet, *Inorg. Chem.*, 2003, **42**, 7086.
26. F. Paul, W. E. Meyer, L. Toupet, H. Jiao, J. A. Gladysz and C. Lapinte, *J. Am. Chem. Soc.*, 2000, **122**, 9405.
27. V. Guillaume, P. Thominot, F. Coat, A. Mari and C. Lapinte, *J. Organomet. Chem.*, 1998, **565**, 75.
28. Y. Zhu and M. O. Wolf, *J. Am. Chem. Soc.*, 2000, **122**, 10121.
29. Y. Zhu, D. B. Millet, M. O. Wolf and S. J. Rettig, *Organometallics*, 1999, **18**, 1930.
30. M. Sato, H. Shintate, Y. Kawata, M. Sekino, M. Katada and S. Kawata, *Organometallics*, 1994, **13**, 1956.
31. M. Sato, Y. Hayashi, M. Katada and S. Kawata, *J. Organomet. Chem.*, 1994, **471**, 179.
32. M. Sato, Y. Hayashi, H. Shintate, M. Katada and S. Kawata, *J. Organomet. Chem.*, 1994, **471**, 179.
33. M. Sato, E. Mogi and M. Katada, *Organometallics*, 1995, **14**, 4837.
34. M. Sato, Y. Hayashi, S. Kumakura, N. Shimizu, M. Katada and S. Kawata, *Organometallics*, 1996, **15**, 721.
35. G.-L. Xu, M. C. DeRosa, R. J. Crutchley and T. Ren, *J. Am. Chem. Soc.*, 2004, **126**, 3728.
36. R. F. Winter, *"Rutheniumkomplexe mit hoch ungesättigten C3- und C4-Liganden aus Diacetylen"*, Verlag Grauer, Beuren Stuttgart, 2002.
37. R. F. Winter and S. Záliš, *Coord. Chem. Rev.*, 2004, **248**, 1565.
38. R. F. Winter, *Chem. Commun.*, 1998, 2209.
39. S. Hartmann, R. F. Winter, T. Scheiring and M. Wanner, *J. Organomet. Chem.*, 2001, **637–639**, 240.
40. H. Fischer and N. Szesni, *Coord. Chem. Rev.*, 2004, **248**, 1659.

41. B. Buriez, I. D. Burns, A. F. Hill, A. J. P. White, D. J. Williams and J. D. E. T. Wilton-Ely, *Organometallics*, 1999, **18**, 1504.
42. S. Hartmann, R. F. Winter, B. M. Brunner, B. Sarkar, A. Knödler and I. Hartenbach, *Eur. J. Inorg. Chem.*, 2003, 876.
43. J. Maurer, R. F. Winter, B. Sarkar, J. Fiedler and S. Záliš, *Chem. Commun.*, 2004, 1900.
44. C. J. Adams, S. L. James, X. Liu, P. R. Raithby and L. J. Yellowlees, *J. Chem. Soc. Dalton Trans.*, 2000, 63.
45. C. J. Adams, L. E. Bowen, M. G. Humphrey, J. P. L. Morrall, M. Samoc and L. J. Yellowlees, *Dalton Trans.*, 2004, 4130.
46. D. L. Lichtenberger, S. K. Renshaw and R. M. Bullock, *J. Am. Chem. Soc.*, 1993, **115**, 3276.
47. D. L. Lichtenberger, S. K. Renshaw, A. Wong and C. D. Tagge, *Organometallics*, 1993, **12**, 3522.
48. J. N. Louwen, R. Hengelmolen, D. M. Grove, A. Oskam and R. L. DeKock, *Organometallics*, 1984, **3**, 908.
49. J. E. McGrady, T. Lovell, R. Stranger and M. G. Humphrey, *Organometallics*, 1997, 4004.
50. C. D. Delfs, R. Stranger, M. G. Humphrey and A. M. McDonagh, *J. Organomet. Chem.*, 2000, **607**, 208.
51. K. D. John, T. C. Stoner and M. D. Hopkins, *Organometallics*, 1997, **16**, 4948.
52. R. Denis, L. Toupet, F. Paul and C. Lapinte, *Organometallics*, 2000, **19**, 4240.
53. F. Paul, J.-Y. Mevellec and C. Lapinte, *Dalton Trans.*, 2002, **1**, 1783.
54. K. Costuas, F. Paul, L. Toupet, J.-F. Halet and C. Lapinte, *Organometallics*, 2004, **23**, 2053.
55. F. Paul, K. Costuas, I. Ledoux, S. Deveau, J. Zyss, J.-F. Halet and C. Lapinte, *Organometallics*, 2002, **21**, 5229.
56. J. Manna, K. D. John and M. D. Hopkins, *Adv. Organomet. Chem*, 1995, **38**, 79.
57. M.-Y. Choi, M. C.-W. Chan, S. Zhang, K.-K. Cheung, C.-M. Che and K.-Y. Wong, *Organometallics*, 1999, 2074.
58. M. Sato and M. Watanabe, *Chem. Commun.*, 2002, 1574.
59. M. Sato, Y. Kawata, H. Shintate, Y. Habata, S. Akabori and K. Unoura, *Organometallics*, 1997, **16**, 1693.
60. M. Sato, A. Iwai and M. Watanabe, *Organometallics*, 1999, **18**, 3208.
61. S. Kheradmandan, K. Venkatesan, O. Blacque, H. W. Schmalle and H. Berke, *Chem. Eur. J.*, 2004, **10**, 4872.
62. M. I. Bruce, P. J. Low, K. Costuas, J.-F. Halet, S. P. Best and G. A. Heath, *J. Am. Chem. Soc.*, 2000, **122**, 1949.
63. M. I. Bruce, B. G. Ellis, P. J. Low, B. W. Skelton and A. H. White, *Organometallics*, 2003, **22**, 3184.
64. M. B. Robin and P. Day, *Adv. Inorg. Chem. Radiochem.*, 1967, **10**, 247.

65. S. Kheradmandan, K. Heinze, H. W. Schmalle, H. Berke, *Angew. Chem*, 1999, **111**, 2412; *Angew. Chem. Int. Ed. Engl.* 1999, **38**, 2270.
66. F. J. Fernández, O. Blacque, M. Alfonso and H. Berke, *Chem. Commun.*, 2001, 1266.
67. F. J. Fernández, K. Venkatesan, O. Blacque, M. Alfonso, H. W. Schmalle and H. Berke, *Chem. Eur. J.*, 2003, **9**, 6192.
68. J. W. Seyler, W. Weng, Y. Zhou and J. A. Gladysz, *Organometallics*, 1993, **12**, 3802.
69. Y. Zhou, J. W. Seyler, W. Weng, A. M. Arif and J. A. Gladysz, *J. Am. Chem. Soc.*, 1993, **115**, 8509.
70. M. Brady, W. Weng, Y. Zhou, J. W. Seyler, A. J. Amoroso, A. M. Arif, M. Böhme, G. Frenking and J. A. Gladysz, *J. Am. Chem. Soc.*, 1997, **119**, 775.
71. B. E. Woodworth, P. S. White and J. L. Templeton, *J. Am. Chem. Soc.*, 1997, **119**, 828.
72. N. Le Narvor and C. Lapinte, *J. Chem. Soc., Chem. Commun.*, 1993, 357.
73. N. Le Narvor, L. Toupet and C. Lapinte, *J. Am. Chem. Soc.*, 1995, **117**, 7129.
74. M. Guillemot, L. Toupet and C. Lapinte, *Organometallics*, 1998, **17**, 1928.
75. F. Coat and C. Lapinte, *Organometallics*, 1996, **15**, 477.
76. M. C. B. Colbert, J. Lewis, N. J. Long, P. R. Raithby, M. Younus, A. J. P. White, D. J. Williams, N. N. Payne, L. Yellowlees, D. Beljonne, N. Chawdhury and R. H. Friend, *Organometallics*, 1998, **17**, 3034.
77. L. D. Field, A. V. George, F. Laschi, E. Y. Malouf and P. Zanello, *J. Organomet. Chemistry*, 1992, **435**, 347.
78. N. Le Narvor and C. Lapinte, *Organometallics*, 1995, **14**, 634.
79. S. Le Stang, F. Paul and C. Lapinte, *Organometallics*, 2000, **19**, 1035.
80. T. Weyland, C. Lapinte, G. Frapper, M. J. Calhorda, J.-F. Halet and L. Toupet, *Organometallics*, 1997, **16**, 2024.
81. T. Weyland, K. Costuas, A. Mari, J.-F. Halet and C. Lapinte, *Organometallics*, 1998, **17**, 5569.
82. T. Weyland, K. Costuas, L. Toupet, J.-F. Halet and C. Lapinte, *Organometallics*, 2000, **19**, 4228.
83. R. F. Winter, K.-W. Klinkhammer and S. Záliš, *Organometallics*, 2001, **20**, 1317.
84. R. F. Winter, S. Hartmann, S. Záliš and K. W. Klinkhammer, *Dalton Trans.*, 2003, 2342.
85. R. F. Winter, *Eur. J. Inorg. Chem.*, 1999, 2121.
86. S. Rigaut, O. Maury, D. Touchard and P. H. Dixneuf, *Chem. Commun.*, 2001, 373.
87. S. Rigaut, J. Perruchon, S. Guesmi, C. Fave, D. Touchard and P. H. Dixneuf, *Eur. J. Inorg. Chem.*, 2005, 447.
88. N. Mantovani, M. Brugnati, L. Gonsalvi, E. Grigiotti, F. Laschi, L. Marvelli, M. Peruzzini, G. Reginato, R. Rossi and P. Zanello, *Organometallics*, 2005, **24**, 405.

89. A. Rabier, N. Lugan and G. L. Geoffroy, *Organometallics*, 1994, **13**, 4676.
90. A. Rabier, N. Lugan and R. Mathieu, *J. Organomet. Chem.*, 2001, **617–618**, 681.
91. V. Guillaume, V. Mahias, A. Mari and C. Lapinte, *Organometallics*, 2000, **19**, 1422.
92. B. Etzenhouser, M. D. B. Cavanough, H. N. Spurgeon and M. B. Sponsler, *J. Am. Chem. Soc.*, 1994, **116**, 2221.
93. B. A. Etzenhouser, Q. Chen and M. B. Sponsler, *Organometallics*, 1994, **13**, 4176.
94. M. B. Sponsler, *Organometallics*, 1995, **14**, 1920.
95. M.-C. Chung, X. Gu, B. A. Etzenhouser, A. M. Spuches, P. T. Rye, S. K. Seetharaman, D. J. Rose, J. Zubieta and M. B. Sponsler, *Organometallics*, 2003, **22**, 3485.
96. J. Maurer, B. Sarkar, S. Záliš and R. F. Winter, *J. Solid State Electrochem.*, 2005, **9**, 738.
97. M. Sato, A. Kudo, Y. Kawata and H. Saitoh, *Chem. Commun.*, 1996, 25.
98. M. Sato, Y. Kawata, A. Kudo, A. Iwai, H. Saitoh and S. Ochiai, *J. Chem. Soc., Dalton Trans.*, 1998, 2215.
99. M. Sato, T. Nagata, A. Tanemura, T. Fujihara, S. Kumakura and K. Unoura, *Chem. Eur. J.*, 2004, **10**, 2166.
100. A.-C. Ribou, J.-P. Launey, M. L. Sachtleben, H. Li and C. W. Spangler, *Inorg. Chem.*, 1996, **35**, 3735.
101. Y. J. Chen, D.-S. Pan, C.-F. Chiu, J.-X. Su, S. J. Lin and K. S. Kwan, *Inorg. Chem.*, 2000, **39**, 953.
102. S. Barlow and S. R. Marder, *Chem. Commun.*, 2000, 1555.
103. E. M. McGale, B. H. Robinson and J. Simpson, *Organometallics*, 2003, **22**, 931.
104. A. Aviram and M. A. Ratner, *Chem. Phys. Lett.*, 1974, **29**, 277.
105. F. Paul and C. Lapinte, *Coord. Chem, Rev.*, 1998, **178–180**, 431.
106. J. M. Tour, *Acc. Chem. Res.*, 2000, **33**, 791.
107. J. E. Sutton and H. Taube, *Inorg. Chem.*, 1981, **20**, 3125.
108. J.-P. Launay and C. Coudret, in *"Electron Transfer in Chemistry"*, ed. A. P. De Silva, V. Balzani, VCH, Weinheim, 2001, Vol. **5**, pp. 3–47.
109. F. Barrière, N. Camire, W. E. Geiger, U. T. Mueller-Westerhoff and R. Sanders, *J. Am. Chem. Soc.*, 2002, **124**, 7262.
110. A. B. P. Lever, *in "Comprehensive Organometallic Chemistry II"*, Elsevier, Amsterdam, 2003, Vol. **2**, pp. 435–438.
111. C. G. Atwood, W. E. Geiger and A. L. Rheingold, *J. Am. Chem. Soc.*, 1993, **115**, 5310.
112. C. G. Atwood and W. E. Geiger, *J. Am. Chem. Soc.*, 2000, **122**, 5477.
113. N. S. Hush, *Coord. Chem. Rev.*, 1985, **64**, 135.
114. T. Bartik, B. Bartik, M. Brady, R. Dembinski, J. A. Gladysz, *Angew. Chem.*, 1996, **35**, 467; *Angew. Chem. Int. Ed. Engl.*, 1996, **108**, 414.
115. R. Dembinski, T. Bartik, B. Bartik, M. Jaeger and J. A. Gladysz, *J. Am. Chem. Soc.*, 2000, **122**, 810.
116. S. Szafert and J. A. Gladysz, *Chem. Rev.*, 2003, **103**, 4175.

117. S. Szafert, P. Haquette, S. B. Falloon and J. A. Gladysz, *J. Organomet. Chem*, 2000, **604**, 52.

118. H. Jiao, K. Costuas, J. A. Gladysz, J.-F. Halet, M. Guillemot, L. Toupet, F. Paul and C. Lapinte, *J. Am. Chem. Soc.*, 2003, **125**, 9511.

119. F. Coat, M.-A. Guillevic, L. Toupet, F. Paul and C. Lapinte, *Organometallics*, 1997, **16**, 5988.

120. G.-L. Xu, G. Zou, Y.-H. Ni, M. C. DeRosa, R. J. Crutchley and T. Ren, *J. Am. Chem. Soc.*, 2003, **125**, 10057.

121. L.-B. Gao, L.-Y. Zhang, L.-X. Shi and Z.-N. Chen, *Organometallics*, 2005, **24**, 1678.

122. N. S. Hush, *Prog. Inorg. Chem.*, 1967, **8**, 391.

123. C. Creutz, *Progr. Inorg. Chem.*, 1983, **30**, 1.

124. J. A. Reimers and N. S. Hush, *Chem. Phys.*, 1990, **146**, 89.

125. J. R. Reimers and N. S. Hush, *J. Photochem. Photobiol. A: Chem.*, 1994, **82**, 31.

126. A. Ceccon, S. Santi, L. Orian and A. Bisello, *Coord. Chem. Rev.*, 2004, **248**, 683.

127. J. H. K. Yip, J. Wu, K.-Y. Wong, K.-W. Yeung and J. J. Vittal, *Organometallics*, 2002, **21**, 1612.

128. J. H. K. Yip, J. Wu, K.-Y. Wong, K. P. Ho, C. S.-G. Pun and J. J. Vittal, *Organometallics*, 2002, **21**, 5292.

129. S. C. Jones, V. Coropceanu, S. Barlow, T. Kinnibrugh, T. Timofeeva, J.-L. Brédas and S. R. Marder, *J. Am. Chem. Soc.*, 2004, **126**, 11782.

130. H. P. Xia, R. C. Y. Yeung and G. Jia, *Organometallics*, 1998, **17**, 4762.

131. S. H. Liu, H. Xia, T. B. Wen, Z. Zhou and G. Jia, *Organometallics*, 2003, **22**, 737.

132. S. H. Liu, Y. Chen, K. L. Wan, T. B. Wen, Z. Zhou, M. F. Lo, I. D. Williams and G. Jia, *Organometallics*, 2002, **21**, 4984.

133. T. Ito, T. Hamaguchi, H. Nagino, T. Yamaguchi, H. Kido, I. S. Zavarine, T. Richmond, J. Washington and C. P. Kubiak, *J. Am. Chem. Soc.*, 1999, **121**, 4625.

134. T. Yamaguchi, N. Imai, T. Ito and C. P. Kubiak, *Bull. Chem. Soc. Jpn*, 2000, **73**, 1205.

135. C. Patoux, C. Coudret, J.-P. Launay, C. Joachim and A. Gourdon, *Inorg. Chem.*, 1997, **36**, 5037.

136. S. Fraysse, C. Coudret and J.-P. Launay, *J. Am. Chem. Soc.*, 2003, **125**, 5880.

137. H. Xia, T. B. Wen, Q. Y. Hu, X. Wang, X. Chen, L. Y. Shek, I. D. Williams, K. S. Wong, G. K. L. Wong and G. Jia, *Organometallics*, 2005, **24**, 562.

138. C.-C. Huang, Y.-C. Lin, S.-L. Huang, Y.-H. Liu and Y. Wang, *Organometallics*, 2003, **22**, 1512.

139. B. Bildstein, *Coord. Chem. Rev.*, 2000, **206–207**, 369.

140. W. Skibar, H. Kopacka, K. Wurst, C. Salzmann, K.-H. Ongania, F. Fabrizi de Biani, P. Zanello and B. Bildstein, *Organometallics*, 2004, **23**, 1024.

141. B. Bildstein, O. Loza and Y. Chizhov, *Organometallics*, 2004, **23**, 1825.
142. T. Bartik, W. Weng, J. A. Ramsden, S. Szafert, S. B. Falloon, A. M. Arif and J. A. Gladysz, *J. Am. Chem. Soc.*, 1998, **120**, 11071.
143. B. E. Woodworth and J. L. Templeton, *J. Am. Chem. Soc.*, 1996, **118**, 7418.
144. R. Dembinski, S. Szafert, P. Haquette, T. Lis and J. A. Gladysz, *Organometallics*, 1999, **18**, 5438.
145. R. D. Dewhurst, A. F. Hill and A. C. Willis, *Organometallics*, 2004, **23**, 1646.
146. R. D. Dewhurst, A. F. Hill, M. K. Smith, *Angew. Chem.*, 2004, **116**, 482; *Angew. Chem. Int. Ed. Engl.*, 2004, **43**, 476.
147. R. D. Dewhurst, A. F. Hill and A. C. Willis, *Organometallics*, 2004, **23**, 5903.
148. P. Belanzoni, N. Re and A. Sgamellotti, *J. Organomet. Chem.*, 2002, **656**, 156.
149. H. Jiao and J. A. Gladysz, *New. J. Chem.*, 2001, **25**, 551.
150. C. Hartbaum, G. Roth and H. Fischer, *Chem. Ber./Recueil*, 1997, **130**, 479.
151. C. Hartbaum and H. Fischer, *Chem. Ber./Recueil*, 1997, **130**, 1063.
152. C. Hartbaum, G. Roth and H. Fischer, *Eur. J. Inorg. Chem.*, 1998, 191.
153. G. Jia, H. P. Xia, W. F. Wu and W. S. Ng, *Organometallics*, 1996, **15**, 3634.
154. H. P. Xia, W. F. Wu, W. S. Ng, I. D. Williams and G. Jia, *Organometallics*, 1997, **13**, 2940.
155. H. P. Xia, W. S. Ng, J. S. Ye, X.-L. Li, W. T. Wong, Z. Lin, C. Yang and G. Jia, *Organometallics*, 1999, **18**, 4552.
156. W. Weng, T. Bartik, M. T. Johnson, A. M. Arif and J. A. Gladysz, *Organometallics*, 1995, **14**, 889.
157. M. S. Morton, J. P. Selegue and A. Carrillo, *Organometallics*, 1996, **15**, 4664.
158. H. Fischer, F. Leroux, R. Stumpf and G. Roth, *Chem. Ber.*, 1996, **129**, 1475.
159. H. Fischer, F. Leroux, G. Roth and R. Stumpf, *Organometallics*, 1996, **15**, 3723.
160. H. Fischer, O. Podschadly, G. Roth, S. Herminghaus, S. Kiewitz, J. Heck, S. Houbrechts and T. Meyer, *J. Organomet. Chem.*, 1997, **541**, 321.
161. F. Leroux, R. Stumpf and H. Fischer, *Eur. J. Inorg. Chem.*, 1998, 1225.
162. B. Fuss, M. Dede, B. Weibert and H. Fischer, *Organometallics*, 2002, **21**, 4425.
163. S. Rigaut, L. Le Pichon, J.-C. Daran, D. Touchard and P. H. Dixneuf, *Chem. Commun.*, 2001, 1206.
164. S. Guesmi, D. Touchard and P. H. Dixneuf, *Chem. Commun.*, 1996, 2773.
165. S. Rigaut, J. Massue, D. Touchard, J.-L. Fillaut, S. Golhen, P. H. Dixneuf, *Angew. Chem.*, 2002, **114**, 4695; *Angew. Chem. Int. Ed. Engl.*, 2002, **41**, 4513.

166. S. Rigaut, K. Costuas, D. Touchard, J.-Y. Saillard, S. Golhen and P.H. Dixneuf, *J. Am. Chem. Soc.*, 2004, **126**, 4072.

167. Y. Shi, G. T. Yee, G. Wang and T. Ren, *J. Am. Chem. Soc.*, 2004, **126**, 10552.

168. D. S. Frohnapfel, B. E. Woodworth, H. H. Thorp and J. L. Templeton, *J. Phys. Chem.*, 1998, **102**, 5665.

169. M. Glöckle, W. Kaim and J. Fiedler, *Organometallics*, 1998, **17**, 4923.

170. N. J. Long, C. K. Williams, *Angew. Chem.*, 2003, **115**, 2690; *Angew. Chem. Int. Ed. Engl.*, 2003, **42**, 2586.

171. I. R. Whittall, A. W. McDonagh and M. G. Humphrey, *Adv. Organomet. Chem.*, 1998, **42**, 291.

172. I. R. Whittall, A. M. McDonagh, M. G. Humphrey and M. Samoc, *Adv. Organomet. Chem.*, 1999, **43**, 349.

173. S. Di Bella, *Chem. Soc. Rev.*, 2001, **30**, 355.

174. M. G. Humphrey and C. P. Powell, *Coord. Chem. Rev.*, 2004, **248**, 725.

175. V. W.-W. Yam, *Chem. Commun.*, 2001, 789.

176. V. W.-W. Yam and K. M. C. Wong, *Top. in Curr. Chem.*, 2005, **257**, 1.

177. K.-L. Cheung, S.-K. Yip and V. W.-W. Yam, *J. Organomet. Chem.*, 2004, **689**, 4451.

178. C. W. Baxter, T. C. Higgs, A. C. Jones, S. Parsons, P. J. Bailey and P. A. Tasker, *Dalton Trans.*, 2002, 4395.

179. P. A. Jelliss, K. M. Wampler and A. Siemiarczuk, *Organometallics*, 2005, **24**, 707.

180. B. J. Coe, *Chem. Eur. J.*, 1999, **5**, 2464.

181. C. E. Powell, M. G. Humphrey, M. P. Cifuentes, J. P. Morrall, M. Samoc and B. Luther-Davies, *J. Phys. Chem. A*, 2003, **127**, 11264.

182. B. J. Coe, S. Houbrechts, I. Asselberghs and A. Persoons, *Angew. Chem.*, 1999, **111**, 377; *Angew. Chem. Int. Ed. Engl.*, 1999, **38**, 366.

183. I. Asselberghs, K. Clays, A. Persoons, A. M. McDonagh, M. D. Ward and J. A. McCleverty, *Chem. Phys. Lett.*, 2003, **368**, 408.

184. M. P. Cifuentes, C. E. Powell, M. G. Humphrey, G. A. Heath, M. Samoc and B. Luther-Davies, *J. Phys. Chem. A*, 2001, **105**, 9625.

185. T. Weyland, I. Ledoux, S. Brasselet, J. Zyss and C. Lapinte, *Organometallics*, 2000, **19**, 5235.

186. G. De Santis, L. Fabrizzi, M. Licchelli, C. Mangano and D. Sacchi, *Inorg. Chem.*, 1995, **34**, 2581.

187. G. De Santis, L. Fabrizzi, M. Licchelli, N. Sardone and A. H. Velders, *Chem. Eur. J.*, 1996, **2**, 1243.

188. S. H.-F. Chong, S. C.-F. Lam, V. W.-W. Yam, N. Zhu, K.-K. Cheung, S. Fathallah, K. Costuas and J.-F. Halet, *Organometallics*, 2004, **23**, 4924.

189. Y. Yamamoto, M. Shiotsuka and S. Onaka, *J. Organomet. Chem.*, 2004, **689**, 2905.

190. M. Younus, A. Köhler, S. Cron, N. Chawdhury, M. R. A. Al-Mandhary, M. S. Khan, J. Lewis, N. J. Long, R. H. Friend, P. R. Raithby, *Angew. Chem.*, 1998, **110**, 3180; *Angew. Chem. Int. Ed. Engl*, 1998, **37**, 3036.

191. W.-Y. Wong, K.-H. Choi and K.-W. Cheah, *Dalton Trans.*, 2000, 113.
192. M. S. Khan, M. K. Al-Suti, M. R. A. Al-Mandhary, B. Ahrens, J.K. Bjernemose, M. F. Mahon, L. Male, P. R. Raithby, R. H. Friend, A. Köhler and J. S. Wilson, *Dalton Trans.*, 2003, 65.
193. I. Fratoddi, C. Battocchio, A. Furlani, P. Mataloni, G. Polzonetti and M. V. Russo, *J. Organomet. Chem.*, 2003, **674**, 10.
194. W. Lu, M. C. W. Chan, Z. Hui, C.-M. Che, N. Zhu and S.-T. Lee, *J. Am. Chem. Soc.*, 2004, **126**, 4958.
195. Q.-Z. Yang, Q.-X. Tong, L.-Z. Wu, Z.-X. Wu, L.-P. Zhang and C.-H. Tung, *Eur. J. Inorg. Chem.*, 2004, 1948.
196. V. W.-W. Yam, *Coord. Chem. Rev.*, 2003, **245**, 39.
197. Q.-H. Wei, G.-Y. Yin, L.-Y. Zhang, L.-X. Shi, Z.-M. Mao and Z.-N. Chen, *Inorg. Chem.*, 2004, **43**, 3484.
198. T. E. Müller, S. W.-K. Choi, D. M. P. Mingos, D. Murphy, D. J. Williams and V. W.-W. Yam, *J. Organomet. Chem.*, 1994, **484**, 209.
199. B.-C. Tzeng, W.-C. Lo, C.-M. Che and S.-M. Peng, *Chem. Commun.*, 1996, 181.
200. S.-C. Chan, M. C. W. Chan, Y. Wang, C.-M. Che, K.-K. Cheung and N. Zhu, *Chem. Eur. J.*, 2001, **7**, 4180.
201. W. Lu, H.-F. Xiang, N. Zhu and C.-M. Che, *Organometallics*, 2002, **21**, 2343.
202. V. W.-W. Yam, K.-L. Cheung, E. C.-C. Cheng, N. Zhu, K.-K. Cheung, *Dalton Trans.*, 2003.
203. I. E. Pomestchenko and F. N. Casellano, *J. Phys. Chem. A*, 2004, **108**, 3485.
204. F. Hua, S. Kinayyigit, J. R. Cable and F. N. Castellano, *Inorg. Chem.*, 2005, **44**, 471.
205. C. J. McAdam, J. L. Morgan, B. H. Robinson, J. Simpson, P. H. Rieger and A. L. Rieger, *Organometallics*, 2003, **22**, 5126.
206. D. Astruc, M.-C. Daniel and J. Ruiz, *Chem. Comm.*, 2004, 2637.
207. P. V. Bernhardt and E. G. Moore, *Aust. J. Chem.*, 2003, **56**, 239.
208. K. M.-C. Wong, W.-S. Tang, X.-X. Lu, N. Zhu and V. W.-W. Yam, *Inorg. Chem.*, 2005, **44**, 1492.
209. P. K. M. Siu, S.-W. Lai, W. Lu, N. Zhu and C.-M. Che, *Eur. J. Inorg. Chem.*, 2003, **1**, 2749.
210. X.-X. Lu, C.-K. Li, C.-C. Cheng, N. Zhu and V. W.-W. Yam, *Inorg. Chem.*, 2004, **43**, 2225.
211. R. L. Beddoes, C. Bitchon, R. W. Grime, A. Ricalton and M. W. Whiteley, *J. Chem. Soc., Dalton Trans.*, 1995, 2873.
212. Y. Hayashi, M. Osawa, K. Kobayashi and Y. Wakatsuki, *Chem. Commun.*, 1996, 1617.
213. D. A. Valyaev, O. A. Semeikin and N. A. Ustynyuk, *Coord. Chem. Rev.*, 2004, **248**, 1679.
214. K. Venkatesan, O. Blacque, T. Fox, M. Alfonso, H. W. Schmalle and H. Berke, *Organometallics*, 2004, **23**, 1183.
215. R. S. Iyer and J. P. Selegue, *J. Am. Chem. Soc.*, 1987, **109**, 910.

216. L. N. Novikova, M. G. Peterleitner, K. A. Sevumyan, O. V. Semeikin, D. A. Valyaev, N. A. Ustynyuk, V. N. Khrustalev, L. N. Kuleshova and M. Y. Antipin, *J. Organomet. Chem.*, 2001, **631**, 47.
217. D. Unseld, V. V. Krivykh, K. Heinze, F. Wild, G. Artus, H. Schmalle and H. Berke, *Organometallics*, 1999, **18**, 1525.
218. D. A. Valyaev, M. G. Peterleitner, L. I. Leont'eva, L. N. Novikova, O. V. Semeikin, V. N. Khrustalev, M. Y. Antipin, N. A. Ustynyuk, B. W. Skelton and A. H. White, *Organometallics*, 2003, **22**, 5491.
219. S. Rigaut, F. Mornier, F. Mousset, D. Touchard and P. H. Dixneuf, *Organometallics*, 2002, **21**, 2654.
220. C. J. Adams and S. J. A. Pope, *Inorg. Chem.*, 2004, **43**, 3492.

CHAPTER 7
EPR Spectroelectrochemistry

P. R. MURRAY AND L. J. YELLOWLEES

School of Chemistry, University of Edinburgh, The King's Buildings, West Mains Road, Edinburgh, EH9 3JJ

7.1 Introduction

The *in situ* combination of electrochemistry and EPR spectroscopy make for an ideal partnership since a reversible one-electron-transfer process must result in either a paramagnetic starting material or product species. It is therefore surprising that there are considerably fewer examples of the use of EPR spectroelectrochemistry in the literature compared to its UV–Vis analogue. This chapter will discuss EPR spectroelectrochemical cells, the experiment itself and what information can be gained about the system under study. It will not consider EPR spectra that have been collected as a result of electrogeneration performed *ex situ* and the paramagnetic solution then transferred to an EPR tube. Furthermore, only paramagnetic species in solution will be discussed.

In general, EPR spectroelectrochemical experiments are usually conducted on species that are diamagnetic in their starting redox state and hence the EPR signal appears on electrogeneration. Transition-metal coordination compounds often show rich electrochemical behaviour and it is from this class of compounds that many of the examples considered here are taken.

EPR spectrochemical experiments yield information about (i) the site of redox activity in the compound[1-4] (ii) the contribution of various nuclei to the molecular orbital occupied by the unpaired electron that may then be checked against the results of theoretical calculations and (iii) the half-wave potentials of systems where direct measurement is difficult (usually as a result of slow electron-transfer rates). Examples of each of these have been taken from our investigations and are detailed below. Initially, however, the cell used in these experiments is described followed by a review of some other cell designs.

Spectroelectrochemistry
Edited by Wolfgang Kaim and Axel Klein
© Royal Society of Chemistry, 2008

7.2 Experimental Setup

7.2.1 Design of the *In-Situ*, Variable-Temperature EPR Spectroelectrochemical Cell

One of the challenges of combining electrochemical techniques with EPR spectroscopy is that the solvents most suited to the electrochemical experiment (*e.g.* dimethylformamide or acetonitrile) have a high dielectric constant that makes them less than ideal for study by EPR spectroscopy. This is due to the fact that the radiation required for EPR spectroscopy lies in the microwave region. Solvents with high dielectric constants absorb microwaves and thus a spectrum often cannot be obtained in a cylindrical EPR tube. Such problems can be overcome by using a flat cell with a small pathlength. An additional problem lies in the fact that many redox products are not stable at room temperature. These may be inhibited from decomposing by decreasing the temperature. Therefore, the ability to be able to generate paramagnetic species at depressed temperatures is of enormous benefit to these spectroelectrochemical experiments. The temperature may be decreased further such that a frozen glass is obtained and therefore the cell design must also incorporate the capacity to record both the solution and solid EPR spectra of the same solution. The EPR cavity used for variable-temperature (VT) work in the X-band is narrow and thus the flat cell could be no wider than $\sim$4 mm. EPR spectroelectrochemical results are usually run on an X-band spectrometer (9.5 GHz) and all results given herein are indeed recorded at this field strength.

Thus, the desire was to design a narrow, robust, easily cleaned, flat cell in which the pathlength is short enough such that microwave absorption is minimal. The cell was also required to be spacious enough such that the three electrodes for electrogeneration could be incorporated.

A schematic of the *in-situ* variable-temperature spectroelectrochemical EPR cell used in our laboratory is shown in Figure 7.1.[5] The quartz flat cell is purchased from Wilmad and then customised in-house. The Pt/Rh gauze working electrode is placed in the flat region of the cell with the Ag/AgCl reference electrode and Pt counter-electrode placed as near to the working electrode as the cell geometry will allow. The connecting Pt wire from the gauze electrode is sheathed in PTFE tubing so that electrolysis only occurs in the flat portion of the cell. The total cell volume is 2 ml. The temperature of the flat cell may be varied by placing it inside the variable-temperature EPR cavity and passing prechilled N$_2$ gas over its surface. This cell design has been successfully used over the temperature range 100–350 K.

Variable-temperature work was carried out using a Bruker ER 4111 VT unit. EPR spectra were recorded on an X-band Bruker ER 200 D spectrometer using EPR acquisition system version 2.42 software.[6] All EPR spectra were corrected for DPPH, $g = 2.0036 \pm 0.0003$. EPR solution spectra were simulated using WIN-EPR Simfonia software[7] and frozen glass spectra by the EPR service in Manchester. Spectroelectrochemical electron-transfer reactions were initiated using a BAS CV-27 voltammograph potentiostat.

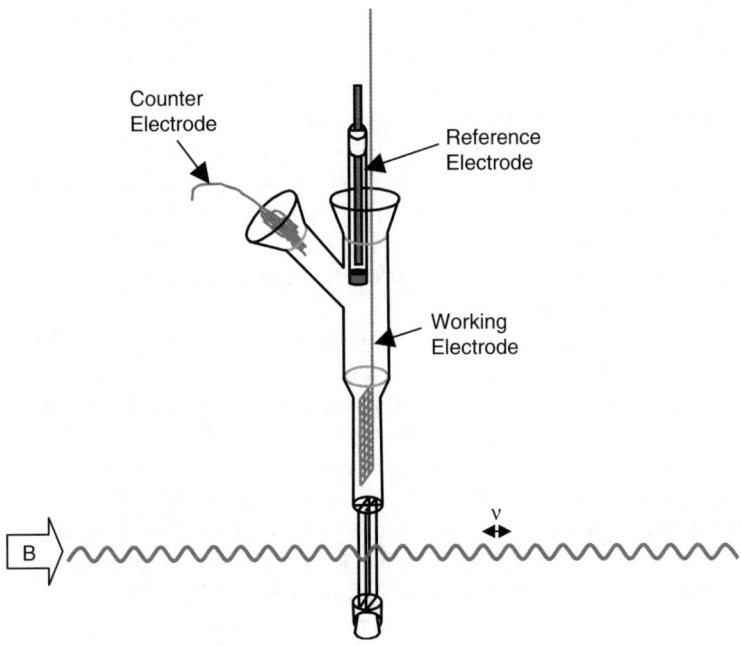

Figure 7.1   Schematic representation of the *in-situ*, variable-temperature flat cell for EPR spectroelectrochemistry.

7.2.2   Other Reported Cell Designs

Hartl and Vlcek[8] describe a low-temperature, air-tight EPR cell based on a round-tube design. The Au working electrode has a helical shape with the Pt counter electrode placed down the middle of the working electrode.[9] The electrochemical circuit is completed by a Ag wire pseudoreference electrode. The small solution volume between the working electrode and the cell wall permits the use of solvents such as CH_3CN. The cell was used to study the one-electron oxidation of 3,6-diphenyl-1,2-dithiine in CH_2Cl_2 and the radical anions of 6-methyl-6-phenyl fulvene and *fac*-[Re(Bn)(CO)$_3$(dmb)]⁻, where Bn = CH$_2$Ph, dmb = 4,4'-dimethyl-2,2'-bipyridine. Hartl and Vlcek used a Pt minigrid electrode in an EPR tube to study the redox chemistry of [Mn(CO)$_3$(DBCat)]⁻, DBCat = 3,5-di-*tert*-butyl catecholate in THF.[8]

Bond and coworkers[10,11] have developed a small-volume (0.2 ml) variable-temperature EPR spectroelectrochemical cell that enables simultaneous rapid-scan voltammetry and EPR measurements to be made. The performance of this cell is compared to that of a flow-through cell designed by Coles and Compton.[12,13] The small-volume cell permits cyclic voltammetric studies at variable temperatures but has significantly lower sensitivity compared to the flow-through cell, which is not amenable to low-temperature work.

In 1990 Kaim *et al.*[14] detailed an *in-situ* two-electrode EPR cell with a Pt tip working electrode and a Pt wire electrode contained in an EPR tube with an internal diameter of 1 mm. Radical species are generated by short electro-lysis times. Recent uses of this cell include the study of [(acac)$_2$Ru(μ-bpytz)Ru(acac)], (acac = acetyacetonate, bpytz = (3,5-dimethylpyrazolyl)-1,2,4,5-tetrazine) in CHCl$_2$ in which the reduction site is confirmed to be the bridging bpytz ligand.[15] The EPR spectroelectrochemical study of the family of com-plexes (abpy)Re(CO)$_3$(Hal) (Hal = F, Cl, Br, I, abpy = 2,2′-azobispyridine) confirms the one-electron reduction to be primarily localised on the abpy lig-and.[16] However, coupling is observed of the unpaired electron to the halide ligand when Hal = F, since fluoride has the largest isotropic hyperfine coupling constant for all the halides.

A laminated platinum-mesh working electrode positioned in an EPR flat cell combined with a Ag pseudoreference electrode and a Pt counter electrode was used by Dunsch and coworkers[17] for the *in-situ* study of C$_{120}$O. The same group also performed simultaneous variable temperature EPR/UV–Vis–NIR experi-ments on Wurster's reagent and thianthrene.[18,19] Similar cells using laminated ITO and gold working electrodes have also been reported.[20,21]

An EPR microspectroelectrochemical low-temperature cell has been detailed by Wilgocki and Rybak[22] with a Pt working electrode and has been used to characterise the unstable cation [O = Re(OEt)Cl$_2$(py)$_2$]$^+$.

7.3   EPR Spectroelectrochemical Experiment

Figure 7.2 shows the growth of the EPR signal as [Pt(bpy)Cl$_2$], bpy = 2,2′-bipyridine, is reduced by one-electron in DMF/0.1M [^{n}Bu$_4$N][BF$_4$] using the cell shown in Figure 7.1[5] at 253 K. The diamagnetic [Pt(bpy)Cl$_2$] is EPR-silent whereas the one-electron reduction product [Pt(bpy)Cl$_2$]$^-$ is paramagnetic and responsible for the growth of the signal observed in Figure 7.2. The EPR signal arises from coupling of the unpaired electron to ^{195}Pt ($I = \frac{1}{2}$, natural abundance 34%). The spectra show several clear isosbestic points that indicate a clean conversion from the neutral starting material to the monoreduced paramagnetic product. Further discussion of the spectrum of [Pt(bpy)Cl$_2$]$^-$ is given later.

It should be noted here that the temperature at which the spectra are re-corded can have a dramatic effect on the resolution of the signal. Figure 7.3 shows the EPR spectra of [Pt(4,4′-(NO$_2$)$_2$-bpy)Cl$_2$]$^{2-}$ generated at –0.75 V in DMF/0.5M [^{n}Bu$_4$N][BF$_4$] at several temperatures.[23] The signal is significantly better resolved at 313 K compared to 233 K. As the temperature of the cell is decreased so the viscosity of the solution is increased, thereby hindering the tumbling of the paramagnetic species and resulting in loss of signal resolution. Thus, the variable-temperature facility of the *in-situ* cell may well enable reactive intermediates to be generated and stabilised, however, low tempera-tures do not necessarily lead to well-resolved EPR spectra. It is also often observed that increasing the temperature of the cell in an attempt to improve the resolution results in collapse of the EPR signal and loss of resolution as the redox product decomposes at the elevated temperature. It is therefore necessary

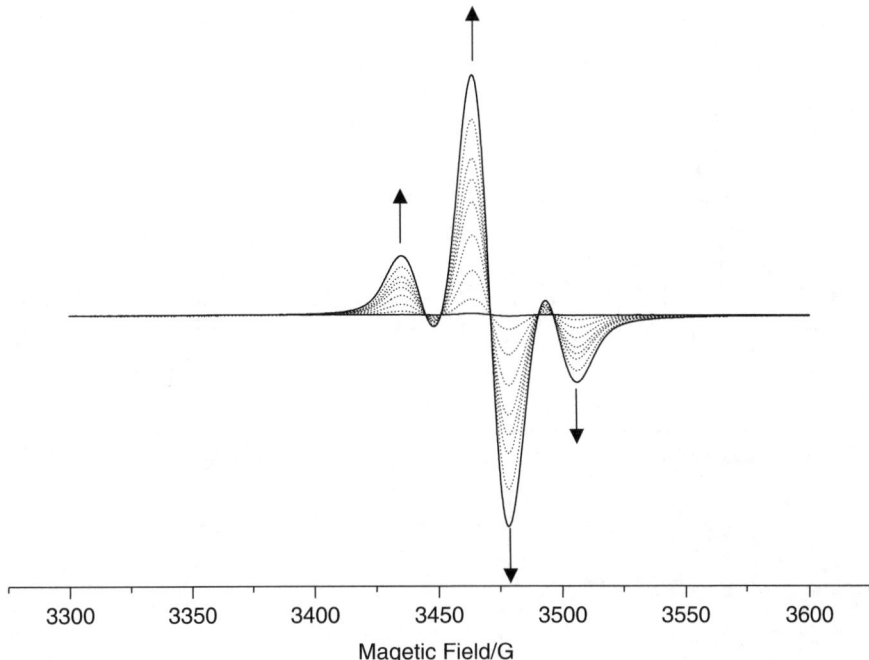

Figure 7.2 EPR spectroelectrochemical study of one-electron reduction of [Pt(bpy)Cl$_2$] ($E_{gen} = -1.6$ V) in DMF/0.1 M [^{n}Bu$_4$N][BF$_4$] at 253 K.

to record any EPR spectrum at several temperatures in order to obtain the optimum result.

The chemical integrity of a redox-active solution is easily established using UV–Vis spectroelectrochemistry since after recording the spectrum of the reduced or oxidised spectrum the applied potential may be reversed such that the starting material is reformed. The UV–Vis spectrum should then change from that of the redox product to that of the starting material, thereby ensuring that the redox product has not undergone a chemical reaction and that the recorded spectrum is indeed that species generated as a result of the electron-transfer process only. Unfortunately, this definitive test is not open to the EPR spectroelectrochemical experiment. The starting material is usually EPR-silent as are most decomposition products. It is therefore best practice to firstly carry out the UV–Vis spectroelectrochemical study and ascertain which temperature is necessary to stabilise the redox product and then to undertake the EPR experiment at this temperature.

The choice and concentration of electrolyte can be important in EPR spectroelectrochemical experiments. Usually, it is assumed that the electrolyte, added to lower the resistance of the solvent and hence aid the electrogeneration process, is inert. This is not always the case, particularly, if the unpaired electron occupies an orbital that is localised on one part of the redox-active molecule. Such a situation will lead to a highly polarised redox product that

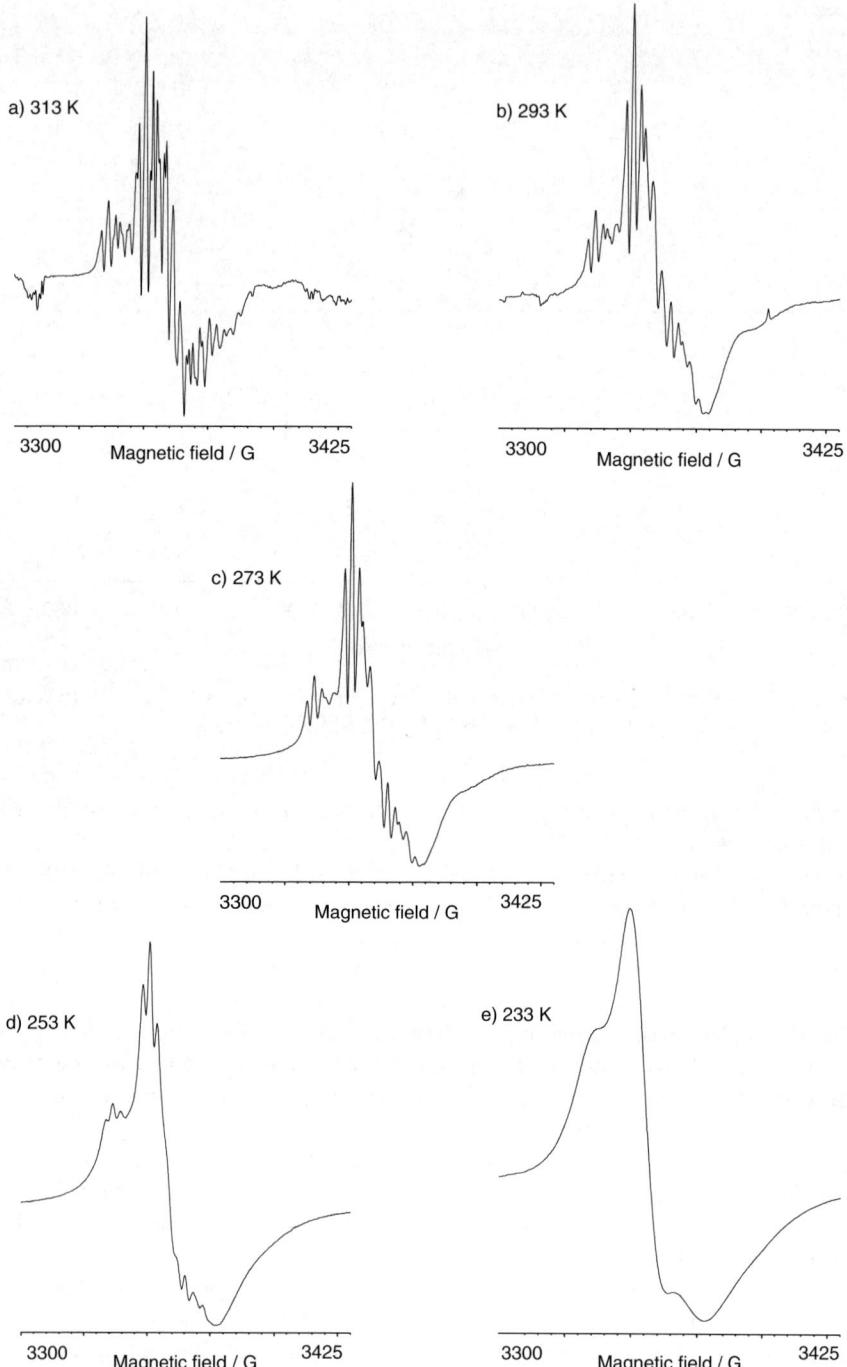

Figure 7.3   EPR spectra of [Pt(4,4'-(NO$_2$)$_2$-bpy)Cl$_2$]$^{2-}$ generated at –0.75 V in DMF/
0.5M [^{n}Bu$_4$N][BF$_4$] at several temperatures.

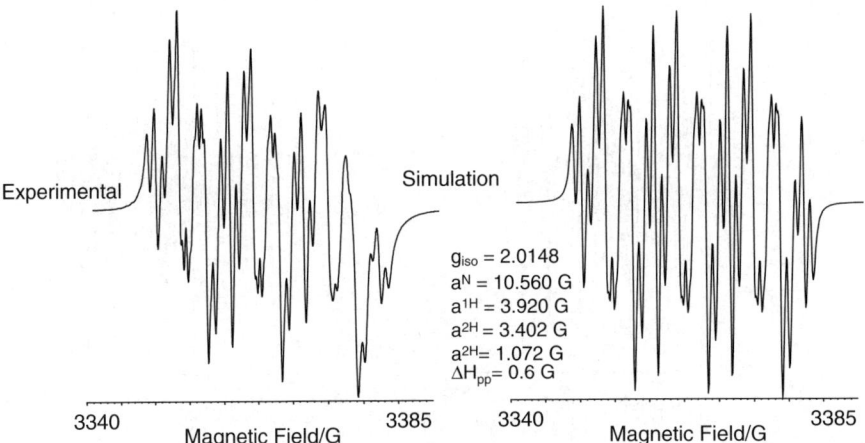

Figure 7.4 Experimental and simulated EPR spectra of $[C_6H_5NO_2]^-$ generated *in situ* at -1.0 V *vs.* Ag/AgCl at 233 K, recorded at 233 K in 0.1 M [^{n}Bu$_4$N][BF$_4$]/ CH$_3$CN.

may strongly interact with the electrolyte (and/or the solvent). A good example of this behaviour is observed for the nitrobenzene anion radical.

Nitrobenzene undergoes a one-electron reversible reduction with a half-wave potential of -1.00 V *vs.* Ag/AgCl in acetonitrile/0.1 M [^{n}Bu$_4$N][BF$_4$] at 298 K.[24–26] The anion radical $[C_6H_5NO_2]^-$ was generated in the EPR spectro-electrochemical cell at 233 K at -1.5 V in CH$_3$CN/0.1 M [^{n}Bu$_4$N][BF$_4$]. The resultant spectrum is shown in Figure 7.4. The experimental spectrum is best simulated (as shown in Figure 7.4) by a large coupling of the unpaired electron to the nitrogen of the NO$_2$ group (10.56 G) and further couplings to the ring protons; 3.92 G (1H), 3.40 G (2H) and 1.07 G (2H). The unique proton is obviously the hydrogen in the *para* position and it is assumed that the larger of the two equivalent proton couplings is to the *ortho* positions with the smallest coupling observed to the *meta* protons.[26] As expected there is only a small deviation of g from that of the free electron.

Repeating the experiment in acetonitrile to which no electrolyte had been added gave a very different EPR spectrum, see Figure 7.5. The EPR spectro-electrochemical experiment may be conducted in the absence of electrolyte, albeit on a longer timescale than if electrolyte is present, since the reduced anion can serve as a charge carrier. We are confident that the EPR spectrum is that of reduced nitrobenzene since the UV–Vis results are indistinguishable in the presence and absence of electrolyte and the parameters required for successful EPR simulation fit those of nitrobenzene. The EPR simulation of nitrobenzene radical formed in acetonitrile alone is shown in Figure 7.5. Comparison of Figure 7.4 and Figure 7.5 shows several similar features but also two important differences. The g value is the same within experimental error. The hyperfine couplings to the ring protons are either identical or very similar; 3.80 G *vs.* 3.92 G (*para*), 3.37 G *vs.* 3.40 G (*ortho*) and 1.08 G *vs.* 1.07 G (*meta*). However, the two

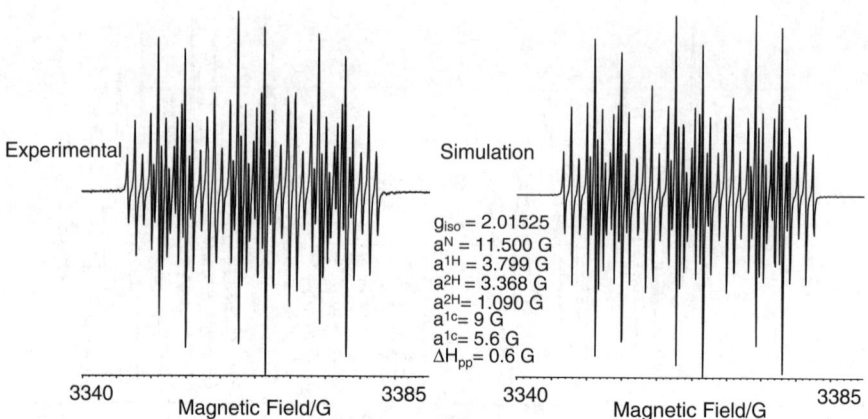

Figure 7.5 Experimental and simulated EPR spectra of $[C_6H_5NO_2]^{1-}$ generated *in situ* at -1.0 V *vs.* Ag/AgCl at 233 K, recorded at 233 K in CH_3CN (no electrolyte).

significant differences are the hyperfine coupling values to the nitrogen nucleus namely 11.60 G in the absence of electrolyte and 10.56 G in 0.1 M [$^nBu_4N$][BF_4], and the linewidth of the signals 0.15 G (no electrolyte) *vs.* 0.6 G (0.1 M [$^nBu_4N$][BF_4]). The very sharp linewidths of the no-electrolyte solution result in observation of hyperfine couplings to ^{13}C (1% natural abundance) of 9.00 and 5.60 G presumably to the NO_2 carbon and the *para* carbon nuclei, respectively. It is remarkable that the ^{14}N hyperfine coupling constant can be affected by more than 1 G simply by changing the amount of electrolyte in the solvent.

The large coupling to the ^{14}N nucleus suggests that the molecular orbital occupied by the unpaired reduction electron has a large contribution from the nitro group, which is as expected for such a strong electron-withdrawing group. Concentration of the electron density on the nitro group in the monoreduced species results in an electrostatic interaction between the negatively charged nitro group and the positively charged tetra butyl ammonium cation, that is, an ion pair is formed. The interaction with the [$^nBu_4N$]$^+$ cation will decrease the electron density on the nitro group and hence a smaller ^{14}N hyperfine coupling is observed. Ion pair formation between the reduced NO_2 group and the [$^nBu_4N$]$^+$-electrolyte cation has little effect on the other nuclei of the nitro-benzene anion as evidenced by the minor changes in the coupling constants to the hydrogen nuclei in the presence and absence of electrolyte.

Altering the electrolyte cation, but leaving the [BF_4]$^-$ unchanged should affect the strength of the ion pair formed and indeed the ^{14}N hyperfine coupling constant does vary with electrolyte cation. A smaller cation should form a tighter ion pair than a larger cation, thus a cation with a large radius should result in a large ^{14}N coupling (smaller interaction) and conversely a small radius cation should give rise to a small ^{14}N coupling. Table 7.1 gives coupling data for the nitrobenzene anion in acetonitrile with 0.1 M concentrations of tetrapropyl-, tetrabutyl- and tetrahexylammonium cations.[27]

Table 7.1   EPR hyperfine coupling constants for $(NO_2–C_6H_5)^-$ in CH_3CN for various electrolyte cations (0.1 M).

Electrolyte cation	a^N	a^{Ho}	a^{Hm}	a^{Hp}
$(C_3H_7)_4N^+$	10.32	3.39	1.09	3.97
$(C_4H_9)_4N^+$	10.56	3.40	1.07	3.92
$(C_6H_{13})_4N^+$	10.90	3.35	1.10	3.86

The EPR spectrum of the nitrobenzene anion radical is also dependent on the solvent. For example, the coupling constants measured in 0.1 M $[C_3H_7N]ClO_4/CH_3CN$ and 0.1 M $[C_3H_7N]ClO_4/DMF$ are $a^N = 10.3$ G vs. 9.70 G, $a^{Ho} = 3.39$ G vs. 3.36 G, $a^{Hm} = 1.09$ G vs. 1.07 G and $a^{Hp} = 3.97$ G vs. 4.03 G.[28] The largest reported value for the nitrogen coupling constant in the nitrobenzene anion radical is 13.3 G measured in H_2O.[29] The EPR coupling parameters for the nitrobenzene radical anion generated electrochemically in the presence of electrolyte have been referred to as "free-ion" values.[30,31] This study calls into question this assignment since even so-called "inert" electrolytes influence the value of the a^N coupling. Most electrogenerated radicals show little or no dependence on solvent or electrolyte. However, if the site of reduction is localised then the EPR study should be repeated in different solvents and electrolytes.

7.4   Uses of EPR Spectroelectrochemistry

7.4.1   Determination of the Site of a Redox Process

The site of redox activity in reducible or oxidisable compounds may be considered to be primarily based on the metal centre, or on one of the ligands or on a molecular orbital spanning both the metal and the ligand. Interrogation of the EPR results can reveal much about the electronic character of the orbital occupied by the unpaired electron. However, if the redox orbital has a significant contribution from the central transition metal then the resultant EPR signal may be dominated by the contribution from the metal since the isotropic coupling constant, a_0, of most transition metals is larger than that of most common ligand nuclei. In addition the spin-orbit coupling constant of the transition metal may be large enough to result in significant broadening of the EPR signal thereby masking the superhyperfine coupling to ligand nuclei.

The compounds $[Pt(4,4'-^tBu_2-bpy)L_2]$ (Figure 7.6) undergo reversible reduction processes at -1.32 V (X = H) and -1.08 V (X = NO$_2$).[32] The cyclic voltammogram for the Pt complex where X = H is shown in Figure 7.7. The ligands $4,4'-^tBu_2$-bpy, C_2-Ph and $C_2(C_6H_4$-p-$NO_2)$ are all redox-active and therefore the question arises as to where the site of redox activity is located. EPR spectroelectrochemical study proved definitive in this regard rather than UV–Vis–NIR spectroelectrochemistry. Figure 7.8(a) gives the experimental and simulated spectra for $[Pt(4,4'-^tBu_2-bpy)(C_2(Ph))_2]^-$ and Figure 7.8(b) shows the spectra for $[Pt(4,4'-^tBu_2-bpy)(C_2(C_6H_4$-$p$-$NO_2))_2]^{2-}$. Note that the spectra in

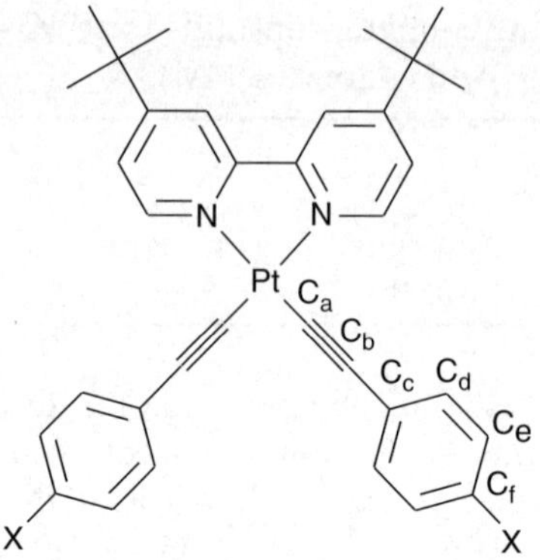

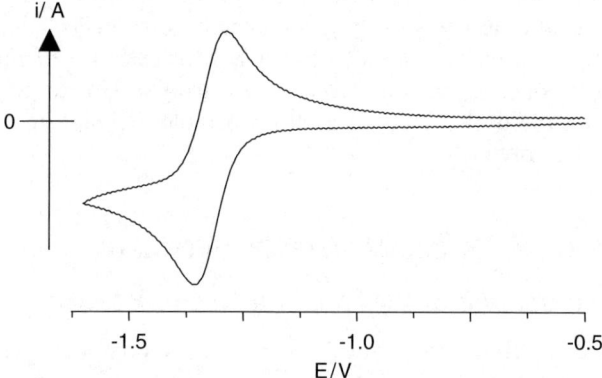

Figure 7.6   [Pt(4,4′-^{t}Bu$_2$-bpy)(C$_2$(C$_6$H$_4$-p-X))$_2$] (X = H, NO$_2$).

Figure 7.7   Cyclic voltammogram of [Pt(4,4′-^{t}Bu$_2$-bpy)(C$_2$(ph)$_2$] in 0.1 M [^{n}Bu$_4$N] [BF$_4$]/DMF, 290 K.

Figure 7.8(a) contain the whole spectral range of 100 G, whereas the spectra in Figure 7.8(b) only contain the low field half of the signal.

The spectra of the monoreduced anion [Pt(4,4′-^{t}Bu$_2$-bpy)(C$_2$(Ph))$_2$]$^-$ may be simulated using the parameters given in Table 7.2. Thus, the unpaired electron couples to the Pt nucleus, two equivalent nitrogen nuclei and two equivalent protons. The spectrum shows a great similarity to the previously reported so-lution EPR spectrum of [Pt(bpy)(CN)$_2$],[1-33] and the coupling constants are in good agreement. Thus the conclusions reached for [Pt(bpy)(CN)$_2$]$^-$ apply to [Pt(4,4′-^{t}Bu$_2$-bpy)(C$_2$(Ph))$_2$]$^-$; that is, that the unpaired electron in [Pt(4,4′-^{t}Bu$_2$-bpy)(C$_2$(Ph))$_2$]$^-$ primarily resides on the coordinated 4,4′-^{t}Bu$_2$-bpy ligand. Table 7.2 shows that two sets of two equivalent protons are required for a successful simulation of the spectrum of [Pt(bpy)(CN)$_2$]$^-$, whereas only one set of two equivalent protons are required for [Pt(4,4′-^{t}Bu$_2$-bpy)(C$_2$(Ph))$_2$]$^-$. Thus, one pair of protons in [Pt(bpy)(CN)$_2$]$^-$ must be attributed to 4,4′ positions since

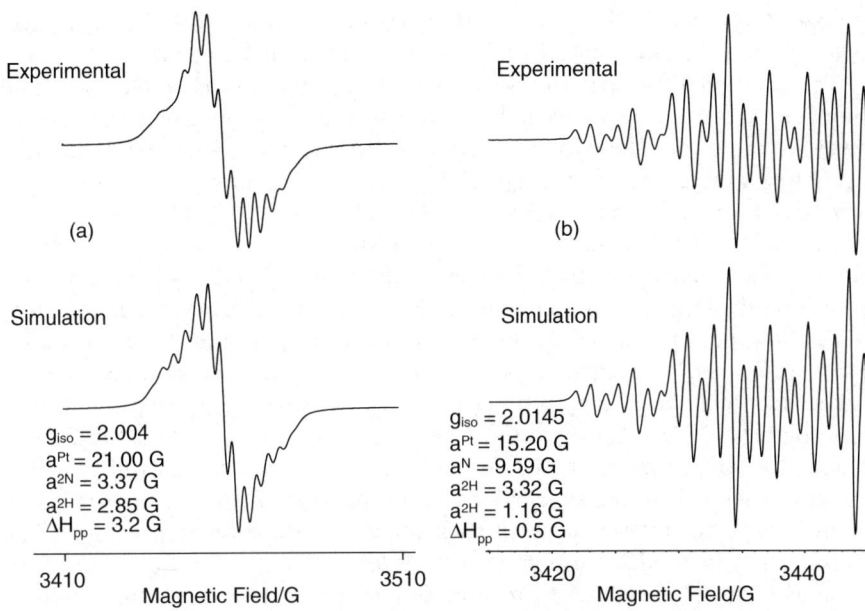

Figure 7.8 Solution EPR spectra and simulations for (a) [Pt(4,4′-^{t}Bu$_2$-bpy)(C$_2$(ph)$_2$)]$^-$ and (b) [Pt(4,4′-^{t}Bu$_2$-bpy)(C$_2$(C$_6$H$_4$-p-NO$_2$))$_2$)]$^{2-}$ in 0.1 M [^{n}Bu$_4$N][BF$_4$]/ DMF.

Table 7.2 Parameters used to model the EPR spectrum of [Pt(4,4′-^{t}Bu$_2$-bpy) (C$_2$(Ph))$_2$)]$^{1-}$, [Pt(4,4′-^{t}Bu$_2$-bpy) (C$_2$(C$_6$H$_4$-p-NO$_2$))$_2$)]$^{2-}$ [Pt(bpy) (CN)$_2$]$^{1-}$ and [Ru(4,4′-^{t}Bu$_2$-bpy)(PPh$_3$)$_2$(C$_2$(C$_6$H$_4$-p-NO$_2$))$_2$)]$^{2-}$.

Compound	$A(^{195}Pt/^{99,101}Ru)/G$	$A(^{14}N)/G$	$A(^1H)/G$	ΔH_{pp}
[Pt(4,4′-^{t}Bu$_2$- bpy)(C$_2$(ph))$_2$]$^-$	21.0	2×3.37	2×2.85	3.2
[Pt(4,4′-^{t}Bu$_2$- bpy)(C$_2$(C$_6$H$_4$-p- NO$_2$))$_2$]$^{2-}$	15.2	2×9.59	2×3.32 2×1.16	0.5
[Pt(bpy)(CN)$_2$]$^-$	20.5	2×3.4	2×2.8 2×3.2	
[Ru(4,4′-^{t}Bu$_2$- bpy)(PPh$_3$)$_2$(C$_2$ (C$_6$H$_4$-p-(NO$_2$)$_2$))$_2$]^2	0.53	2×9.86	2×3.28 2×1.05	0.45

substitution by ^{t}Bu groups removes coupling to this position. Further studies indicate that the 5,5′ position has the most significant contribution to the orbital occupied by the unpaired electron. Therefore, the ^{14}N coupling in [Pt(4,4′-^{t}Bu$_2$-bpy)(C$_2$(Ph))$_2$]$^-$ is assigned to the bpy nitrogens and the ^{1}H coupling to the 5,5′ positions on the derivatised bpy ligand and the monoanion can be formulated as [Pt(II)(4,4′-^{t}Bu$_2$-bpy)$^-$(C$_2$(Ph))$_2$]$^-$. In addition, the similar EPR behaviours of

$[Pt(4,4'-^tBu_2-bpy)(C_2(Ph))_2]^-$ and $[Pt(bpy)(CN)_2]^-$ indicate that the phenylace-tylide ligands are electronically comparable to the cyanide ligands.

The solution EPR spectrum of $[Pt(4,4'-^tBu_2-bpy)(C_2(C_6H_4-p-NO_2))_2]^{2-}$ and simulated spectrum are shown in Figure 7.8(b) and the simulated parameters are given in Table 7.2. The two experimental EPR spectra in Figure 7.8 are obviously very different as are the hyperfine coupling constants given in Table 7.2. The ligand simulation parameters for $[Pt(4,4'-^tBu_2-bpy)(C_2(C_6H_4-p-NO_2))_2]^{2-}$ of one large ^{14}N coupling and two sets of two equivalent protons suggest that the site of redox activity in this complex is the $[C_2(C_6H_4-p-NO_2)]^-$ ligand. Furthermore, the large ^{14}N coupling indicates that the nitro group carries much of the electron density when the compound is in its reduced state. Note that the di-anion experimental spectrum is best simulated to a spin-$\frac{1}{2}$ system, that is, there is no evidence in the EPR spectrum (solution or frozen) of a triplet ground state. The complex $[Pt(4,4'-^tBu_2-bpy)(C_2(C_6H_4-p-NO_2))_2]^{2-}$ therefore has a degenerate pair of LUMOs primarily based on the $C_6H_4-p-NO_2$ groups that will be ortho-gonal to each other and are spatially well separated. Thus, the two reduction electrons are not interacting with each other and the direduced complex can satisfactorily be modeled using a $S = \frac{1}{2}$ system.

Assignment of the coupling to specific nuclei was done using the results of semiempirical molecular orbital calculations at the PM3 level. These calculations reveal that the LUMO of $[Pt(4,4'-^tBu_2-bpy)(C_2(C_6H_4-p-NO_2))_2]$ is based on one $[(C_2(C_6H_4-p-NO_2))]^-$ ligand and that the spin densities in the carbon $2p_z$ orbital contributing to the π system (ρ_C^π) on C_e and C_d, Figure 7.6, are 0.130 and 0.055, respectively. Thus the reduction electron is coupling to the Pt nucleus (15.20 G), the N atom of the NO_2 group (9.59 G), two equivalent protons attached to C_e (3.32 G) and two equivalent protons bonded to C_d (1.16 G). The McConnell equation, eqn (7.1), which relates the hydrogen hyperfine coupling value to the spin density, $\rho_C^{\pi},^{34}$ where Q is a semiempirical parameter of the order of 24 G[4] can be used to check the above assignments.

$$a^H = Q\rho_C^\pi \tag{7.1}$$

Feeding the values obtained above for a^H and ρ_C^π into eqn (7.1) gives values of Q of 25.5 and 21.1 G for C_e and C_d, respectively.

Thus the site of redox activity in $[Pt(4,4'-^tBu_2-bpy)L_2]$ where $L = C_2(C_6H_4-p-X)$ (X = H or NO_2), is critically dependent on X. When X = H, then the derivatised bpy ligand is reduced, but when X = NO_2 the redox activity is located on the organometallic ligand. This is directly attributable to the strong electron-withdrawing nature of the NO_2 group compared to H.

The related complex *trans,cis*-$[Ru(4,4'-^tBu_2-bpy)(PPh_3)_2(C_2(C_6H_4-p-NO_2))_2]$ exhibits a one-electron metal-based oxidation at $+0.44$ V and a two-electron reversible reduction at -0.84 V. The experimental and simulated EPR spectra of the di-reduced species are shown in Figure 7.9[35] with hyperfine coupling constants given in Table 7.2. The couplings indicate that the sites of redox activity in $[Ru(4,4'-^tBu_2-bpy)(PPh_3)_2(C_2(C_6H_4-p-NO_2))_2]^{2-}$ are the non-interacting NO_2 containing ligands.

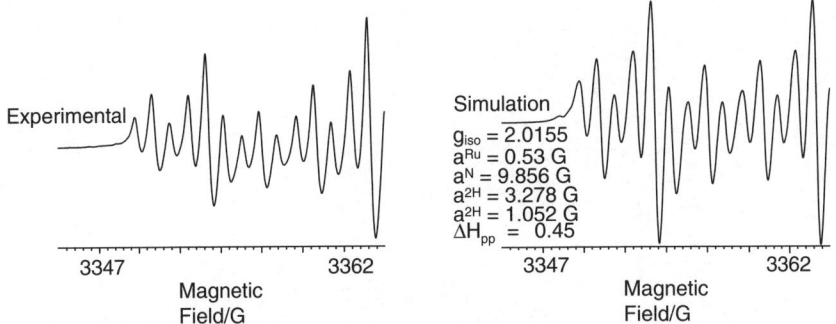

Experimental

Simulation

g_{iso} = 2.0155
a^{Ru} = 0.53 G
a^N = 9.856 G
a^{2H} = 3.278 G
a^{2H} = 1.052 G
ΔH_{pp} = 0.45

3347   3362   3347   3362
Magnetic   Magnetic
Field/G   Field/G

Figure 7.9   Solution EPR spectra and simulations for [Ru(4,4'-^{t}Bu$_2$-bpy)(PPh$_3$)$_2$(C$_2$(C$_6$H$_4$-p-NO$_2$))$_2$]$^{2-}$ in 0.1 M [^{n}Bu$_4$N][BF$_4$]/DMF.

The power of *in-situ* generation of radical species is fully exploited in the study of a redox series. In such series several redox states can be characterised from the starting material that undergoes more than one reversible reduction/oxidation process. The site of redox activity accompanying the stepwise reductions or oxidations is difficult to probe by means other than *in-situ* spectroelectrochemical studies since it is imperative that complete generation of each of the redox partners is achieved and recorded for successful assignment. The use of a large surface area working electrode makes accomplishment of 100% electrogeneration of the sequential redox species possible. In addition, use of a reference electrode such as Ag/AgCl or SCE rather than a pseudo-reference electrode is preferred so that the exact value of E_{app} is known.

The donor–acceptor (diad) molecule 5,10,15,20-[*N*-benzyl-*N'*-(4-benzyl-4,4'-bipyridinium-4-pyridyl)]-triphenylporphyrin tris(hexafluorophosphate, Figure 7.10, comprises a porphyrin ring (donor) covalently attached to a dibenzylviologen unit (acceptor). The diad has a one-electron reduction at –0.30 V followed by two closely spaced one-electron reductions centred at –0.78 V in CH$_3$CN/0.1 M[^{n}Bu$_4$N][BF$_4$].[36]

EPR spectroelectrochemistry at 233 K elucidated which centres on the molecule were reduced. Reduction at –0.61 V results in the EPR spectrum shown in Figure 7.11(a) that is best simulated by coupling of the unpaired electron to three equivalent nitrogens (a(N) = 4.08 G) and two sets of six equivalent protons (a(H) = 1.41, 1.00 G). Consideration of the structure of the diad in Figure 7.10 suggests that the redox site at –0.30 V be attributed to the acceptor part of the molecule that has three equivalent nitrogen sites. It is likely that six H couplings observed in the EPR spectrum are due to methylene protons (2 per nitrogen) and six are aromatic protons (again 2 per nitrogen) adjacent to N$^+$.

On application of a potential of –1.40 V the EPR signal decays as the system becomes diamagnetic. Thus, the second reduction electron probably enters the same orbital as the first on the acceptor part of the compound. A new signal subsequently develops as a third electron is added to the molecule, see Figure 7.11(b). The third electron enters a molecular orbital linked to the

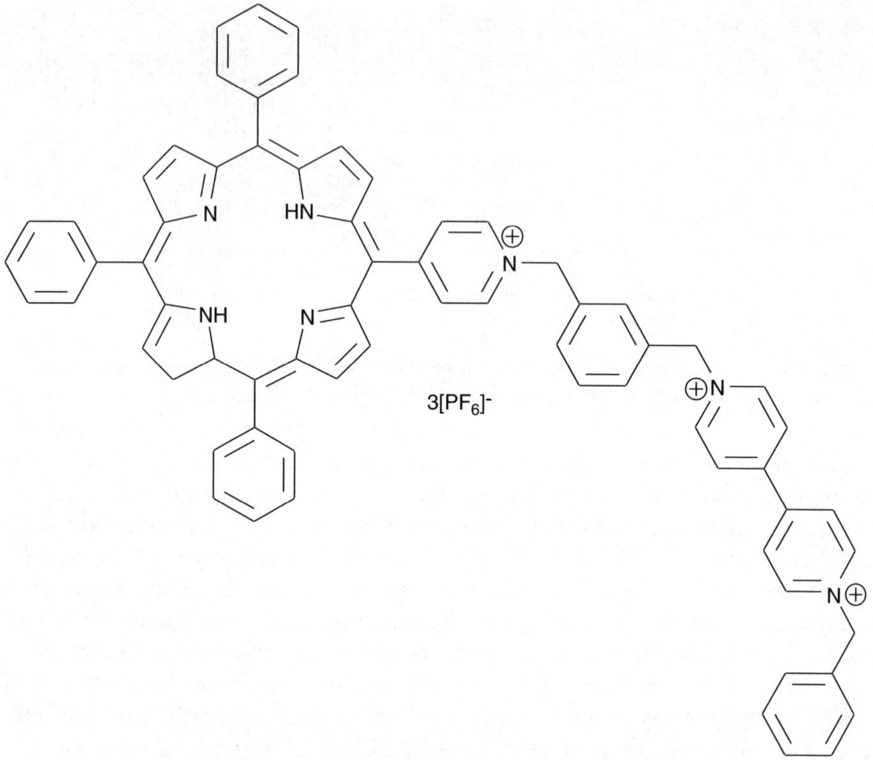

Figure 7.10 Structure of 5,10,15,20-[N-benzyl-N′-(4-benzyl-4,4′-bipyridinium-4-pyridyl)]-triphenylporphyrin tris(hexafluorophosphate).

porphyrin ring since the EPR signal is best simulated using four equivalent nitrogen nuclei (2.65 G) and two equivalent protons (5.40 G) and this fragment is clearly to be found at the heart of the porphyrin ring. Note that the EPR spectra generated on addition of one and three reduction electrons are very different in form and are centred at differing magnetic fields. Reversal of the applied potential to –0.61 V shows collapse of the signal at $g = 2.011$ to give a featureless EPR spectrum and then growth of the first signal at $g = 2.012$. The EPR experiments show that at 233 K the diad is stable in four redox states, two of which are diamagnetic and two are paramagnetic.

Successful interpretation of EPR results can also be used to quantify the amount a given nucleus contributes to the molecular orbital containing the unpaired electron. The solution EPR spectrum of $[Pt(bpy)Cl_2]^{1-}$ (Figure 7.12) may be best explained as arising from the reduction electron interacting with the Pt metal centre since ^{195}Pt is 34% naturally abundant with $I = \frac{1}{2}$ and all other naturally occurring isotopes of Pt have $I = 0$. Thus, a broad singlet resonance is observed at $g_{iso} = 1.998$ with ^{195}Pt satellites ($A_{iso} = -54 \times 10^{-4}\,cm^{-1}$). Any superhyperfine coupling to the ligand nuclei is unresolved. The small shift of g_{iso} from the free-electron value of 2.0023 is indicative of only a small

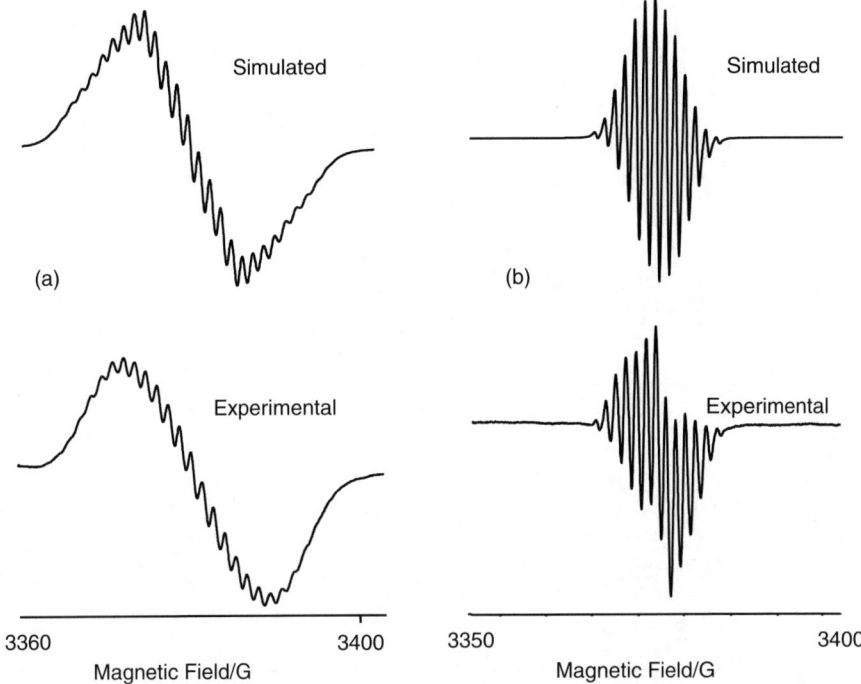

Figure 7.11 EPR, experimental and simulated, spectra for reduced diad in DMF/0.2 M ["Bu$_4$N][BF$_4$] at 233 K; (a) one-electron reduction product $E_{gen} = -0.61$ V, $g = 2.012$; (b) three-electron reduction product $E_{gen} = -1.40$ V, $g = 2.011$. Simulation parameters as in text and linewidth = 1.00 G.

contribution to the semioccupied molecular orbital (SOMO) from the Pt orbitals.

The frozen solution spectrum for [Pt(bpy)Cl$_2$]$^{1-}$ is shown in Figure 7.13. The spectrum is clearly rhombic and the g values are given in Table 7.3.[31] The ^{195}Pt coupling is apparent on g_1 and g_2 and values are given in Table 7.3. The ^{195}Pt hyperfine splitting about the high field g_3 component is not resolved and the value given in Table 7.3 has an uncertainty of $\pm 5 \times 10^{-4}$ cm^{-1}. The SOMO of [Pt(bpy)Cl$_2$]$^{1-}$ is of b$_2$ symmetry (in C_2 symmetry with the axes as defined in Figure 7.14. The Pt 5d$_{ys}$ and 6p$_z$ orbitals are the only metal orbitals that can formally admix to this ground state and thus the metal contribution to the SOMO can be described by eqn (7.2):

$$|SOMO> = \underline{a}|ys> + \underline{b}|z> \tag{7.2}$$

where $\underline{a}$ and $\underline{b}$ are the linear combination of atomic orbital coefficients of the 5d$_{ys}$ and 6p$_z$ orbitals in the SOMO, respectively. Following Rieger's methodology[37] it is possible to analyse the ^{195}Pt hyperfine couplings in order to calculate the unpaired electron spin densities in the 5d$_{ys}$ and 6p$_z$ orbitals (a^2 and b^2

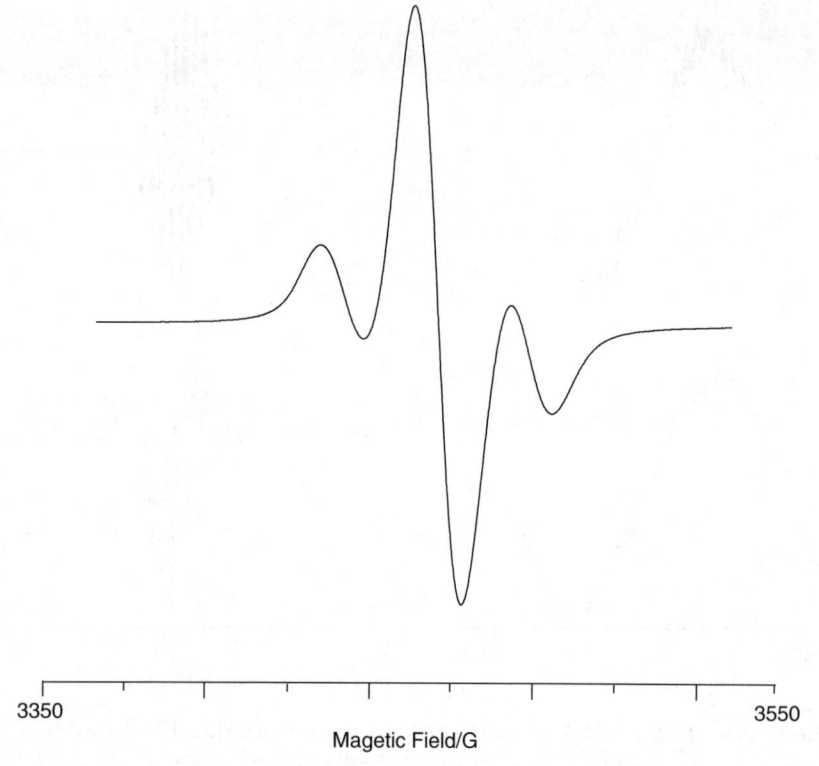

3350   3550

Magetic Field/G

Figure 7.12   Solution X-band EPR spectrum of $[Pt(bpy)Cl_2]^{1-}$ in 0.1 M $[^nBu_4N][BF_4]/$ DMF.

respectively) from eqns (7.3–7.5)

$$A_{xx} = A_s - 4/7P_d a^2 - 2/5P_p b^2 \qquad (7.3)$$

$$A_{yy} = A_s + 2/7P_d a^2 - 2/5P_p b^2 \qquad (7.4)$$

$$A_{zz} = A_s + 2/7P_d a^2 + 4/5P_p b^2 \qquad (7.5)$$

where P_d and P_p are the electron nuclear dipolar coupling parameters for platinum 5_d and 6_p electrons, respectively, and A_s is the isotropic Fermi contact term. This approach ignores the contributions to the hyperfines from mixing in of excited states *via* spin-orbit coupling. In experimental terms this is to ignore deviations of the *g* values from g_e. Since g_1, g_2 and g_3 values for $[Pt(bpy)Cl_2]^{1-}$ are close to g_e this assumption should not have a major effect on the final result.

Solving eqns (7.3)–(7.5) requires experimental A_1 to be mapped onto A_{xx}, A_{yy} or A_{zz} and likewise for A_2 and A_3. The three A values obtained from simulation average well to the experimental solution value and therefore all A values have the same sign, which is negative in this case. Since P_d and P_p are positive quantities, eqns (7.3)–(7.5) suggest that A_{xx} will have the most negative value and A_{zz} the least negative. Hence A_1 is assigned to A_{yy}, A_2 to A_{xx} and A_3

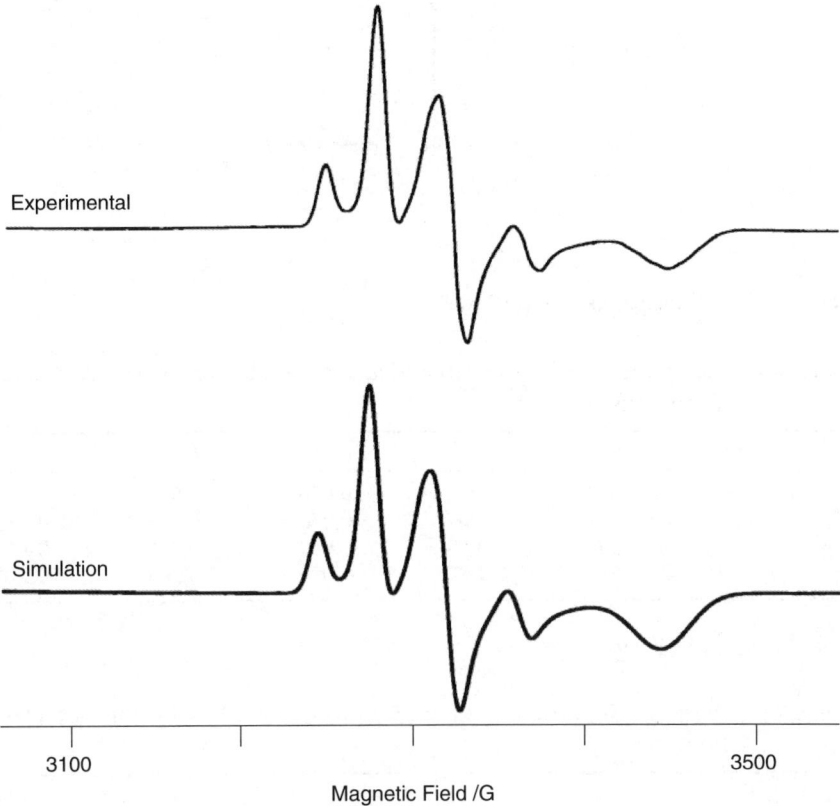

Figure 7.13 Experimental and simulated frozen solution EPR spectra of $[Pt(bpy)_2Cl_2]^{1-}$ in 0.1 M $[^nBu_4N][BF_4]$/DMF.

Table 7.3 Experimental and simulated frozen solution EPR parameters for $[Pt(bpy)_2Cl_2]^{1-}$ in 0.1 M $[^nBu4N][BF4]$/DMF.

	g_{iso}	g_1	g_2	g_3	$A_{iso}(Pt)^c$	A_1^c	A_2^c	A_3^c
Expt.[a]	1.998	2.038	2.009	1.935	−54	−56	−95	−
Sim.[b]	−	2.038	2.011	1.938	−	−57.4	−80.6	−20.8

[a]Parameters estimated from spectra.
[b]Parameters from computer simulations.
[c]$A/10^{-4}\,cm^{-1}$.

to A_{zz}. Equations (7.3)–(7.5) can now be solved for $P_d a^2$ and $P_p b^2$. Rieger[37] has derived equations for calculating the required dipolar coupling parameters P_d and P_p. These parameters represent the expected hyperfine couplings for unit population of the metal valence orbitals and hence are necessarily dependent on the assumed electronic configuration. Table 7.4 lists the calculated values of P_d, P_p, a^2 and b^2 for various assumed configurations.

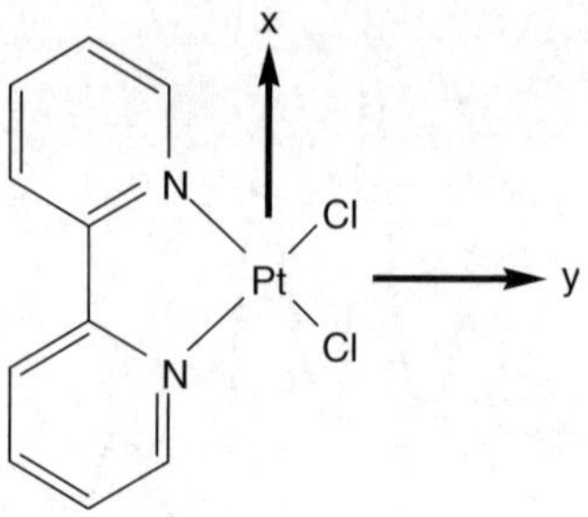

Figure 7.14   Principal axes of [Pt(bpy)Cl$_2$].

Table 7.4   Platinum unpaired electron densities for [Pt(bpy)Cl$_2$]$^{1-}$ calculated from EPR data.

Configuration	$P_d{}^a$	$P_p{}^a$	a^2	b^2	$Total^b$
d^8	549	402	0.050	0.076	0.126
d^{8}s^1	539	268	0.050	0.113	0.163
d^{8}p^1	544	402	0.050	0.076	0.126
d^9	507	183	0.054	0.166	0.220

^{a}10^{-4} cm^{-1}.
^{b}Total Pt unpaired electron density.

Table 7.5   Platinum 5d$_{ys}$ and 6p$_z$ admixtures to the SOMO of [Pt(4,4′-(NO$_2$)$_2$-bpy)Cl$_2$]$^{1-}$.

X	a^2	b^2	Total
OEt	0.015	0.057	0.07
Me	0.033	0.075	0.11
Ph	0.049	0.060	0.11
H	0.050	0.076	0.13
Cl	0.053	0.057	0.11
CO$_2$Me	0.055	0.067	0.12

UV–Vis spectroelectrochemical results for [Pt(bpy)Cl$_2$]$^{1-}$ indicate that the reduction electron is primarily located on the bpy ligand and that the reduced complex should be formulated as [Pt(II)(bpy)$^{1-}$Cl$_2$]$^{1-}$. Thus, the platinum centre should be regarded as having a d^8 configuration. The platinum 5d$_{ys}$ and 6p$_z$ orbitals contribute 5% and 7.6% to the SOMO. Therefore, the Pt centre contributes <13% to the SOMO which is in agreement with the formulation of a predominantly ligand-based LUMO in [Pt(bpy)Cl$_2$].

The electronic influence of substitution in the 4,4′ position of 2,2′-bipyridine in complexes of the type [Pt(4,4′-X$_2$-bpy)Cl$_2$]$^{1-}$ on the metal contribution to the SOMO has also been investigated using this method.[38] Results are tabulated in Table 7.5.

In each case, the contributions from the metal orbitals to the SOMO consist of almost equally small amounts of 5d$_{ys}$ and 6p$_z$, although the latter is always slightly larger. The total metal contributions to the SOMO are in the range 7–13%, thus confirming a predominantly ligand-based orbital in each case. One apparent trend is that the $5d_{ys}$ admixture to the SOMO increases as X becomes a stronger electron withdrawing group. As the energy of the ligand π^* (b_2) orbital is decreased by stronger electron-withdrawing substituents there will be a better energy match with the filled low-lying platinum $5d_{ys}$ (b_2) orbital and hence, a more efficient metal-ligand orbital overlap. This results in a still predominantly ligand-based π^* orbital but with greater metal 5d character.

7.4.2 Determination of the Half-Wave Potential of a Redox Process

The spectroelectrochemical experiments where the EPR spectrum is recorded as a function of applied potential is a useful method for measuring the half-wave potential of a redox process and the number of electrons transferred in that process. It is a particularly good method when the electron-transfer process is slow at an electrode surface and the resultant cyclic voltammogram too broad to analyse. Biological molecules often fall into this category.[39,40] They are usually large and the redox-active centre may be buried within the molecule making direct electrochemical study of the half-wave potential difficult. The solvent system containing the biological species of interest may or may not contain molecular mediators to aid electron transfer.

Thus, as a paramagnetic species is generated so the EPR signal will increase in intensity, with the relative concentration of the EPR active material obtained from the relative resonance intensity.[41] The EPR response as a function of applied potential for a solution of *E. coli* flavodoxin from −150 to −550 mV is presented in Figure 7.15.[5]

A flavodoxin-containing species reveals two redox processes, the first one-electron reduction yielding a paramagnetic radical and the second one-electron reduction resulting in the diamagnetic form. Thus, as the applied potential is stepped from −150 to −550 mV the EPR signal increases in intensity to a maximum value and then decreases. If the applied potential is reversed from −550 to −150 mV, the result is the growth and then collapse of the EPR signal. The profile exactly mirrors that of the response in the forward direction. The half-wave potential is given by the potential at which the relative peak size is half the final value. Thus analysis of the data in Figure 7.15 gives the two formal potentials −300 and −506 mV *vs.* SHE for the first and second electron-transfer processes, respectively. Note that using a large surface area working electrode considerably shortens the electrolysis time and that the solution is allowed to equilibrate fully at each applied potential before the EPR signal is recorded.

7.4.3 Determination of Zero-Point Energy Splitting

The complex *cis*-Ru(bpy)$_2$(CN)$_2$ (bpy = 2,2′-bipyridine) exhibits a reversible one-electron oxidation at +0.93 V and two reversible one-electron reductions

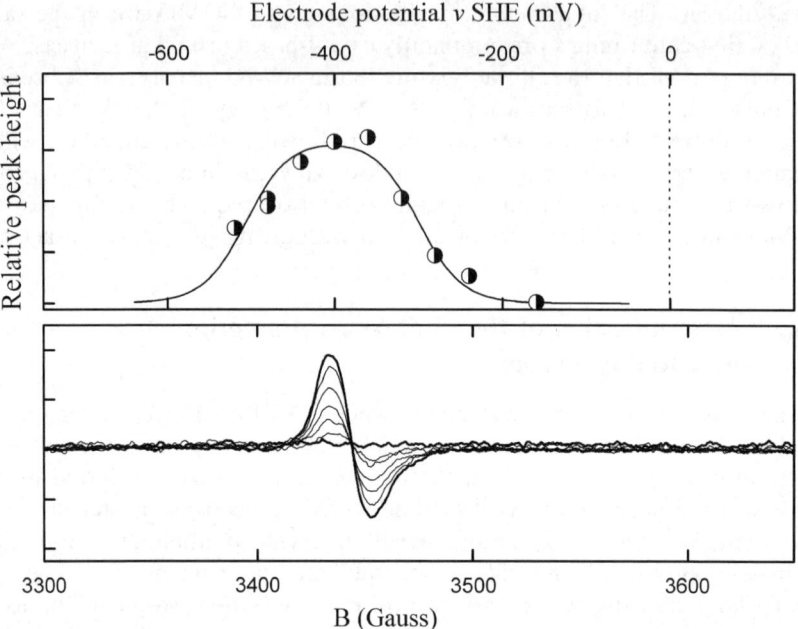

Figure 7.15 EPR signal of reduction of *E. coli* flavodoxin from –150 to –550 mV at 293 K.

at –1.50 V and –1.78 V in 0.1 M [^{n}Bu$_4$N][BF$_4$]/DMF at 240 K. The oxidation is assigned to a metal-based Ru(II)/Ru(III) process whilst the redox waves at negative potentials involve the sequential reduction of the two bpy ligands. The potential of the Ru(II)/Ru(III) couple and the position of the Ru(II) → bpy MLCT band (v_{max} around 20 000 cm^{-1}) both depend on the solvent with both parameters showing a linear correlation with solvent acceptor numbers.[42–44] The frozen glass (100 K) EPR spectrum of electrogenerated [Ru(bpy)$_2$(CN)$_2$]$^{1+}$, is shown in Figure 7.16. The rhombic signal ($g_1 = 2.469$, $g_2 = 2.305$, $g_3 = 1.895$) is typical of that for a low-spin d^6 metal centre. Interestingly, the EPR spectra and the UV–Vis adsorption bands for the mono-oxidised species [Ru(bpy)$_2$(CN)$_2$]$^{1+}$ show minimal solvent dependence.

The frozen glass EPR spectrum exhibits a rhombic signal because the Ru centre has, at best, C_{2V} symmetry. The net result of the distortion from O_h symmetry to C_{2V} is to remove the degeneracy of the "t_2" orbitals. The axial distortion (Δ) splits the t_2 subshell into $e + b$ components, whilst the super-imposed rhombic distortion (V) further splits the e into two nondegenerate components (Figure 7.17). The splitting of the t_2 orbitals will result in two optical transitions E_1 and E_2 that are predicted to have weak molar extinction coefficients.

The EPR spectrum can be analysed by g-tensor theory of low-spin d^5 metal ions.[45–47] Unfortunately, the EPR experiment only yields the absolute magni-tudes of the g values, not their signs. It also gives no information on the

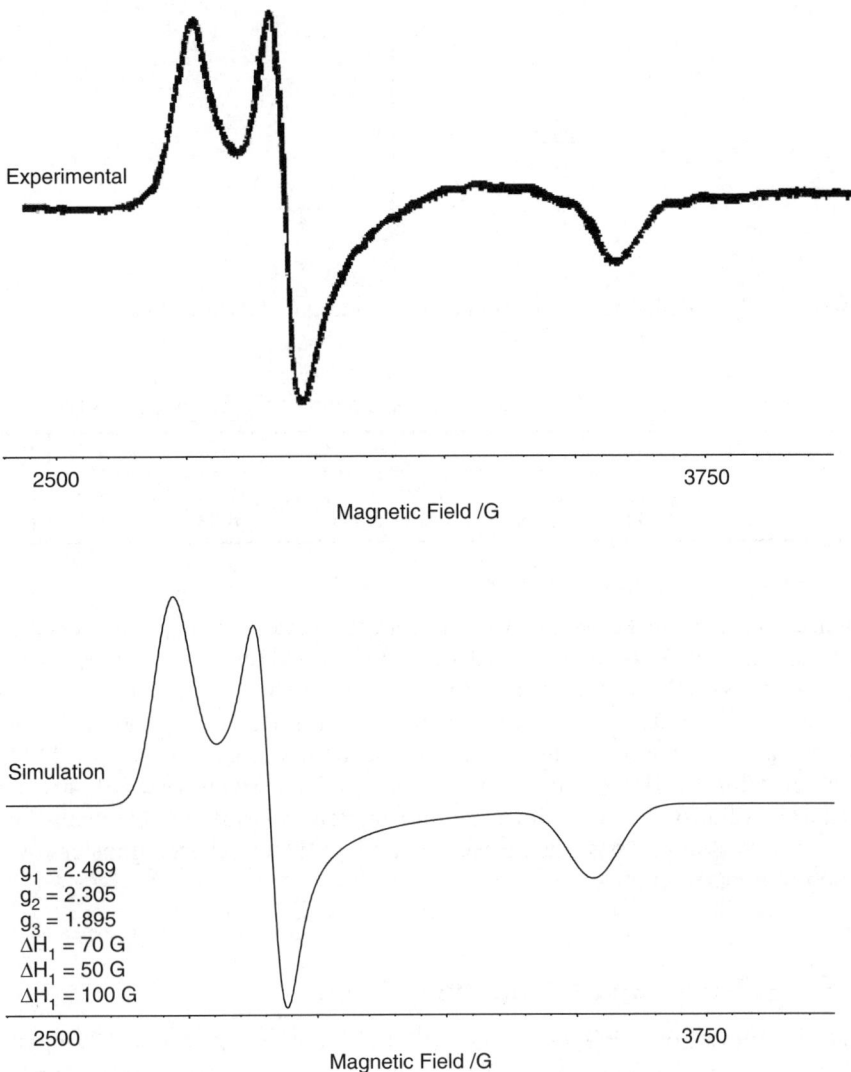

Figure 7.16 Experimental and simulated frozen solution EPR spectra of *cis*-[Ru(bpy)$_2$ (CN)$_2$]$^+$ in 0.1 M [^{n}Bu$_4$N][BF$_4$]/DMF.

correspondence between g_1, g_2 and g_3 with g_x, g_y and g_z. Thus, in analysing the data it is necessary to solve for each combination of $\pm g_1$, $\pm g_2$ and $\pm g_3$ with g_x, g_y and g_z. Of the forty-eight variations, only two solutions gave sensible values of E_1 and E_2 (Table 7.6). Note that the parameters Δ, $\overset{*}{v}$, E_1 and E_2 are given as a function of the spin-orbit coupling constant λ of Ru(III).

The UV–Vis–NIR absorption spectrum of a concentrated solution of [Ru(bpy)$_2$(CN)$_2$]$^{1+}$ has a very weak band ($\varepsilon \sim 40\,M^{-1}\,cm^{-1}$) at 4300 cm^{-1}. The spin-orbit coupling constant of complexed Ru has a value around 1000 cm^{-1}.

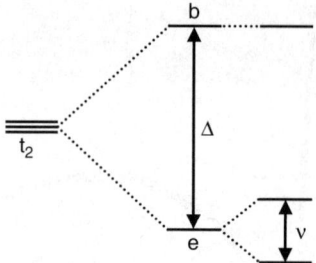

Figure 7.17   Splitting of t_2 orbitals under axial and rhombic distortions.

Table 7.6   Assignment of g values and parameters for $[Ru(bpy)_2(CN)_2]^{1+}$.

Solution no.	g_x	g_y	g_z	Δ/λ	v/λ	E_1/λ	E_2/λ
1	−2.500	2.300	1.789	4.10	1.50	3.46	5.17
2	−2.500	2.300	−1.789	0.21	0.08	1.44	1.58

Hence, solution 1 in Table 7.6 gives the preferred values for the parameters Δ/λ, v/λ, E_1/λ and E_2/λ. If the observed transition at $4300 \, cm^{-1}$ is assigned to E_1, then λ for Ru(III) is computed to be $1240 \, cm^{-1}$, whereas the alternative assignment of $E_2 = 4300 \, cm^{-1}$ gives a value of λ for Ru(III) of $840 \, cm^{-1}$. The larger value of λ is in better agreement with the reported values of 1180–$1250 \, cm^{-1}$ for Ru(III).[48] Table 7.6 gives $E_2/\lambda = 5.17$, which for $\lambda = 1240 \, cm^{-1}$ predicts a value of $7400 \, cm^{-1}$ for E_2. Unfortunately, it is not possible to observe this transition directly as it is masked by a more intense charge-transfer transition that peaks at $9000 \, cm^{-1}$.

7.5   Solvents and Compounds Used

The syntheses of $[Pt(4,4'-{}^tBu_2\text{-}bpy)(C_2(C_6H_4\text{-}p\text{-}NO_2))_2]$, bpy $= 2,2'$-bipyridine,[49] 5,10,15,20-[N-benzyl-N'-(4-benzyl-4,4'-bipyridinium-4-pyridyl]-triphenylporphyrin tris(hexafluorophosphate),[36] $[Pt(4,4'-X_2\text{-}bpy)Cl_2]$, X $=$ NO$_2$,[50] H,[51] are described in the literature.

$[Ru(bpy)_2(CN_2)]$ was prepared using a literature method.[52]

Dichloromethane was pretreated with KOH pellets and distilled over P_2O_5 immediately prior to use. All other solvents and compounds were used as obtained without further purification. $[^nBu_4N][BF_4]$ was synthesised from the neutralisation of $[^nBu_4N]OH \, H[BF_4]$, recrystalised from water/methanol and dried at 350 K under vacuum for three weeks before electrochemical use. All electrochemical studies were performed using a Dell GX110 personal computer with general-purpose electrochemical system (GPES) version 4.8 software connected to an Autolab system containing either a PSTAT 20 or PSTAT 30 potentiostat. Cyclic volumetric experiments employed a Pt microworking electrode (we, 0.5 mm diameter), a Pt counter electrode (ce) and a Ag/AgCl reference electrode (re) in a 0.45 M [$^nBu_4N$][BF_4]/0.05M [$^nBu_4N$]Cl/dichloromethane solution. The ferrocinium/ferrocene couple was measured at +0.55 V *vs.* the Ag/AgCl reference electrode. The exact composition of the solution under study is detailed as appropriate in the text. Dissolved oxygen is removed from the solvent by N_2 saturation prior to study.

References

1. T. M. McKinney, *Electroanalytical Chemistry Vol. 10: Electron Spin Resonance and Electrochemistry*, A. J. Bard (ed.), Marcel Dekker, New York, 1977, p. 97.
2. A. J. Bard and L. R. Faulkner, *Electrochemical Methods: Fundamentals and Applications*, John Wiley and Sons, New York, 1980, p. 615.
3. R. G. Compton and A. M. Waller, *Spectroelectrochemistry:Theory and Practise*, R. J. Gale (ed.), Plenum Press, New York, 1980, p. 615.
4. I. B. Goldberg and T. M. McKinney, *Laboratory Techniques in Electroanalytical Chemistry*, P. T. Kissinger and W. R. Heineman (ed.), Marcel Dekker, New York, 1996, p. 901.
5. E. Alessio, S. Daff, M. Elliot, E. Iengo, L. A. Jack, K. G. Macnamara, J. M. Pratt and L. J. Yellowlees, *Chapter 11, Spectroelectrochemical Techniques in Trends in Molecular Electrochemistry*, A. J. L. Pombeiro and C. Amotore (ed.), Fontis Media and Marcel Dekker, Netherlands, 2004.
6. P. Morse, Scientific Software Services, Bloomington, USA, 1994.
7. WIN-EPR SimFonia, version 1.25, Brucker, Billerica, USA, 1996.
8. F. Hartl and A. Vlcek Jr., *Inorg. Chem.*, 1991, **30**, 3048.
9. R. D. Allendoerfer, G. A. Martinchek and S. Bruckenstein, *Anal. Chem.*, 1975, **47**, 890.
10. D. A. Fiedler, M. Koppenol and A. M. Bond, *J. Electrochem. Soc.*, 1995, **142**, 862.
11. A. M. Bond, P. J. Dyson, D. G. Humphrey, G. Lazarev and P. Suman, *J. Chem. Soc., Dalton Trans.*, 1999, **3**, 443.
12. B. A. Coles and R. G. Compton, *J. Electroanalytical Chem.*, 1983, **144**, 87.
13. R. D. Webster, A. M. Bond, B. A. Coles and R. G. Compton, *J. Electroanal. Chem.*, 1996, **404**, 303.
14. W. Kaim, S. Ernst and V. Kasak, *J. Am. Chem. Soc.*, 1990, **112**, 173.
15. S. Patra, B. Sarkar, S. Ghumaan, J. Fiedler, W. Kaim and G. K. Lahiri, *Inorg. Chem.*, 2004, **43**, 6108.
16. S. Frantz, J. Fiedler, I. Hartenbach, T. Schleid and W. Kaim, *J. Organomet. Chem.*, 2004, **689**, 3031.
17. P. Rapta, A. Staŝko, A. V. Gromov, A. Bartl and L. Dunsch, *Electrochem. Soc. Proc.*, 2000, **10**.
18. P. Rapta, A. L. Kress, P. Hapiot and L. Dunsch, *Phys. Chem. Chem. Phys.*, 2002, **4**, 4181.
19. P. Rapta and L. Dunsch, *J. Electroanalytical Chem.*, 2001, **507**, 287.

20. P. Rapta, R. Fàber, L. Dunsch, A. Neudeck and O. Nuyke, *Spectrochim. Acta Part A*, 2000, **56**, 357.
21. A. Neudeck, A. Petrr and L. Dunsch, *Synth. Met.*, 1999, **107**, 143.
22. M. Wilgocki and W. K. Rybak, *Port. Electrochim. Acta*, 1995, **13**, 211.
23. E. J. L. McInnes, A. J. Welch and L. J. Yellowlees, *Chem. Commun.*, 1996, 2393.
24. N. Sutin, M. J. Weaver and E. L. Yee, *Inorg. Chem.*, 1980, **19**, 1096.
25. M. F. Bento, M. J. Medeiros and M. I. Montenegro, *J. Electroanalytical Chem.*, 1993, **345**, 273.
26. A. H. Maki and D. H. Geske, *J. Am. Chem. Soc.*, 1960, **82**, 267.
27. D. H. Geska and A. H. Maki, *J. Am. Chem. Soc.*, 1960, **82**, 2671.
28. P. H. Rieger and G. Fraenkel, *J. Chem. Phys.*, 1963, **39**, 609.
29. P. L. Loker and W. A. Waters, *Proc. Chem. Soc.*, 1963, **55**.
30. G. R. Stevenson and L. Echegoyen, *J. Phys Chem.*, 1973, **77**, 2339.
31. N. M. Atherton and A. S. Shalabi, *J. Chem. Soc., Faraday Trans. 2*, 1980, **76**, 822.
32. C. J. Adams, S. L. James, X. M. Liu, P. R. Raithby and L. J. Yellowlees, *J. Chem. Soc., Dalton, Trans.*, 2000, **63**.
33. E. J. L. McInnes, R. D. Farley, S. A. MacGregor, K. J. Taylor, L. J. Yellowlees and C. C. Rowlands, *J. Chem. Soc. Faraday Trans.*, 1998, **94**, 2985–2991; D. Collinson, F. E. Mabbs, E. J. L. McInnes, K. J. Taylor, A. J. Welch and L. J. Yellowlees, *J. Chem. Soc. Dalton Trans.*, 1996, 329.
34. H. M. McConnell, *J. Chem. Phys.*, 1956, **24**, 764.
35. C. J. Adams, L. E. Bowen, M. G. Humphrey, J. P. L. Morrall, M. Samoc and L. J. Yellowlees, *J. Chem. Soc. Dalton Trans.*, 2004, 4130.
36. M. T. Barton, N. M. Rowley, P. R. Ashton, C. J. Jones, N. Spencer, M. S. Tolley and L. J. Yellowlees, *New J. Chem.*, 2000, **24**(7), 555.
37. P. H. Rieger, *J. Magn. Reson.*, 1997, **124**, 140.
38. E. J. L. McInnes, R. D. Farley, C. C. Rowlands, A. J. Welch, L. Rovatti and L. J. Yellowlees, *J. Chem. Soc. Dalton Trans.*, 1999, 4203.
39. K. Niki, O. Vrána and V. Brabec, *Experimental Techniques in Bio-chemistry*, V. Brabec, D. Waltz and G. Milazzo (ed.), Birkhäuserr Verlag, Basel, 1996.
40. M. Comtat and H. Durliat, *Biosens.r Bioelectron.*, 1994, **9**, 663.
41. Y. -T. Long, Z. -H. Yu and H. -Y. Chen, *Electrochem. Commun.*, 1999, **1**, 194.
42. E. Y. Fung, A. C. M. Chua and J. C. Curtis, *Inorg. Chem.*, 1988, **27**, 1294.
43. J. N. Demas, T. F. Turner and G. A. Crosby, *Inorg. Chem.*, 1969, **8**, 674.
44. V. Gutmann, *Electrochim. Acta.*, 1976, **21**, 661.
45. J. S. Griffith, *The Theory of Transition Metal Ions*, Cambridge University Press, London, 1961, p. 364.
46. B. Bleany and M. C. O'Brien, *Proc. Phys. Soc. London, Sect. B.*, 1956, **69**, 1216.
47. G. K. Lahir, S. Bhattacharya, B. K. Ghosh and A. Chakravorthy, *Inorg. Chem.*, 1987, **26**, 4324.

48. R. E. DeSimone, *J. Am. Chem. Soc.*, 1973, **95**, 6238.
49. C. J. Adams, S. L. James, X. M. Liu, P. R. Raithby and L. J. Yellowlees, *J. Chem. Soc., Dalton Trans.*, 2000, 63.
50. E. J. L. McInnes, A. J. Welch and L. J. Yellowlees, *Chem. Commun.*, 1996, 2393.
51. E. J. L. McInnes, A. J. Welch and L. J. Yellowlees, *J. Chem. Soc. Dalton Trans.*, 1999, 4203.
52. A. A. Schilt, *J. Am. Chem. Soc.*, 1963, **85**, 904.

Subject Index